National Aeronautics and Space Administration (NASA):
Background, Issues, Bibliography

National Aeronautics and Space Administration (NASA): Background, Issues, Bibliography

C.V. Anderson (Editor)

Nova Science Publishers, Inc.
New York

Senior Editors: Susan Boriotti and Donna Dennis
Coordinating Editor: Tatiana Shohov
Office Manager: Annette Hellinger
Graphics: Wanda Serrano
Editorial Production: Jennifer Vogt, Matthew Kozlowski and Maya Columbus
Circulation: Ave Maria Gonzalez, Indah Becker, Raymond Davis and Vladimir Klestov
Communications and Acquisitions: Serge P. Shohov
Marketing: Cathy DeGregory

Library of Congress Cataloging-in-Publication Data
Available Upon Request

ISBN 1-59033-165-6.

Copyright © 2002 by Nova Science Publishers, Inc.
400 Oser Ave, Suite 1600
Hauppauge, New York 11788-3619
Tele. 631-231-7269   Fax 631-231-8175
e-mail: Novascience@earthlink.net
Web Site: http://www.novapubishers.com

All rights reserved. No part of this book may be reproduced, stored in a retrieval system or transmitted in any form or by any means: electronic, electrostatic, magnetic, tape, mechanical photocopying, recording or otherwise without permission from the publishers.

The publisher has taken reasonable care in the preparation of this book, but makes no expressed or implied warranty of any kind and assumes no responsibility for any errors or omissions. No liability is assumed for incidental or consequential damages in connection with or arising out of information contained in this book.

This publication is designed to provide accurate and authoritative information with regard to the subject matter covered herein. It is sold with the clear understanding that the publisher is not engaged in rendering legal or any other professional services. If legal or any other expert assistance is required, the services of a competent person should be sought. FROM A DECLARATION OF PARTICIPANTS JOINTLY ADOPTED BY A COMMITTEE OF THE AMERICAN BAR ASSOCIATION AND A COMMITTEE OF PUBLISHERS.

Printed in the United States of America

CONTENTS

PART I.	BACKGROUND	1
Chapter 1	A Brief History of the National Aeronautics and Space Administration *Stephen J. Garber and Roger D. Launius*	3
Chapter 2	The National Aeronautics and Space Administration (NASA): History and Organization *Erin C. Hatch*	13
PART II.	ISSUES	43
Chapter 3	NASA's Earth Science Enterprise *Erin C. Hatch*	45
Chapter 4	Space Launch Vehicles: Government Activities, Commercial Competition, and Satellite Exports *Marcia S. Smith*	53
Chapter 5	Space Stations *Marcia S. Smith*	69
Chapter 6	Military Space Activities: Highlights of the Rumsfeld Commission Report and Key Organization and Management Issues *Marcia S. Smith*	87
Chapter 7	Commonly used Acronyms and Program Names in the Space Program *Marcia S. Smith and David P. Radzanowsk*	95
Chapter 8	An Overview of NASA's Mission to Planet Earth (MTPE) *David P. Radzanowski and Stephen J. Garber*	119

PART III.	BIBLIOGRAPHY	**125**
Books		**127**
Websites		**229**
Author Index		**237**
Title Index		**243**
Subject Index		**253**

Part I. Background

Chapter 1

A Brief History of the National Aeronautics and Space Administration

Stephen J. Garber and Roger D. Launius

Launching NASA

"**An Act to provide for research into the problems of flight within and outside the Earth's atmosphere, and for other purposes.**" With this simple preamble, the Congress and the President of the United States created the national Aeronautics and Space Administration (NASA) on October 1, 1958. NASA's birth was directly related to the pressures of national defense. After World War II, the United States and the Soviet Union were engaged in the Cold War, a broad contest over the ideologies and allegiances of the nonaligned nations. During this period, space exploration emerged as a major area of contest and became known as the space race.

During the late 1940s, the Department of Defense pursued research and rocketry and upper atmospheric sciences as a means of assuring American leadership in technology. A major step forward came when President Dwight D. Eisenhower approved a plan to orbit a scientific satellite as part of the International Geophysical Year (IGY) for the period, July 1, 1957 to December 31, 1958, a cooperative effort to gather scientific data about the Earth. The Soviet Union quickly followed suit, announcing plans to orbit its own satellite.

The Naval Research Laboratory's Project Vanguard was chosen on 9 September 1955 to support the IGY effort, largely because it did not interfere with high-priority ballistic missile development programs. It used the non-military Viking rocket as its basis while an Army proposal to use the Redstone ballistic missile as the launch vehicle waited in the wings. Project Vanguard enjoyed exceptional publicity throughout the second half of 1955, and all of 1956, but the technological demands upon the program were too great and the funding levels too small to ensure success.

A full-scale crisis resulted on October 4, 1957 when the Soviets launched *Sputnik 1*, the world's first artificial satellite as its IGY entry. This had a "Pearl Harbor" effect on American public opinion, creating an illusion of a technological gap and provided the impetus for increased spending for aerospace endeavors, technical and scientific educational programs, and the chartering of new federal agencies to manage air and space research and development.

More immediately, the United States launched its first Earth satellite on January 31, 1958, when *Explorer 1* documented the existence of radiation zones encircling the Earth. Shaped by the Earth's magnetic field, what came to be called the Van Allen Radiation Belt, these zones partially dictate the electrical charges in the atmosphere and the solar radiation that reaches Earth. The U.S. also began a series of scientific missions to the Moon and planets in the latter 1950s and early 1960s.

A direct result of the Sputnik crisis, NASA began operations on October 1, 1958, absorbing into itself the earlier National Advisory Committee for Aeronautics intact: its 8,000 employees, an annual budget of $100 million, three major research laboratories-Langley Aeronautical Laboratory, and Lewis Flight Propulsion Laboratory-and two smaller test facilities. It quickly incorporated other organizations into the new agency, notably the space science group of the Naval Research Laboratory in Maryland, the Jet Propulsion Laboratory managed by the California Institute of Technology for the Army, and the Army Ballistic Missile Agency in Huntsville, Alabama, where Wernher von Braun's team of engineers were engaged in the development of large rockets. Eventually NASA created other Centers and today it has ten located around the country.

NASA began to conduct space missions within months of its creation, and during its first twenty years NASA conducted several major programs:

- Human space flight initiatives-Mercury's single astronaut program (flights during 1961-1963) to ascertain if a human could survive in space; Project Gemini (flights during 1965-1966) with two astronauts to practice space operations, especially rendezvous and docking of spacecraft and extravehicular activity (EVA); and Project Apollo (flights during 1968-1972) to explore the Moon.
- Robotic missions to the Moon (Ranger, Surveyor, and Lunar Orbiter), Venus (*Pioneer Venus*), Mars (*Mariner 4, Viking 1* and *2*), and the outer planets (*Pioneer 10* and *11, Voyager 1* and *2*).
- Aeronautics research to enhance air transport safety, reliability, efficiency, and speed (X-15 hypersonic flight, lifting body flight research, avionics and electronics studies, propulsion technologies, structures research, aerodynamics investigations).
- Remote-sensing Earth satellites for information gathering (Landsat satellites for environmental monitoring).
- Applications satellites for communications (*Echo 1, TIROS*, and *Telstar*) and weather monitoring.
- An orbital workshop for astronauts, *Skylab*.
- A reusable spacecraft for traveling to and from Earth orbit, the Space Shuttle.

EARLY SPACEFLIGHTS: MERCURY AND GEMINI

NASA's first high-profile program involving human spaceflight was Project Mercury, an effort to learn if humans could survive the rigors of spaceflight. On May 5, 1961, Alan B. Shepard Jr. became the first American to fly into space, when he rode his Mercury capsule on a 15-minute suborbital mission. John H. Glenn Jr. became the first U.S. astronaut to orbit the Earth on February 20, 1962. With six flights, Project Mercury achieved its goal of putting piloted spacecraft into Earth orbit and retrieving the astronauts safely.

Project Gemini built on Mercury's achievements and extended NASA's human spaceflight program to spacecraft built for two astronauts. Gemini's 10 flights also provided NASA scientists and engineers with more data on weightlessness, perfected reentry and splashdown procedures, and demonstrated rendezvous and docking in space. One of the highlights of the program occurred during Gemini 4, on June 3, 1965, when Edward H. White, Jr., became the first U.S. astronaut to conduct a spacewalk.

The singular achievement of NASA during its early years involved the human exploration of the Moon, Project Apollo. Apollo became a NASA priority on May 25 1961, when President John F. Kennedy announced "I believe that this nation should commit itself to achieving the goal, before this decade is out, of landing a man on the Moon and returning him safely to Earth." A direct response to Soviet successes in space, Kennedy used Apollo as a high-profile effort for the U.S. to demonstrate to the world its scientific and technological superiority over its cold war adversary.

Going to the Moon - Project Apollo
Edwin E. "Buzz" Aldrin, Jr. descends from the Apollo 11 Lunar Module to become the second human to walk on the Moon. Neil A. Armstrong, who took this photograph, was the commander of the mission and the first to walk on the lunar surface.

In response to the Kennedy decision, NASA was consumed with carrying out Project Apollo and spent the next 11 years doing so. This effort required significant expenditures, costing $25.4 billion over the life of the program, to make it a reality. Only the building of the Panama Canal rivaled the size of the Apollo program as the largest nonmilitary technological endeavor ever undertaken by the United States; only the Manhattan Project was comparable in a wartime setting. Although there were major challenges and some failures - notably a January 27, 1967 fire in an Apollo capsule on the ground that took the lives of astronauts Roger B. Chaffee, Virgil "Gus" Grissom, and Edward H. White Jr. - the program moved forward inexorably.

Less than two years later, in October 1968, NASA bounced back with the successful Apollo 7 mission, which orbited the Earth and tested the redesigned Apollo command module. The Apollo 8 mission, which orbited the Moon on December 24-25, 1968, when its crew read from the book of Genesis, was another crucial accomplishment on the way to the Moon.

"That's one small step for [a] man, one giant leap for mankind." Neil A. Armstrong uttered these famous words on July 20, 1969, when the Apollo 11 mission fulfilled Kennedy's challenge by successfully landing Armstrong and Edwin E. "Buzz" Aldrin, Jr. on the Moon. Armstrong dramatically piloted the lunar module to the lunar surface with less than 30 seconds worth of fuel remaining. After taking soil samples, photographs, and doing other tasks on the Moon, Armstrong and Aldrin rendezvoused with their colleague Michael Collins in lunar orbit for a safe voyage back to Earth.

Five more successful lunar landing missions followed. The Apollo 13 mission of April 1970 attracted the public's attention when astronauts and ground crews had to improvise to end the mission safely after an oxygen tank burst midway through the journey to the Moon. Although this mission never landed on the Moon, it reinforced the notion that NASA had a remarkable ability to adapt to the unforeseen technical difficulties inherent in human spaceflight.

With the Apollo 17 mission of December 1972, NASA completed a successful engineering and scientific program. Fittingly, Harrison H. "Jack" Schmitt, a geologist who participated on this mission, was the first scientist to be selected as an astronaut. NASA learned a good deal about the origins of the Moon, as well as how to support humans in outer space. In total, 12 astronauts walked on the Moon during 6 Apollo lunar landing missions.

In 1975, NASA cooperated with the Soviet Union to achieve the first international human spaceflight, the Apollo-Soyuz Test Project (ASTP). This project successfully tested joint rendezvous and docking procedures for spacecraft from the U.S. and the U.S.S.R. After being launched separately from their respective countries, the Apollo and Soyuz crews met in space and conducted various experiments for two days.

After a gap of six years, NASA returned to human spaceflight in 1981, with the advent of the Space Shuttle. The Shuttle's first mission, STS-1, took off on April 12, 1981, demonstrating that it could take off vertically and glide to an unpowered airplane-like landing. On STS-6, during April 4-9, 1983, F. Story Musgrave and Donald H. Peterson conducted the first Shuttle EVA, to test new spacesuits and work in the Shuttle's cargo bay. Sally K. Ride became the first American woman to fly in space when STS-7 lifted off on June 18, 1983, another early milestone of the Shuttle program.

On January 28, 1986 a leak in the joints of one of two Solid Rocket Boosters attached to the *Challenger* orbiter caused the main liquid fuel tank to explode 73 seconds after launch, killing all 7 crew members. The Shuttle program was grounded for over two years, while NASA and its contractors worked to redesign the Solid Rocket Boosters and implement management reforms to increase safety. On September 29, 1988, the Shuttle successfully returned to flight. Through mid-1998, NASA has safely launched 65 Shuttle missions since the return to flight. These have included a wide variety of scientific and engineering missions. There are four Shuttle orbiters in NASA's fleet: *Atlantis, Columbia, Discovery*, and *Endeavour*.

Space Shuttle
This rare view of two Space Shuttle orbiters simultaneously on launch pads at the Kennedy Space Center was taken on September 5, 1990. The orbiter *Columbia* is shown in the foreground on pad 39A, where it was being prepared for a launch (STS-35) the next morning. This launch ended up being delayed until December 1990. In the background, the orbiter *Discovery* sits on pad 39B in preparation for an October liftoff on STS-41.

TOWARD A PERMANENT HUMAN PRESENCE IN SPACE

The core mission of any future space exploration will be humanity's departure from Earth orbit and journeying to the Moon or Mars, this time for extended and perhaps permanent stays. A dream for centuries, active efforts to develop both the technology and the scientific knowledge necessary to carry this off are now well underway. The next generation of launch vehicles taking us from the Earth into orbit are being developed right now. The X-33, X-34, and other hypersonic research projects presently underway will help to realize routine, affordable access to space in the first decades of the twenty-first century.

An initial effort in this area was NASA's Skylab program in 1973. After Apollo, NASA used its huge Saturn rockets to launch a relatively small orbital space workshop. There were three human Skylab missions, with the crews staying aboard the orbital workshop for 28, 59, and then 84 days. The first crew manually fixed a broken meteoroid shield, demonstrating that humans could successfully work in space. The Skylab program also served as a successful experiment in long-duration human spaceflight.

In 1984, Congress authorized NASA to build a major new space station as a base for further exploration of space. By 1986, the design depicted a complex, large, and multipurpose facility. In 1991, after much debate over the station's purpose and budget, NASA released plans for a restructured facility called Space Station Freedom. Another redesign took place after the Clinton administration took office in 1993 and the facility became known as Space Station Alpha.

Then Russia, which had many years of experience in long-duration human spaceflight, such as with its *Salyut* and *Mir* space stations, joined with the U.S. and other international partners in 1993 to build a joint facility that became known formally as the International Space Station (ISS). To prepare for building the ISS starting in late 1998, NASA participated in a series of Shuttle missions to *Mir* and seven American astronauts lived aboard *Mir* for extended stays.

The Science of Space
The Sojourner rover and undeployed ramps aboard the Mars Pathfinder spacecraft are shown shortly after landing on the Martian surface on July 4, 1997. Partially deflated airbags are also clearly visible.

In addition to major human spaceflight programs, there have been significant scientific probes that have explored the Moon, the planets, and other areas of our solar system. In particular, the 1970s heralded the advent of a new generation of scientific spacecraft. Two similar spacecraft, Pioneer 10 and Pioneer 11, launched on March 2, 1972 and April 5, 1973, respectively, traveled to Jupiter and Saturn to study the composition of interplanetary space. Voyagers 1 and 2, launched on September 5, 1977 and August 20, 1977, respectively, conducted a "Grand Tour" of our solar system.

In 1990, the Hubble Space Telescope was launched into orbit around the Earth. Unfortunately, NASA scientists soon discovered that a microscopic spherical aberration in the polishing of the Hubble's mirror significantly limited the instrument's observing power. During a previously scheduled servicing mission in December, 1993, a team of astronauts performed a dramatic series of spacewalks to install a corrective optics package and other hardware. The hardware functioned like a contact lens and the elegant solution worked perfectly to restore Hubble's capabilities. The servicing mission again demonstrated the unique ability of humans to work in space, enabled Hubble to make a number of important astronomical discoveries, and greatly restored public confidence in NASA.

Several months before this first HST servicing mission, however, NASA suffered another major disappointment when the Mars Observer spacecraft disappeared on August 21, 1993, just three days before it was to go into orbit around the red planet. In response, NASA began developing a series of "better, faster, cheaper" spacecraft to go to Mars.

Mars Global Surveyor was the first of these spacecraft; it was launched on November 7, 1996, and has been in a Martian orbit mapping Mars since 1998. Using some innovative technologies, the Mars Pathfinder spacecraft landed on Mars on July 4, 1997 and explored the surface of the planet with its miniature rover, Sojourner. The Mars Pathfinder mission was a scientific and popular success, with the world following along via the Internet.

Over the years, NASA has continued to look for life beyond our planet. In 1975, NASA launched the two Viking spacecraft to look for basic signs of life on Mars; the spacecraft arrived on Mars in 1976 but did not find any indications of past or present biological activity there. In 1996 a probe from the Galileo spacecraft that was examining Jupiter and its moon, Europa, revealed that Europa may contain ice or even liquid water, thought to be a key component in any life-sustaining environment. NASA also has used radio astronomy to scan the heavens for potential signals from extraterrestrial intelligent life. It continues to investigate whether any Martian meteorites contain microbiological organisms and in the late 1990s, organized an "Origins" program to search for life using powerful new telescopes and biological techniques.

The "First A in NASA:" Aeronautics Research
The rocket-powered X-15 aircraft set a number of altitude and speed records. Its flights during the 1960s also provided engineers and scientists with much useful data for the Space Shuttle program.

Building on its roots in the National Advisory Committee for Aeronautics, NASA has continued to conduct many types of cutting-edge aeronautics research on aerodynamics, wind shear, and other important topics using wind tunnels, flight testing, and computer simulations. In the 1960s, NASA's highly successful X-15 program involved a rocket-powered airplane that flew above the atmosphere and then glided back to Earth unpowered. The X-15 pilots helped researchers gain much useful information about supersonic aeronautics and the program also provided data for development of the Space Shuttle. NASA also cooperated with the Air Force in the 1960s on the X-20 Dyna-Soar program, which was designed to fly into orbit. The Dyna-Soar was a precursor to later similar efforts such as the National Aerospace Plane, on which NASA and other Government agencies and private companies did advanced hypersonics research in such areas as structures, materials, propulsion, and aerodynamics.

NASA has also done significant research on flight maneuverability on high speed aircraft that is often applicable to lower speed airplanes. NASA scientist Richard Whitcomb invented the "supercritical wing" that was specially shaped to delay and lessen the impact of shock waves on transonic military aircraft and had a significant impact on civil aircraft design. Beginning in 1972, the watershed F-8 digital-fly-by-wire (DFBW) program laid the groundwork for electronic DFBW flight in various later aircraft such as the F/A-18, the Boeing 777, and the Space Shuttle. More sophisticated DFBW systems were used on the X-29 and X-31 aircraft, which would have been uncontrollable otherwise.

From 1963 to 1975, NASA conducted a research program on "lifting bodies," aircraft without wings. This valuable research paved the way for the Shuttle to glide to a safe unpowered landing, as well as for the later X-33 project, and for a prototype for a future crew return vehicle from the International Space Station.

Applications Satellites
This dramatic view of Earth was taken by the crew of Apollo 17. The Apollo program put into perspective for many people just how small and fragile our planet is. Over its forty-year existence, NASA has been involved in many meteorological and Earth science missions that help us better understand our Earth.

NASA did pioneering work in space applications such as communications satellites in the 1960s. The Echo, Telstar, Relay, and Syncom satellites were built by NASA or by the private sector based on significant NASA advances.

In the 1970s, NASA's Landsat program literally changed the way we look at our planet Earth. The first three Landsat satellites, launched in 1972, 1975, and 1978, transmitted back to Earth complex data streams that could be converted into colored pictures. Landsat data has been used in a variety of practical commercial applications such as crop management and fault line detection, and to track many kinds of weather such as droughts, forest fires, and ice floes. NASA has been involved in a variety of other Earth science efforts such as the Earth Observation System of spacecraft and data processing that have yielded important scientific results in such areas as tropical deforestation, global warming, and climate change.

CONCLUSION

Since its inception in 1958, NASA has accomplished many great scientific and technological feats. NASA technology has been adapted for many non-aerospace uses by the private sector. At its 40th anniversary, NASA remains a leading force in scientific research and in stimulating public interest in aerospace exploration, as well as science and technology in general. Perhaps more importantly, our exploration of space has taught us to view the Earth, ourselves, and the universe in a new way. While the tremendous technical and scientific accomplishments of NASA demonstrate vividly that humans can achieve previously inconceivable feats, we also are humbled by the realization that Earth is just a tiny "blue marble" in the cosmos.

FOR FURTHER READING

Roger E. Bilstein, *Orders of Magnitude: A History of the NACA and NASA, 1915-1990.* (NASA SP-4406) Washington, D.C.: Government Printing Office, 1989.

John M. Logsdon, et. al., *Exploring the Unknown: Selected Documents in the History of the U.S. Civil Space Program* (NASA SP-4407). Washington, D.C.: Government Printing Office. *Volume 1: Organizing for Exploration* (1995). *Volume 2: External Relationships* (1996). *Volume 3: Using Space* (1998).

Chapter 2

THE NATIONAL AERONAUTICS AND SPACE ADMINISTRATION (NASA): HISTORY AND ORGANIZATION

Erin C. Hatch

ABSTRACT

The National Aeronautics and Space Administration (NASA) was created by the *National Aeronautics and Space Act of 1958* (P.L. 85-568) to undertake civilian research, development, and flight activities in aeronautics and space. During its more than forty years of existence, NASA has undertaken a wide variety of programs and projects. Notable achievements include the Apollo landings on the Moon, the development of communications and weather satellellites, the sending of planetary probes to all the planets except Pluto, and research in aeronautics that has improved aircraft performance and safety and assisted the competitive stance of the U.S. aeronautics industry. The agency is currently undertaking ambitious programs such as the International Space Station to provide a permanently inhabited international space station in Earth orbit, and a series of Mars exploration probes to assist in laying the groundwork for future human missions to Mars.

In the early 1990s, the downsizing of the aerospace industry due to the end of the Cold War and the large federal budget deficit caused many observers inside and outside of NASA to reexamine the space agency's goals. The agency responded by pushing for the "reinvention" of NASA, and stressing, "cheaper, faster, better" missions as the key to more efficient programs that still achieve their technical and scientific goals. However, recent agency failures and near failures are causing some policy makers to question the agency's management and structure. Independent review teams have assessed NASA management practices, and the agency is planning to consolidate and respond to these recommendations. Although NASA's FY2001 budget request represents its first budgetary increase in more than seven years, it is yet to be seen what impact the agency's recent problems will have on the likelihood of Congress approving this request.

NASA is organized into four strategic enterprises: Earth Science, Human Exploration and Development of Space, Space Science, and Aero-Space Technology. Earth Science supports programs that focus on the effects of natural and human-induced changes on the global environment. Human Exploration and Development of Space supports programs that provide transportation to and from space for people and payloads, and develop and operate habitable space facilities. The Space Science enterprise is responsible for NASA's programs related to the study of the universe, the solar system and the sun and its interaction with Earth. Aero-Space Technology coordinates NASA science and technology efforts in air and space transportation.

NASA's approximate $14 billion annual budget supports an organization consisting of NASA Headquarters and 16 research and space flight field centers and associated facilities located throughout the United States. NASA Headquarters offices manage the space flight centers, research centers, and associated installations. The 10 main field centers are Ames Research Center, Dryden Flight Research Center, Glenn Research Center, Goddard Space Flight Center, Jet Propulsion Laboratory, Johnson Space Center, Kennedy Space Center, Langley Research Center, Marshall Space Flight Center, and Stennis Space Center. Each of these functions as a "center of excellence" for the agency, and has an identified primary management mission.

INTRODUCTION

The National Aeronautics and Space Administration (NASA) was created in 1958 to lead the United States civilian space effort. Since that time, the agency has sought to advance America's exploration, use, and development of space. NASA funds research and development in advanced aeronautics, space, and related technologies, and enables the transfer of those technologies to other governmental agencies and the commercial sector. The agency's science programs attempt to advance and communicate scientific understanding of the Earth, Solar System, and Universe.

NASA efforts have included a wide variety of programs and projects. The Apollo program resulted in the landing of Americans on the Moon. The agency assisted in developing the first communications and weather satellites. NASA has sent planetary probes to all the planets except Pluto. The agency's research in aeronautics has improved aircraft performance and safety and assisted the competitive stance of the U.S. aeronautics industry.

Today NASA supports a number of programs that seek to: understand how the universe, galaxies, stars, and planets formed and evolved and the origin and existence of life beyond Earth; provide data relevant to the study of the globally interactive Earth environment; understands the role of gravity and cosmic radiation in biological, physical, and chemical systems in space, on other planetary bodies, and on Earth, and how we might apply this knowledge toward establishing a permanent human presence in space; and provide air and space transportation with higher safety, lower costs, less environmental impact, and better business opportunities. In order to accomplish these endeavors, NASA programs are organized into four strategic enterprises: Space Science, Earth Science, Human Exploration and Development of Space, and Aero-Space Technology. The current organization consists of NASA Headquarters and several research and space flight filed centers, and associated facilities located throughout the United States. These facilities and their associate workforce are supported by an annual budget of approximately $14 billion.

THE NATIONAL AERONAUTICS AND SPACE ACT OF 1958

The National Aeronautics and Space Administration (NASA) was created by the *National Aeronautics and Space Act of 1958* (P.L. 85-568) to conduct civilian research, development, and flight activities in aeronautics and space. The Act, also known as the *NASA Act of 1958* (P.L. 85-568) to conduct civilian research, development, and flight activities in aeronautics and space. The Act, also known as the *NASA Act*, established the agency and outlined its objectives: NASA is to undertake civilian research, development, and flight activities in aeronautics and space to maintain preeminence of the United States in those areas. The Act stipulated, however, that the Department of Defense (DOD) is responsible for all military space activities. NASA incorporated a predecessor agency, the National Advisory Committee for Aeronautics (NACA), whose staff and facilities were transferred to NASA by the Act. NASA's objectives for aeronautical and space activities as outlined in the *NASA Act* are:

- To expand human knowledge of the Earth and of phenomena in the atmosphere and space;
- To improve the usefulness, performance, speed, safety, and efficiency of aeronautical and space vehicles;
- To develop and operate vehicles capable of carrying instruments, equipment, supplies, and living organisms through space;
- To establish long-range studies of the potential benefits to be gained from, the opportunities for, and the problems involved in the utilization of aeronautical and space activities for peaceful and scientific purposes;
- To preserve the role of the United States as a leader in aeronautical and space science and technology and in the application thereof to the conduct of peaceful activities within and outside the atmosphere;
- To make available the agencies directly concerned with national defense discoveries that have military value or significance, and to furnish such agencies, from the civilian agency established to direct and control nonmilitary aeronautical and space activities, with information concerning discoveries which have value or significance to that agency;
- To coordinate the cooperation of the United States with other nations and groups of nations in work done pursuant to this Act and in the peaceful application of the results thereof;
- To effectively utilize the scientific and engineering resources of the United States, with close cooperation among all interested agencies of the United States in order to avoid unnecessary duplication of effort, facilities, and equipment; and
- To preserve the United States' preeminent position in aeronautics and space through research and technology development related to associated manufacturing processes.[1]

---

[1] 42 U.S.C. § 2451.

A Brief History of Activities

In October 1957, the Soviet Union surprised the world by launching the first artificial satellite, Sputnik 1. In response, the United States launched its first satellite, Explorer 1, in January 1958, symbolizing the entrance of the United States into the space age. Later that year, the *NASA Act* created NASA. Since its creation, NASA has undertaken a wide variety of programs and projects such as:

- Human spaceflight, including trips to the Moon;
- Satellites for communications, weather, and land and ocean remote sensing;
- Scientific satellites and probes to examine processes that affect the Earth and its environment, explore the Moon and other planets, and provide data on stars, comets, and interstellar space; and
- Aeronautical research in aerodynamics, materials and propulsion, in addition to flight tests with advanced aircraft, to improve aircraft performance and improve safety in the nation's airspace.

The following sections provide a brief history of NASA's activities in each decade, including successes and failures, and budgetary and organizational changes. This is concluded by a brief summary of NASA's recent activities, including the current controversy over its management practices.

The 1960s

The 1960s are best known for the landing of Americans on the Moon, programs leading up to the Moon landing, and development of scientific, communications, and meteorological satellites. In his May 1961 address to Congress, President Kennedy called for the United States to commit itself to landing an astronaut on the Moon before the end of the decade. To prepare for this endeavor, two programs were developed. The Mercury program (1961-1963) increased knowledge about human reaction to the space environment. The Gemini program (1956-66 expanded the knowledge of humans in space, and developed expertise in space vehicle rendezvous and docking, and in Earth orbit extra vehicular activity. The Apollo program succeeded in meeting President Kennedy's goal with the July 20, 1969, landing of Apollo 11 on the Moon with astronauts Armstrong and Aldrin.[2]

The 1960s also saw the development of scientific, communications and meteorological satellites. Once NASA developed communications and weather satellites, operational responsibility was often turned over to the private sector or to another government agency. This was the case when meteorological spacecraft were turned over to the National Oceanic and Atmospheric Administration (NOAA). Many scientific probes and satellites provided information about the interaction between the Sun and Earth as well as data on other stars and the cosmos. As part of this research effort, the agency began to launch a series of Earth orbiting observatories: the Orbiting Solar Observatory (OSO) (1962-1975), the Orbiting

---

[2] The Apollo program was not without its failures. On January 27, 1967, three astronauts (Grisson, White, and Chaffee) died of asphyxiation during a fire in the first Apollo spacecraft during pre-launch tests.

Astronomy Observatory (OAO) (1966-1972), and the Orbiting Geophysical Observatory (OGO) (1964-1969). Robotic probes were launched to the Moon and other planets. By the end of the decade, Venus and Mars had been visited by Mariner probes in flybys.

The 1970s

The 1970s were characterized by significantly lower budgets for NASA. In the early 1970s, five more Apollo lunar landings were accomplished, but the last three of the eight planned missions were canceled. A near tragedy occurred en route to the Moon on April 13, 1970, when an oxygen tank exploded aboard Apollo 13. The crew made it back safely to Earth four days later after improvising with an ingenious plan. In 1972, President Nixon approved the start of the Space Shuttle program to develop the first reusable launch vehicle for taking people and cargo into space. From 1973 to 1974, three crews visited a space station called Skylab. The third crew remained on Skylab for 84 days, a record at the time. Skylab reentered the Earth's atmosphere in 1979.[3] In 1975, the Apollo-Soyuz Test Project, in which a two-man Soviet crew docked in orbit with a three-man U.S. crew for the first such international space flight, heralded the end of the Apollo era. No U.S. human spaceflights occurred for the remained of the decade.

NASA's success in scientific and application satellites, and planetary probes continued in the 1970s. The Landsat program was initiated to provide environmental information on such topics as land resources and pollution on the Earth's surface. In 1973, Pioneer 10 became the first probe to fly by Jupiter, and in 1979, Pioneer 11 was the first probe to fly by Saturn. N 1974, Mariner 10 became the first probe to fly by Mercury. In 1976, Vikings 1 and 2 wee the first probes to land on the surface of Mars.

NASA undertook other astronomical and space physics projects, such as two Helios spacecraft, which were joint U.S. –German efforts to examine the Sun, launched in 1974 and 1976. The High-Energy Astronomy observatory 1 (HEAO 1), launched in 1977, was the first in a series of three satellite observatories designed to continue X-ray and gamma-ray studies initiated by the OAO and OSO series. NASA also continued a series of Explorer spacecraft probes such as the International Ultraviolet Explorer (IUE), which was launched in 1978 and conducted a variety of astrophysical observations in the ultraviolet spectra.

The 1980s

The 1980s are best known for the beginning of flights of the Space Shuttle[4], and for the tragedy of the *Challenger* accident. After six years of hiatus following the Apollo program, American astronauts once again returned to space with the first Space Shuttle launch on April 12, 1981. A total of 24 successful Shuttle flights occurred before tragedy struck the agency on January 28, 1986. On this day, the Space Shuttle *Challenger* exploded shortly after launch, killing all seven crewmembers. After the cause of the accident was discovered and fixes were made, the Space Shuttle returned to flight on September 29, 1988.

---

[3] Debris from Skylab that did not burn up in the atmosphere rained over the Indian Ocean and parts of Western Australia.
[4] The Space Shuttle is also known simply as "the Shuttle." For more information, see CRS IB93062: *Space Launch Vehicles: Government Activities, Commercial Competition, and Satellite Exports*, by Marcia S. Smith.

In his 1984 State of the Union address, President Reagan directed NASA to develop and launch a permanently occupied Space Station within ten years.[5] This proposed space station, named Freedom in 1988, was redesigned several times I the 1980s. These redesigns were due to Station program changes following the *Challenger* accident, rising costs, and lower than expected funding.

NASA and DOD were relying on the Space Shuttle fleet to launch most of their major spacecraft. However, limited budgets in the late 1970s and early 1980s, the *Challenger* accident, and the development of fewer, large spacecraft with longer development times than those build previously, led t fewer launchings of space science probes and application satellites. The accident convinced the White House and Congress that a mixed fleet, consisting of both the Shuttle and expendable launch vehicles (ELVs), would be necessary to assure continued access to space. Thus, commercialization of the U.S. launch vehicle industry began to increase markedly as the private sector began to build ELVs to satisfy the needs for the military and NASA, as well as its own commercial customers.

In 1989, the Space Shuttle launched the Magellan and Galileo space probes. Magellan's mission, which was completed successfully, was to map Venus's surface. Galileo flew by the asteroids Gaspra and Ida in 1991 and 1993, respectively. Its primary mission began in December 1995, when it went into orbit around Jupiter. Voyager 2, which was launched in 1977, became the first probe to fly by Uranus (January 1986) and Neptune (August 1989), leaving Pluto the only planet in our solar system not visited by a probe.

In addition to planetary probes, a few astronomical and Earth science spacecraft were launched in the early 1980s. The Solar Max satellite, launched in 1980, was the first spacecraft to study specific phenomena of the Sun using coordinated instrumentation. Landsats 4 and 5 were launched in 1982 and 1984, respectively, to continue operation of the Landsat system. The Infrared Astronomical Satellite (IRAS) was launched in 1983 to make the first all-sky infrared survey. Launched in 1989, the Cosmic Background Explorer (COBE) spacecraft examined subtle differences in background radiation to provide clues about the universe's formation after the Big Bang.

The 1990s

The early 1990s brought significant changes to NSAS: the agency gained a news Administrator, Daniel Goldin, just as NASA's overall budgets began to decline.[6] These budget constraints had been preceded by increased budgets in the late 190s, causing many observers inside and outside of NASA to begin to reexamine the space agency's goals and methods of doing business. Administrator Goldin and other administration officials responded by following "reinvention" trends within the federal government as a whole.[7] These individuals stressed "faster, cheaper, and better" missions as the key to more efficient programs that still achieve their scientific and technical goals. While developing indigenous technology for missions has traditionally been NASA's focus, in this decade the agency's

---

[5] Skylab, the first U.S. space station, was never intended to be permanently occupied.
[6] Daniel S. Goldin, appointed by President Bush, became the ninth Administrator of NASA on April 1, 1992.
[7] Vice President Gore's National Partnership for Reinventing Government, otherwise known as the National Performance Review (NPR), directed agencies to reengineer their business processes with the goal of providing better services to the American taxpayer. For more information on the NPR and "reinventing" government, see: CRS IB93026: *Executive Branch Reorganization*, by Harold C. Relyea.

administration began to also focus attention on utilizing commercially available technologies and establishing partnerships with industry. The development of smaller, more efficient spacecraft through the use of advanced technologies was now emphasized. Administrator Goldin stressed the importance of forging new partnerships between NSAS and industry to share the financial burdens of space and aeronautics programs.

Many observers agreed that NASA needed both to adapt existing commercial technologies for its own purposes, and to improve the "spinoff: of its relevant technologies to the private sector.[8] Some, however, questioned whether the Nation was getting its money's worth from NASA's budget in this regard. While most technology analysts agreed on the difficulty of accurately assessing the economic benefits from the transfer of government-developed technology to the private sector, a few analysts attempted this endeavor. Some of these resulting studies supported the transfer of NASA-developed technologies, and found that the amount of funding that the agency invests in research and development is usually much less than the value of technological spinoffs.[9] Other studies were not as supportive, concluding that the spinoff method had never been an effective approach to enriching commercial technology, and was, at best, a weak and expensive substitute for more direct support of the commercial technology base.[10]

In addition to questioning the validity of NASA's technology transfer programs reinventing government processes provided more dramatic influences on the future of the agency. Especially significant were Presidential directions to cut the agency's budget and to decrease its workforce. In January 1995, in step with the FY1996 budget process, the White House and the Office of Management and Budget (OMB) directed NASA to cut $5 billion from the agency's 5-year budget plan. NASA's response, known as the Zero Base Review (ZBR), identified $4 billion in savings from FY 1997 through FY2000 by cutting jobs and facilities through significant agency restructuring. The review recommended accomplishing these reductions, without any program cuts or closure of NASA centers, by increasing program management responsibilities for NASA centers, and by reducing the agency's workforce by approximately 4,000 full-time equivalent (FTE) civil servants and 25,000 contractor workers.[11] Driven by the need to formulate more realistic budgets and achieve a streamlined workforce, in early 1996 Administrator Goldin followed ZBR recommendations

---

[8] Since its creation, NASA has produced a wide range of products with consumer and industrial application sin non-aerospace fields such as medicine, agriculture, construction, and the environment. NASA refers to these products as "spinoffs." NSAS produces an annual volume entitled *Spinoffs* that covers a variety of offshoot technologies from the agency's research. For more information, see the Spinoff Center for Aero-Space Information website [http://www.sti.nasa.gov/tto/] o the NASA Commercial Technology Network website [http://ncn.hq.nasa.gov/.].

[9] Studies by Chase Econometrics, Inc., and the Midwest Research Institute in the late 1980s determined that every NASA technology transfer activities created significant economic gains. There have been some criticisms of these studies, however.

[10] See, for example: John A. Alic et al, *Beyond Spinoff: Military and Commercial Technologies in a Changing World* (Boston: Harvard Business School Press, 1992).

[11] The ZBR proposed streamlining functions at the NASA centers and encouraging centers to concentrate only on specific aspects of NASA's missions. In this way, each center would become a "center of excellence" in one area of NASA's activities and duplication of center activities would be reduced. At the same time, the review proposed changes to reduce overlap and consolidate administration and program functions across the agency. Te ZBE also recommended transferring headquarters program management responsibility to the field centers and increasing the involvement of outside entities in NASA's science programs by pursuing the establishment of science institutes. Most significantly, the reviews recommended reducing NASA's total civil-service workforce to approximately 17, 500 full-time equivalents (FTEs) by 2000. This reduction would have been the lowest level of civil servants at NASA since 1961.

in transferring program management responsibilities and some support functions to the field centers. The agency established "lead centers" to coordinate major programs, and utilized buyouts and restrictive hiring practices to reduce its civil servant workforce from approximately 25,000 FTEs to today's approximate 18,000 FTEs.[12] In this manner, the agency managed to accomplish these reductions without resorting to personnel reductions-in-force (RIFs).

An equally controversial issue for NASA at the time was development of the International Space Station (ISS). Since 1991, twenty-one attempts in NSAS funding bills and three attempts in broader legislation were made to terminate ISS, all of which failed.[13] NASA responded by restructuring the station program a number of times, including adding Russia to the list of participating countries in 1993. It was originally hoped that Russia's participation, based on its then twenty-two years of operating space stations, would help create significant cost savings.[14] However, delays in the program attributed to lack of Russian funding diminished the anticipated savings. Part of the controversy concerned the basic rationale for building and operating the ISS. Later in the decade, concerns centered on Russia's ability to fund its commitment to the program. While NASA had no specific plans for humans to return to the Moon or to visit Mars at this time,[15] station supporters argued that valuable scientific data on materials research and the long-term effects of weightlessness on the human body could be obtained from a permanent outpost in space. On the other hand, critics contended that there was no urgent need for a space station, and that NASA's program was too expensive and not well justified.

During the 1990s, congressional and media attention also focused on the successes and failures of NASA's space science programs. Shortly after the Shuttle put the much-heralded Hubble Space Telescope (HST) into orbit in 1990, scientists discovered a spherical aberration in the spacecraft's main mirror. During a Shuttle servicing mission in December 1993, astronauts installed corrective optics on HST. A second HST Shuttle servicing mission took place in 1997. However, no such recovery was possible for another NASA spacecraft. In August 1993, contact was lost with the Mars Observer spacecraft just prior to entering into a Martian orbit. To recover from this large disappointment, NASA instituted the Mars Surveyor program, a series of low-cost spacecraft to explore Mars. Built and launched quickly, some of these missions were to collect some of the data that would have been obtained by Mars Observer. Missions to Mars began to receive increased interest in 1996 when NASA scientists announced evidence that life may have once existed on ancient Mars. Two NASA missions to Mars were launched in 1996, Mars Pathfinder and Mars Global Surveyor. Mars Pathfinder landed on the surface of Mars on July 4, 1997, and successfully carried out its program plan. Mars Global Surveyor went into orbit around Mars in September 1997. (NASA lost two Mars spacecraft in 1999. These and other recent failures are discussed in the next section.)

Other significant space science programs included astronomical, physics, and planetary exploration missions. The Compton Gamma Ray Observatory (CGRO) was launched in 1989

---

[12] Information supplied to CRS by NASA on March 23, 2000.
[13] For more information on these attempts to cancel the Space Station, see CRS IB93017: *Space Stations*, by Marcia S. Smith.
[14] Russia's current operational space station, Mir, is its seventh since 1971.
[15] In 1989, the twentieth anniversary of the first Apollo landing on the Moon, President Bush outlined a strategy called the Space Exploration Initiative (SEI). Under this plan, the Space Station would be the initial step for lunar and Martian exploration. The SEI program was canceled by Congress in FY1993 because of budget constraints.

to examine gamma ray emissions from around the universe and learn more about the possible cause of black holes, quasars, and supernovae. CGRO began returning data in early 1990. In the Global Geospace Science (GCS) program for space physics research, one spacecraft-Wind-was launched in 1994 and another-Polar-in February 1996. The Solar and Heliospheric Observatory (SOHO), a cooperative project between the European Space Agency (ESA) and NASA, was successfully launched in December 1995 and provided the first long term, uninterrupted view of the Sun. The Cassini spacecraft to explore Saturn, probably one of the last large space probes that NASA will build for the foreseeable future, was launched in October 1997. The first of the New Millennium, technology development missions, Deep Space 1, launched from Cape Canaveral in October 1998, and tested 12 advanced technologies in space during its primary mission.

Space-based environmental programs grew rapidly during the 1990s. The Upper Atmosphere Research Satellite (UARS), a shuttle-launched, Earth-orbiting observatory and center piece of NASA's upper atmosphere research program, was launched from the Shuttle in September 1991. The second and third Total Ozone Mapping Spectrometer (TOMS) instruments to measure the long-term changes in total ozone and verify chemical models of the stratosphere were launched in 1991 and total ozone and verify chemical models of the stratosphere were launched in 1991 and 1996, respectively. The Ocean Topography Experiment (TOPEX/Posideon) was launched in 1992 in a partnership between the U.S. and France to monitor global ocean circulation, discover ties between the U.S. and France to monitor global ocean circulation, discover ties between the oceans and atmosphere, and improve global climate predictions. TOPEX/Posideon mission data has been widely used to better understand El Nino, La Nina, and Pacific Decadal Oscillation phenomena. At the same time, congressional interest was sparked regarding the impact of the Earth science program in the emerging commercial remote sensing industry. This interest led t the passage of the *Land Remote Sensing Policy Act of 1992* (P.L. 102-555).[16] NASA spent the latter part of the decade further developing its Earth science resources, including creating the Earth Observation System (EOS), a three-part system consisting of scientific research, a data archiving and distribution system, and a series of polar-orbiting and low-inclination satellites for long-term global observations of the Earth. (The EOS program is explained in more detail later in this chapter.) This substantial growth came under scrutiny in the 104th Congress. Much of the criticism stemmed from delays in the Earth science program and the controversial nature of many of the subjects being studied, such as global climate change, leading some to question the value of the Earth science program as a whole. At the same time, the agency was experiencing significant problems in its Small Spacecraft Technology Initiative (SSTI) Program.[17] A primary objective of the SSTI was to demonstrate a significantly reduced

---

[16] 15 U.S.C. § 5601.
[17] The Small Satellite Technology Initiative (SSTI) was developed by NASA's then Office of Space Access and Technology to advance technology and reduce costs associated with the design, integration, launch, and operation of small satellites. In July 1994, the agency awarded contracts to TRW and CTA Space Systems to design and launch the small Earth-observing satellites named Lewis and Clark, respectively. Though Lewis was delayed a year before its Athena 1 launch vehicle was deemed flight ready, the spacecraft was ultimately launched in August 1997 to demonstrate advanced spacecraft instruments and subsystem technologies for measuring changes in Earth's land surface. Shortly after launch, the spacecraft entered a flat spin that resulted in a loss of solar power and a fatal battery discharge. Contact with the spacecraft was lost later in August, and then Lewis re-entered the atmosphere and was destroyed in September. At the same time, Clark suffered excessive schedule delays and projected cost growth. In February 1998, NASA announced the termination of Clark due to

government presence by relying mainly on the contractors' own management and mission assurance processes. The failure of two SSTI spacecraft-Lewis and Clark –caused many policymakers to begin to question NASA's "faster, better, cheaper" philosophy.

In 1995, NASA initiated a new effort, the Reusable Launch Vehicle (RLV) program, to flight-test technologies for a new reusable launch vehicle that one day might replace the Space Shuttle. The program involves significant participation by the private sector. Its goal is to validate technologies that would allow for the development of an operational RLV that would be fully funded and operated by the private sector.

End of the Decade, Beginning of the 2000s

The end of the decade was marked by several successes and some notable failures in NASA's programs.[18] The Chandra X-ray Observatory was successfully launched and deployed by Space Shuttle Columbia in July 1999, representing the most sophisticated X-ray observatory ever built. Terra (formerly EOS AM-1), the flagship of EOS, was successfully launched in December 1999. Six spacecraft were also lost in 1999. In March, the agency lost use of the Wide-Field Infrared Explorer (WIRE) spacecraft shortly after its launch.[19] Communication with the student-built Terriers satellite was lost in May.[20] In September, the Mars Climate Orbiter failed while attempting to descend into Mars' orbit.[21] The mission's companion spacecraft, Mars Polar Lander, and its two accompanying probes, Deep Space 2, were lost in December for unknown reasons. In addition, NASA experienced two serious failures during a July Shuttle launch (STS-93).[22] As a result, NASA delayed Shuttle launches for five-months. The agency also continued to experience delays in the Space Station program and significant problems with developing new launch vehicles.

The combination of these problems in such a short period of time has caused some policymakers to question the agency's management and its "smaller, faster, cheaper" philosophy. NASA has responded to these concerns by requesting several independent reviews of its management practices.[23] Agency officials have stated that they plan to

---

cost limits and the agency's expectation that the contractor could not succeed within acceptable schedule and costs.

[18] The commercial launch industry and Department of Defense also experienced an unusually high number of failures during 1999. For more information on these failures, see CRS RS20248: *Space Launch Vehicles: Issues Arising from the Unusually High Failure Rate of U.S. Launch Vehicles from August 1998-May 1999*, by Marcia S. Smith, and CRS IB93062: *Space Launch Vehicles: Government Activities, Commercial Competition, and Satellite Exports*, by Marcia S. Smith.

[19] Shortly after launch, WIRE's primary telescope cover was prematurely released, causing the spacecraft to spin out of control and lose all of its cooling hydrogen. WIRE's estimated mission cost was $54 million, including launch and spacecraft.

[20] After launch, WIRE's primary telescope cover was prematurely released, causing the spacecraft to spinout of control and lose all of its cooling hydrogen. WIRE's estimated mission cost was $54 million, including launch and spacecraft.

[21] NASA and its contractor, Lockheed Martin, used measurement systems for calculating and sending navigation commands to the spacecraft. NASA used metric units, while its contractor, Lockheed Martin, used English units.

[22] During launch, maintenance-related damage to electrical wiring caused a short circuit in two of the three Shuttle main engines, and holes in the liquid hydrogen tubes that feed into the main engines were caused by the dislocation of a pin.

[23] Five independent review have been conducted: (1) Space Shuttle Independent Assessment Team Report to the Associate Administrator for the Office of Space Flight, known as the "McDonald Report" [http://www.hq.nasa.gov/osf/siat/pdf]; (2) Project Management in NASA by the Mars Climate Orbiter Mishap Investigation Board, the "Stephenson Report" [ftp://ftp/hq.nasa.gov/pub/pao/reports/2000/

consolidate the recommendations from these reviews so that NSAS can further respond. At the same time, the agency has requested an increase in its budget for the first time in seven years. The agency's FY2001 request is for $14 billion, up from $13.6 billion in FY2000, and the out-year projections show NASA's budget rising to $15.6 billion by FY2005.[24] It is unclear what impact NASA's recent failures will have on Congress's review of this request.

NASA ORGANIZATION

NASA's current organization consists of NASA headquarters and several research and space flight field centers, and their associated facilities located throughout the United States. The administrator of NASA is appointed by the President and confirmed by the Senate. The agency currently employs approximately 18,000 civil servants. After significant reductions in budget and workforce in the 1990s, NASA is now requesting budgetary and workforce increases. Recent failures and delays within the agency's programs have caused many policymakers to question NASA's current management structure. Responding to these concerns, Administrator Daniel Goldin has announced agency plans to hire 328 additional people in FY2001, with the long-term goal of increasing NASA's workforce by about 2000 people. Some analysts anticipate the announcement of other organizational and management changes in the agency after NASA responds to several independent reviews of the agency's management practices.

The following sections discuss NASA's organizational structure, including its strategic planning process, four enterprises, headquarters, and field centers.

The Strategic Planning Process

In 1992, after the appointment of Daniel Goldin as administrator, NASA began to recognize a need for a new effective strategic management process. At that time, the agency had questionable budget expectations and lacked consensus on priorities and goals. Externally, there also was the perception that the agency was directionless as a result of the end of the Cold War. Previously-developed agency strategic plans were seen as "wish lists" that did not respond to budget realities.[25] Vice President Gore's *National Performance Review* and the *Government Performance and Results Act (GPRA) of 1993* (P.L. 103-62) added further impetus for the development of an agency strategic planning process.[26]

MCO_MIB-Report.pdf]; (3) NASA Faster, Better, Cheaper Task Final Report, the "Spear Report" [ftp://ftp.hq.nasa.gov/pub/pao/reports/2000/fbctask.pdf]; (4) Mars Program Independent Assessment Team Report, the "Young Report" [http://www.nasa.gov/newsinfo/mpiat_report_1.pdf]; and (5) Report of the Loss of the Mars Polar Lander and Deep Space 2 Missions, the "Casani Report" [http://www.nasa.gov/newsinfo/mpl_report_1.pdf]. All were completed in March 2000. These five reviews are also discussed in CRS IB92011: *U.S. Space Programs: Civilian, Military, and Commercial,* by Marcia S. Smith.

[24] For more information on NASA's FY2001 request, see CRS RL30493: *The National Aeronautics and Space Administration's FY2001 Budget Request: Description and Analysis,* by Richard E. Rowberg and Erin C. Hatch.

[25] During its existence, NASA has made use of many advisory committees and planning processes. For example, the agency produced several "Long Range Plans" between 1959 and 1989. However, these plans were usually conducted with relatively sort term foci and at lower organizational levels, such as at the program or project level, than the current strategic planning process.

[26] The *Government Performance and Results Ac* (GPRA) is intended to encourage greater efficiency, effectiveness, and accountability in federal spending, and requires agencies to set goals and use performance measures for

From October 1993 through May 1994, NASA held a series of retreats for senior managers and established several employee working groups to develop an agency strategic plan. The foundation for the plan was based on the *NASA Act of 1958*, the 1989 *National Space Policy*, and employee inputs. During this time, the agency identified its strategic enterprises and strategic functions and agreed on an agency vision and mission with associated values and goals. The first strategic plan form this process was published in May 1994.[27] Revised versions of this plan were published in February 1995 and February 1996.

To comply with *GRPA*, NASA submitted a second 5-year strategic plan to Congress on September 30, 1997.[28] In the fall of 1998, NASA resubmitted its 1998 strategic plan with 1999 interim revisions. The agency still operates under this 1999 revised strategic plan. The plan included agency organizational changes, the most significant of which was the creation of a new Chief Technologist and the elimination of the Space Technology Enterprise. Programs in the former enterprise, related to reducing space launch costs were combined with the Aeronautics Enterprise, and the new combined enterprise was called Aero-Space Technology. In this way, technology development became the responsibility of each of the four enterprises, with guidance and oversight provided by the Chief Technologist. The plan also reiterated NASA's "Safety first" objective, outlined the agency's three-part mission-Scientific Research, Space Exploration, and Technology Development and Transfer-and contained a roadmap defining the agency's near-, mid-, and long-term goals. In order to accomplish these goals over the next 25 years, the document contained detailed plans for each of the enterprises.

GPRA also requires agencies to transmit to OMB and to Congress annual performance plans, outlining the agency's process to evaluate its achievement of strategic plan goals. Pursuant to the law, NASA sent its first annual performance plan to Congress in the FY1999 budget request.[29] Following internal and external guidance, NASA revised and resubmitted its FY1999 plan, including additional linkages to the budget, more information on the reasonableness of performance targets, and details about performance verification process. Since this time, NASA has submitted performance plans with its budget requests for FY2000 and FY2001.

In March 2000, NASA submitted to Congress its FY1999 performance report that compared actual performance to goals.[30] In this document, the agency asserted that it met 81% of its performance targets, and that successes "far outweighed" failures. However, NASA maintained that the unpredictable nature of science and technology activities made it difficult to comply with *GPRA* requirements to quantify outcomes. In addition, the agency stated that its multi-year development activities proved especially troublesome when relating current outcomes to current fiscal expenditures. Some analysts expect Congress to review this

---

management and, ultimately for budgeting. For more information on GPRA, see CRS RS205257: *Government Performance and Results Act: Brief History and Implementation Activities During the First Session of the 106th Congress,* by Genevieve J. Knezo, CRS 98-726 GOV: *Government Performance and Results Act and the Appropriations Process,* by Sandy Streeter, or CRS 97-382 GOV: *Government Performance and Results Act: Implications for Congressional Oversight,* by Frederick M. Kaiser and Virginia A. McMurttry.

[27] NASA's current and past strategic plans are located on the NASA Headquarters website [http://www.hq.nasa.gov/office/dodez/plans.html].

[28] GPRA requires that agencies update their strategic plans at least every 3 years.

[29] NASA'sFY1999 Performance Report is available on the NASA website [http://ifmp.nasa.gov/codeb/docs/1999_perf_report.pdf].

[30] NASA's FY1999 Performance Report is available on the NASA website [http://ifmp.nasa.gov/codeb/docs/1999_perf_report.df].

and other agencies' performance reports in conjunction with the appropriations and authorization process.

The Four Enterprises

As part of its strategic planning process, NASA established four strategic enterprises to function as primary business areas for implementing its mission. The four enterprises Aero-Space Technology (AST-formerly Aeronautics and Space Transportation Technology), Space Science Enterprise (SSE). Earth Science Enterprise (ESE-formerly Mission to Planet Earth), and Human Exploration and Development of Space (HEDS). Each enterprise has unique strategic goals, objectives, and implementation strategies, and is headed by an enterprise associate administrator. The agency provides capabilities-such as space operations and communications, technology development, facilities construction and commercialization activities-required for each enterprise to achieve its goals. These agency-level activities are cross-agency and their functions are driven primarily by the strategic plans of the enterprises. The four enterprises are discussed below.

Aero-Space Technology [31]

The Aero-Space Technology (AST) enterprise coordinates NASA science and technology efforts in air and space transportation, and is managed by the Office of Aero-Space Technology (OAST). The Enterprise's programs consist of basic research and technology efforts that address fundamental and long-term opportunities, and more focused technology development programs that seek to capitalize on the products of the first. The Enterprise also has agency-wide responsibility for technology transfer and commercialization. This last responsibility requires increasing awareness in non-aerospace industries of NASA-developed technologies, and pro-actively assisting the transfer of technological knowledge into diverse applications.

The Research and Technology (R&T) Base program supports a wide range of aeronautical vehicle classes. The Airframe Systems and Information Technology programs develop advanced technology concepts and methods, and provide advanced tools and techniques. The Propulsion Systems program develops propulsion technologies. The Aviation Operations System program addresses technologies in communication, navigation, and surveillance systems; air traffic management; cockpit systems; operations human factors; and weather and hazardous environment characterization and avoidance systems. The Rotorcraft program develops rotorcraft technologies for the aerospace manufacturing industry, helicopter operators, selected DoD organizations, and the FAA. The Flight Research program addresses Advanced Space Transportation program develops technologies with launch cost reduction potential, and for in-space transportation systems to reduce costs, system mass, and trip time. The agency examines the nature of each specific technology to determine whether it then transitions to the focused technology programs or directly to the customer. In this way, the agency maintains that the R & T Base program provides the basis for future focused technology programs.

---

[31] The factual information in this section is largely drawn from the Aerospace Technology website [http://www.aero-space.nasa.gov/].

The Focused Technology Programs (FTPs) develop promising technologies that help meet enterprise and agency goals and selected national needs. The program seeks to develop products to a readiness level appropriate for transfer to other government agencies or industry. The "X" vehicles-X-33, X-34, and X-37—have been the most visible of the FTP programs. The X-33 is an integrated technology effort to flight-demonstrate "single-stage-to-orbit" (SSTO) technologies for a reusable space transportation vehicle."[32] The X-34 program seeks to demonstrate technologies for a reusable (but not commercially viable) space transportation vehicle. The X-37 program, also known as "Future-X," is a joint effort between the U.S. Air Force and NASA to test key design, manufacturing, structures, and thermal protection technologies on a reusable, in-space demonstrator. The vehicle is designed to be carried into orbit either by the Shuttle or by an expendable launch vehicle; NASA's current plans call for two orbital flights of an X-37 vehicle on the Shuttle. While the decision on how and whether to replace the Shuttle has been delayed until 2005, NASA plans to end its financial contributions to the "X" programs by the end of FY2002.[33] Rather than continuing with the "X"-vehicle programs, the agency is refocusing its efforts on a new competitive program to reduce the risk of RLV development. NASA plans to spend $4.4 billion between FY2001 and FY2001 and FY2005 on a Second Generation Reusable Launch Vehicle (RLV) Initiative. Instead of supporting a single design concept, such as the SSTO X-33 space plane, this program is designed to encourage development of a variety of technology concepts. In this way, the agency hopes to have more than one space transportation technology from which to chose in 2005.

Another FTP program is Aviation Systems Capacity, which is intended to safely increase the capacity, productivity, access, and flexibility of the National Airspace System Subsonic Technology program develops technology for environmentally compatible, operationally efficient subsonic aircraft, in cooperation with the FAA, U.S. aeronautics industry, and academia. The Aviation Safety Program (AvSP) develops and demonstrates technologies and strategies to improve aviation safety, and thereby reduce aircraft accident and fatality rates. The AvSP is conducted in partnership with the FAA, and in coordination with the Department of Defense, other government agencies, and the rotorcraft and general aviation industries. The Aero-Space Technology Enterprise manages of NASA component to the cross-agency High Performance Computing and Communication (HPCC) program, which is coordinated by the National Science Foundation.[34] The federal HPCC program supports projects that accelerate the development, application, and transfer of high-performance computing and communications technologies. Specifically, NASA's HPCC program seeks to meet engineering and science needs of the aeronautics, Earth and space science, spaceborne research, and education communities, and to accelerate the distribution of these technologies to the public.

---

[32] The term "single-stage-to-orbit" describes a rocket that can attain orbit with only one stage, instead of two or more as is common today, while carrying people or cargo.

[33] After NASA ends its financial contribution to the "X" plane programs, the agency's industrial partners must decide whether or not to proceed on their own. For more information about the X-33 or X-34 programs or other space launch vehicles, see CRS OIB93062: *Space Launch Vehicles: Government Activities, Commercial Competition, and Satellite Exports,* by Marcia S. Smith.

[34] For more information on the High Performance Computing and Communication (HPCC) program, see CRS Short Report 97-521: *Next Generation Internet and Related Initiatives.*

Earth Science Enterprise[35]

The Earth Science Enterprise (ESE) supports programs that focus on the effects of natural and human-induced changes on the global environment. The ESE is the largest federal agency program studying the Earth and its environment, and is managed by NASA's Office of Earth Science (OES). The program aids scientific understanding of environmental issues, such as global climate change, through the use of space-based, airborne, and ground-based instruments to acquire long-term data on the Earth climate system. The Enterprise supports research and analysis programs that assist scientists in converting these data into knowledge of the Earth system. At the same time, ESE manages a data and information management system to capture, process, archive, and distribute data to the scientific community and the public. A final cross-cutting objective of the Enterprise is the developed of remote sensing technologies, which can be used to reduce the cost and increase the reliability of future missions. A significant goal of the ESE is to enhance predictive capabilities about potential global environmental risks. In support of this, NASA is a significant contributor to the United States Global Change Research Program (USGCRP), the International Geosphere-Biosphere Program (IGBP), and the World Climate Research Program (WCRP).

The centerpiece of the Enterprise's array of instruments is the Earth Observing System (EOS) spacecraft series. Preceding EOS were a number of individual satellite and Shuttle-based missions, some of which are still providing information about basic Earth system processes.[36] The EOS series will continue this effort by obtaining coordinated, long-term, global observations of the land surface, biosphere, solid Earth, atmosphere, and oceans. The series consists of several polar-orbiting and low inclination satellites of various sizes, many of which include international contributions. The EOS program also supports research designed to analyze data and develop models that might explain the spacecraft's observations. The first EOS satellite launch was in 1999, and NASA plans to continue them through 2003.[37] The Enterprise is in the process of developing a science implementation plan that will drive the selection of follow-on missions to this first phase of EOS spacecraft. The agency has indicated that the National Polar-orbiting Operational Environmental Satellite System (NPOESS)-a joint venture between NASA, the National Oceanic and Atmospheric Administration (NOAA), and the Department of Defense (DoD)-will likely be key among these follow-on missions.

NASA has created and EOS Data Information System (EOSDIS) in order to process EOS flight data into useful information. The agency describes EOSDIS as evolutionary, including the phased deployment of the EOS satellites and their enabling data transmission technology. An initial version of EOSDIS (Version 0) became operational in 1994, and was implemented

---

[35] The factual information in this section is largely drawn from the Earth Science Enterprise website [http://www.earth.nasa.gov/].

[36] Previous missions include: Upper Atmosphere Research Satellite (UARS) (1919); Total Ozone Mapping Spectrometer (TOMS) instruments (1978, 1991, and 1996); Ocean Topography Experiment (TOPEX/Posideon) (1992); Tropical Rainfall Measuring Mission (TRMM) (1997); Sea-viewing Wide Field-of-view Sensor (SEASTR/SeaWifs) (1997); and QuickScat (1999).

[37] According to NASA's *1999 EOS Reference Handbook*, the first three launched spacecraft in the first EOS series are: (1) the seventh spacecraft in the Landsat Program (Landsat-7) (April 1999); (2) Terra (formerly known as EOS-AM-1 (December 1999); and (3) Active Cavity Radiometer Irradiance Monitors (ACRIMSAT) (December 1999). For more information, see: U.S. National Aeronautics and Space Administration (NASA), Earth Observing System (EOS) Project Science Office, *1999WOS Reference Handbook: A Guide to NASA's Earth Science Enterprise and the Earth Observing System* (Washington: 1999) [http://eospso.gsfc.nasa.gov/ftp_docs/andbook99.pdf].

through cooperation with NOAA, the U.S. Geological Survey (USGS), and partnering international space agencies. Significant technical difficulties delayed the deployment of Versions 1 and 2, through the agency reports that both are now performing successfully. The next phase of deployment (Version 3) has a revised deadline of December 2000, and the last phase (Version 4) is due to be implemented by April 2001.

Complementing EOS is the Earth Probes program designed to investigate processes requiring special orbits or short development cycles of one to three years. Within this program are small and rapidly-developed Earth System Science Pathfinder (ESSP) missions, the first two of which are scheduled for launch in 2000-2001.[38] The next round of ESSP missions were selected in December 1998, and two of these are scheduled for launch in 2003.[39] Also contained within the Earth Probes program are the Total Ozone Mapping Spectrometer (TOMS) instruments, which are designed to measure the long-term changes in total ozone levels, and verifying the chemical models of the stratosphere used to predict future trends.[40] In addition, Experiments of Opportunity projects include short duration flights of instruments on the Space Shuttle and other platforms. The University Earth System Science (UnESS) program is classified by NASA as spaceborne investigations of modest science scope, led by U.S. university investigators with significant student involvement. Yet another Earth Probes project, is Triana, a spacecraft that would be located at the Earth-Sun LaGrange-1 (L1) point.[41] The remaining Earth Probe project is the Shuttle Radar Topography Mission (SRTM), and Earth-mapping mission conducted aboard the Shuttle in February 2000. The SRTM was supported jointly by NASA, the National Imagery and Mapping Agency (NIMA), the German Aerospace Center (DLR), and the Italian Space Agency (ASI).

Human Exploration and Development of Space[42]

The Human Exploration and Development of Space (HEDS) enterprise supports programs that provide transportation to and from space for people and payloads, and develop and operate habitable space facilities. The Enterprise is managed by the Office of Space Flight (OSF) and the Office of Life and Microgravity Science and Applications (OLMSA). The Enterprise's stated goals include achieving routine space travel and human exploration of the solar system, and ultimately supporting people living and working in space. To accomplish these goals, HEDS manages and supports the Space Shuttle and International Space Station programs.

The Space Transportation System (STS)-NASA's term for the entire Shuttle program-consists of an orbiter spacecraft, two Solid Rocket Boosters (SRB), and external tank, and three main engines. The Office of Space Flight coordinates and manages the Shuttle system, including launch and space flight, intergovernmental agency requirements, and international and joint projects. However, in 1996, NASA turned much of the ground operations of the

---

[38] These two missions are Vegetation Canopy LIDAR (VCL) and Gravity Recovery and Climate Experiment (GRACE).

[39] The first ESSP mission is Pathfinder for Instruments for Cloud and Aerospace Spaceborne Observations-*Climatologie Etendue des Nuages et des Aerosols* (PICASSO)CENA). The second is CloudSat.

[40] According to NASA officials, the TOMS Flight Model-5 (FM-5) was completed and scheduled to fly as a cooperative mission with Russia in late 2000. However, Russia has indicated that it cannot meet that launch date. The agency now plans to fly FM-5 as QuickToms on U.S. vehicle and spacecraft in August 2000.

[41] For more information on the Trina project, see CRS RS20252: *NASA's Triana Spacecraft: An Overview of Congressional Issues*, by Erin C. Hatch.

[42] The factual information in this section is largely drawn from the Human Exploration and Development of Space website [http://www.hq.nasa.gov/osf/heds/].

shuttle over to a single contractor, United Space Alliance (USA), with the goal of reducing shuttle costs and beginning to privatize the Shuttle fleet.[43] The Space Shuttle orbiter has carried a flight crew of up to eight persons, and missions have lasted up to 16 or 17 days in space.[44] The Space Shuttle launches form Kennedy Space Center (KSC). Landing sites are located at KSC, Edwards Air Force Base I California, and White Sands Test Facility. The agency also has contingency landing sites around the world in the event the orbiter must return to Earth in an emergency.[45]

The United States, Canada, Japan, Russia, Brazil, and several European nations are collectively building the International Space Station (ISS) to enable the long-term exploration and utilization of space.[46] The United States is the coordinator of the project, and NASA's Office of Space Flight is managing ISS development. Because of the enormous size and complexity of the completed Station, individual segments of the Station will be lifted into space n more than 40 separate launches.[47] Though the partners originally planned to complet ethe ISS by 2002, that date has now slipped to between 2005 and 2006. The first component of the ISS is the U.S.-funded Russian-built Zarya control module, or Functional Cargo Block (Russian acronym FGB). The U.S.-build Unity connecting module, sometimes referred to as Node 1, is the second component of ISS. In 1998, these two components were launched and connected in space n two separate missions. The next ISS component is planned to be the Avezda Service Module, the first fully Russian contribution, that will provided for the first human habitation of the station and other early critical systems. However, the service module's launch has been delayed from an originally planned launch in June 199 to the current planned launch in July 2000.

NASA and its international partners plan to utilize the Station to perform long-duration research in a "microgravity" environment in materials and life sciences.[48] In addition to the Station, there are other methods to simulate a microgravity environment, including drop tubes, parabolic flights of the KC-135 aircraft, and the Space Shuttle. The Office of Life and Microgravity Sciences and Applications (OLMSA) supports these basic and applied microgravity research programs. Currently, there are three divisions in OLMSA. The first, the Aerospace Medicine and Occupational Health Division (AMOHD), is responsible for programs ensuring the health, safety, and productivity of astronauts in space, and for protecting and promoting the health and safety of NASA employees in general.[49] The Life Sciences Division (LSD) supports research in the fields of gravitational biology, biomedical

---

[43] United Space Alliance (USA) is a joint venture between Lockheed Martin and Boeing. NASA's Space Flight Operations Contract (SFOC) with USA could total $7 billion over six years.

[44] A total of 10 persons could be carried under emergency conditions.

[45] For more information on the Space Shuttle program, see CRS IB93062: *Space Launch Vehicles: Government Activities, Commercial Competition ,and Satellite Exports,* by Marcia S. Smith.

[46] Beside the United Stations, the ISS international partners include Russia, Japan, Canada and 11 of the European Space Agency's (ESA) member states. The ESA partners are: Brazil, Denmark, France, Germany, Italy, the Netherlands, Norway, Spain Sweden, Switzerland, and the United Kingdom. Brazil is also participating , but not as a partner. For more information about the International Space Station and other space stations, see CRS IB93017: *Space Stations*, by Marcia S. Smith.

[47] The completed Station will have a mass of over 450,000kg.

[48] The term "microgravity" refers to a condition that is a nearly gravity-free environment, and is sometimes described as "weightlessness."

[49] On May 1, 2000, Administrator Goldin announced the creation of a new office to coordinate the agency's programs in health and safety on the ground and in space. Beginning at the end of June 2000, programs currently supported by OLMSA's Aerospace Medicine and Occupational Health Division (AMOHD) will now be supported by the new health office.

research and counter measures, and advanced human support technology. The Microgravity Research Division (MRD) manages research programs in the disciplines of biotechnology, combustion science, fluid physics, fundamental physics, and material science. Finally, the Space Product Development office forms alliances between industry and academia through z, which support various commercial research programs. This last office has the stated goal of generating a demand for doing business in space by the private sector.

Space Science Enterprise[50]

The Space Science Enterprise (SSE) is responsible for all of NASA's programs related to the study of the universe, the solar system, and the sun and its interaction with Earth. It is managed by the Office of Space Science (OSS), and supports space-based missions, ground-based programs, scientific research, education and public outreach, and technology development and transfer. Scientific research in the Space Science Enterprise requires space-based, ground-based, and airborne observations of natural phenomena in space. The Enterprise develops and supports missions for these purposes. The SSE also sponsors research to analyze and interpret results from current and past missions' synthesize these analyses with related airborne, suborbital, and ground-based observations; and develop theory, which yields more scientific questions that motivate subsequent missions. Additional programs develop technologies from the conceptual stage to the point where they are ready to be incorporated in the full-scale development of science mission spacecraft. The Enterprise also seeks to enhance public awareness of space science by incorporating educational and public outreach activities as integral parts of their science investigations. The goals of the Enterprise include exploring the solar system, discovering planets around other stars, searching for life beyond Earth, and solving the mysteries of the universe. These four goals map directly with the Enterprise's four "science themes": Solar System Exploration, Sun-Earth Connection, Astronomical Search for Origins, and Structure and Evolution of the Universe.

The Solar System Exploration (SSE) theme supports NASA's missions to planets within the solar system. These missions seek to understand the nature and history of our Solar System, origin and evolution of life on Earth, external forces (including comet and asteroid impacts) that affect Earth's life and habitability, and origin and existence of life beyond Earth. SSE programs also compare Earth to its planetary neighbors, and identify locales and resources for future human habitation within the solar system. The Mars missions, such as Mars Pathfinder and Mars Global Surveyor, support the Solar System Exploration theme, as do other planetary probes, such as Galieleo and the proposed Europa Orbiter.

The Mars Exploration Program, coordinated by NASA Headquarters and the Jet Propulsion Laboratory, relies on partnerships with universities, industry, and international partners to conduct series of robotic missions to Mars. In support of the Solar System Exploration theme, these missions have the goal of understanding the planet's climate and potential for harboring past or present life, and laying the groundwork for future human missions to Mars. The Mars Global Surveyor, Mars Climate Orbiter, and Mars Polar Lander missions were all part of the Mars Exploration Program. Due to the recent Mars mission failures, the program is currently undergoing a major review within the agency.

---

[50] The factual information in this section is largely drawn from the Space Science Enterprise website [http://spacescinece.nasa.gov/].

The Sun-Earth Connection (SEC) theme supports programs that seek to understand why the Sun varies, how the planets respond to this solar variability, how Sun and galaxy interact, and how solar variability affects life and society. The theme supports missions such as the Japanese-led Solar-B, the Magnetospheric Constellation (MagCon), and the Magnetospheric Multiscale mission.

The Solar Terrestrial Probe (STP) theme supports programs that seek to understand why the Sun varies, how the planets respond to this solar variability, giw Sun and galaxy interact, and how solar variability affects life and society. The theme supports missions such as the Japanese-led Solar-B, the Magnetospheric Constellation (MagCon), and the Magnetospheric Multiscale mission.

The Solar Terrestrial Probe (STP) program provides in-situ and remote sensing observations of the Sun from jmultiple platforms for science investigations. The program supports the Sun-Earth Connection theme by trying to identify, develop, infuse, and transfer technologies that enable or enhance opportunities for frequent scientific investigations. The current series includes missions such as the Thermosphere, Ionosphere, Mesosphere Energetics and Dynamics (TIMED) mission, which will investigate the energetics of the mesosphere and lower thermosphere/ionosphere region of the Earth's atmosphere.

Also in support of the SEC theme, the Living With a Star (LWS) program is a new initiative for FY2001, coordinated by the Space Science Enterprise. The program will seek to quantify the physics, dynamics, and behavior of the Sun-Earth system over an 11-year solar cycle. At the same time, an Earth science component will try to understand the effects of solar variability and disturbances on terrestrial climate change. To aid in human exploration in space, LWD will attempt to provide advanced warnings of solar energetic particle showers that affect human safety. An Aero-Space Technology component will seek to characterize the radiation environment for use in designing more reliable electronic components for air and space transportation, and for communication satellites. The project will establish a Solar Dynamics Observatory, the Solar Sentinals mission, and the Geospace Dynamics Network and will initiate data analysis and modeling targets for space weather objectives. NASA plans to develop the LWS program in partnership with other federal agencies, including the National Science Foundation, U.S. Air Force, and National Oceanic and Atmospheric Administration.

The Astronomical Search for Origins (ASO) theme supports projects that seek to discover the origin of the first galaxies, stars, and planetary systems; the location of life-sustaining planets outside our solar system; the origin of Earth's life; and whether or not primitive or evolved life exists outside our solar system. The theme, also known simply as "Origins," supports many existing and planned space-based missions, including the Hubble Space Telescope (HST), the Next Generation Space Telescope (NGST), and the Terrestrial Planet Finder (TPF). An important element of Origins is the search for planets that have the potential for, or in fact currently are, harboring life. Astrobiology is becoming increasingly important in this endeavor.[51]

---

[51] NASA defines astrobiology as the multi-disciplinary study of the origin, evolution, distribution, and destiny of life in the universe. In 1997, NASA established the Astrobiology Institute at Ames Research Center with members from geographically-distributed research institutions, linked through advanced telecommunications. Astrobiology-related research institutions, linked through advanced telecommunications. Astrobiology-related discoveries in recent years have included: planets found outside our solar system, the existence of life in extreme environments on Earth (suck as temperatures beyond the boiling point of water or under extreme radiation

The Structure and Evolution of the Universe (SEU) theme supports programs that help to explain the structure of the Universe, explore the Universe's cycles of matter and energy, and examine the limits of the Universe's gravity and energy. Past SEU theme missions include the Cosmic Background Explorer (COBE) and the Roentgen Satellite (ROSAT), while current missions include the Compton Gamma Ray Observatory (CGRO) and the Chandra X-ray Observatory (formerly called the Advanced X-ray Astrophysics Facility, or AXAF.).

The Space Science Enterprise contains several mission development programs. The Explorer Program seeks to accomplish frequent space science investigations by utilizing innovative, streamlined, and efficient management approaches. It seeks to substantially reduce mission cost through the control of design, development, and operations costs, as well as to reduce cost and improve performance through the use of new technology. The program develops scientific missions of modest programmatic scope in support of the Astronomical Search for Origins, Sun-Earth Connection, and Structure and Evolution of the Universe science themes.[52] Investigations selected for the Explorer program are usually of a survey nature, or have specific objectives not requiring the capabilities of a major observatory.

The Discovery program launches many smaller missions which cost less than $200 million and take no more than three years to launch from start of development. The goal of the program is to do focused science with fast turn-around times by utilizing consortia of industry, small businesses, and universities. The Near Earth Asteroid Rendezvous (NEAR) Mars Pathfinder, and Lunar Prospector missions are all part of the Discovery Program.

The New Millennium Program (NMP) identifies and tests advanced technologies that could provide spacecraft with additional capabilities needed to address future agency goals. Through a series of deep space and Earth orbiting flights, the program seeks to demonstrate promising but risky technologies and lay the groundwork for their use by new missions. NMP is coordinated by the Solar System Exploration Theme, but also incorporates technology that benefits astronomical and Earth-observing mission spacecraft. Deep Space 1 and Earth Observer 1 are both part of the NMP.[53]

In addition, there are three operational divisions within the Enterprise. Space Science's Research Program Management Division is responsible for the management of all grant programs for Research and Analysis (R & A) and for Mission Operations and Data Analysis (MO&DA) for operating space science missions. Other responsibilities of this division include management of the Enterprise's activities in the suborbital program (sounding rockets, balloons, and shuttle payloads), information systems, and the high performance computing program. The Advanced Technology and Mission Studies Division is responsible for developing the critical technologies required to enhance and enable future space science missions. This division also funds future mission design studies, and oversees the

---

conditions), the past presence of liquid water on the surface of Mars, and an icy crust on Jupiter's moon, Europa (and the potential for a liquid ocean below the surface). For more information, see the Astrobiology Institute's website [http://nai.arc.nasa.gov/].

[52] These missions are categorized by size, starting with large-class projects moving down through Medium-class (MIDEX), Small-class (SMEX) and University-class (UNEX) missions. For example, the Far Ultraviolet Spectroscopy Explorer (FUSE) mission was a large-class project, while the Imager for magnetopause-to-Aurora Global Exploration (IMAGE) mission was a MIDEX project. Contained in the SMEX class are Missions of Opportunity (MO), which cost no more than $21 million (in FY 1998 dollars), and are flown as part of a non-NASA space mission, usually in cooperation with a foreign space agency.

[53] The Deep Space 2 mission failed in December 1999. The mission's two miniature probes, which were piggybacked aboard the Mars Polar Lander, were to slam into the Martian soil on December 3, 1999. However, the lander and probes failed to respond to communication efforts by NASA engineers after they were to reach the planet's surface.

development of technologies that will benefit the other NASA enterprises, as well as the non-NASA aerospace industry. Finally, the Mission and Payload Development (M&PD) Division oversees the building the space science mission spacecraft. M & PD maintains management responsibility for these spacecraft until they are launched and declared operational or, in the case of deep space missions, until the spacecraft reaches its primary science target.

NASA Headquarters

NASA headquarters exercises management over the four enterprises, and the agency's space flight centers, research centers, and associated installations. NASA headquarters is responsible for long-range strategic planning, program formulation, external advocacy, and resource allocation for the agency's science and technology programs. Headquarters also established policy defines requirements and objectives, and assesses performance of the agency's programs. It is divided into several program offices which plan, direct, and manage distinct research and development programs. Headquarters also has several functional and staff offices, which advise the NASA administrator, oversee agency workforce activities and facilities, set agency-wide standards, and assist in compliance with statutory, regulatory, and fiduciary responsibilities. Figure 1 shows NASA Headquarters organization, Table 1 lists the current program offices and managers, and Table 2 shows the functional/staff offices and managers.

Table 1. NASA Headquarters Program Offices

Office	Manager(s)
Office of the Administrator (Code A)*	Administrator: Daniel S. Goldin
	Deputy Administrator: Vacant
	Associate Deputy Administrator: Dr. Daniel R. Mulville
	Assistant for Commercialization: Daniel C. Tam
	White House Liaison: Leslie Tagg
	Chief Engineer: W. Brian Keegan
	Chief Information Officer: Lee B. Holcomb
	Chief Scientist: Dr. Kathie L. Olsen
	Chief Technologist: Samuel L. Venneri**
	Chief of Staff: Edward J. Heffernan***
Office of Space Flight (Code M)	Joseph H. Rothnberg
Office of Aero-Space Technology** (Code R)	Samuel L. Venneri
Office of Space Science (Code S)	Dr. Edward J. Weiler
Office of Life and Microgravity Sciences And Applications (Code U)	Arnauld E. Nicogossian, MD ****
Office of Earth Science (Code Y)	Dr. Ghassem Asrar

Source: NASA. Prepared by CRS. As of March 20, 2000.

Notes: *NASA offices are often referred to by their letter mail code. For example, the Office of Space Science is often referred toa s "Code S."

**According to the NASA Headquarters website [http://www.hq.nasa.gov/hq/org.html], the Office of the Chief Technologist and the Office of Aero-Space Technology are "soon to be combined."

***According to a NASA news release [ftp://ftp.hq.nasa.gov/pub/pao/pressrel/2000/00-057.txt], Edward J. Heffernan has been named Chief of Staff at NASA Headquarters by Administrators Goldin. As Chief of Staff, Heffernan will continue in his roles as head of the Agency's legislative affairs and will also coordinate all staff activities in the immediate office of the Administrator

****According to officials at NASA Headquarters, beginning at the end of June 2000, Dr. Arnauld Nicogossian will serve as the new health office's Chief Health and Medical Officer. Dr. Nicogossian will also continue to serve in the capacity of Associate Administrator for Life and Microgravity Sciences and Applications pending the selection of his replacement.

Table 2. NASA Headquarters Functional/Staff Offices

Office	Manager
Office for the Chief Financial Officer (Code B)*	Arnold G. Holz
Office of Headquarters Operations (Code C)	Michael D. Christensen
Office of Equal Opportunity Programs (Code E)	George E. Reese
Office of Human Resources and Education (Code F)	Vicki A. Novak
Office of the General Counsel (Code G)	Edward A. Frankle
Office of Procurement (Code H)	Thomas S. Luedtke
Office of External Relations (Code I)	John D. Schumacher
Office of Management Systems (Code J)	Jeffrey E. Sutton
Office of Small and Disadvantaged Business Utilization (Code K)	Ralph C. Thomas, III
Office of Legislative Affairs (Code L)	Edward J. Heffernan
Office of Public Affairs (Coe P)	Peggy C. Wilhide
Office of Safety and Mission Assurance (Code Q)	Frederick D. Gregory
Office of the Inspector General (Code W)	Roberta L. Gross
Office of Policy and Plans (Code Z)	Lori B. Garver

Source: NASA. Prepared by CRS. Current as of March 20, 2002. Notes: *NASA offices are often referred to by their letter mail code. For example, the Office of Policy and Plans is often referred to as "Code Z."

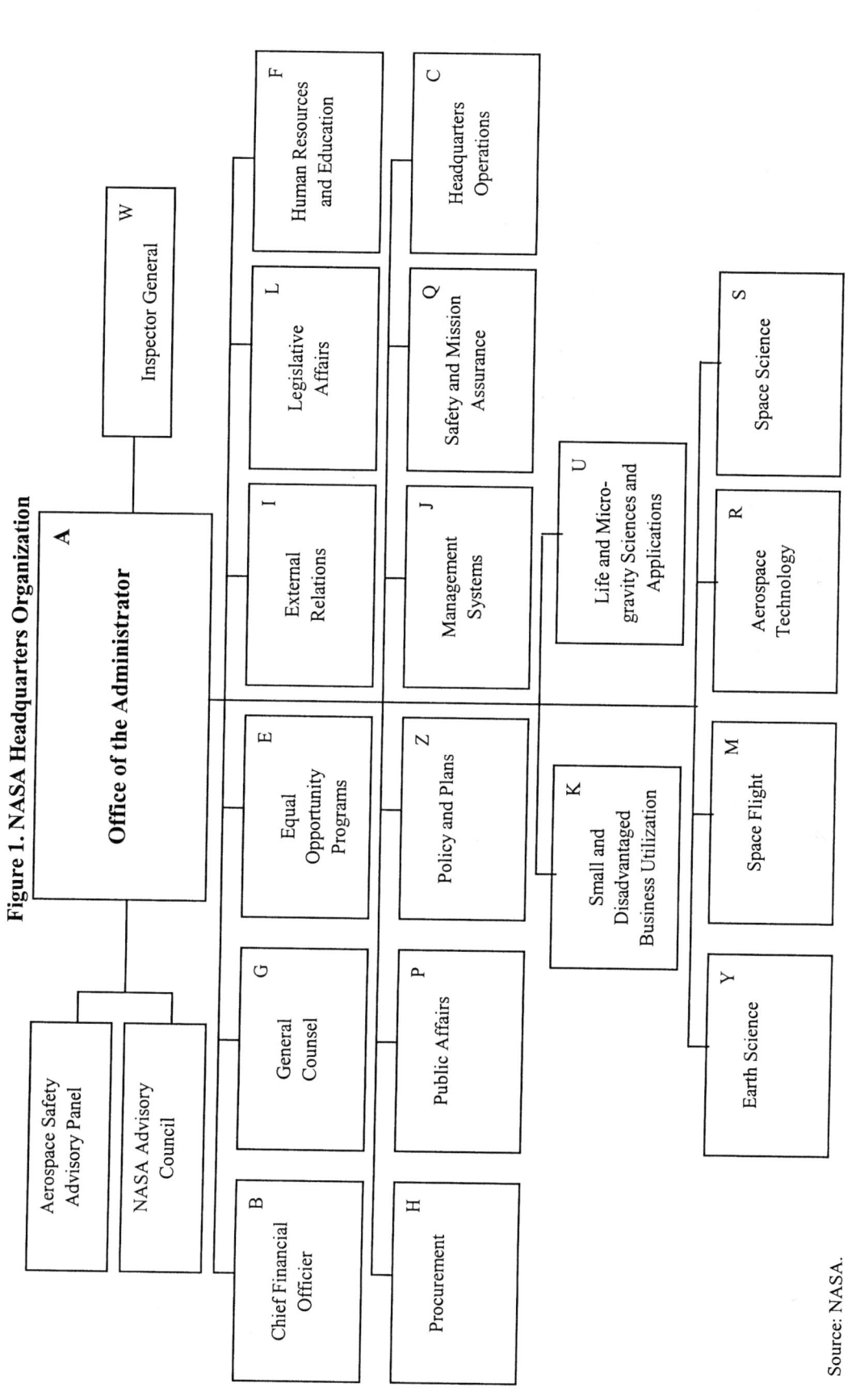

Figure 1. NASA Headquarters Organization

Source: NASA.

NASA Field Centers

NASA's principal technical strength lies in its nine main field centers, five affiliates, and one federally funded research and development center (FFRDC) facility.[54] Each center has its own mission and roles that carry out NASA's many programs and projects. Four of the centers are principally concerned with aero-space technology: Ames Research Center, Dryden Flight Research Center, Langley Research Center, and Glenn Research Center. Four of NASA's centers deal primarily with human space flight Center operations and systems: Johnson Space Center, Kennedy Space Center, Marshall Space Flight center, and Stennis Space Center. Two centers are the principal space and Earth sciences centers: Goddard Space Flight Center and Jet Propulsion Laboratory (JPL). Except for JPL, all of these centers are federally owned and operated facilities. Figure 2 shows the location of the centers, and Table 3 contains NASA's funding distribution by center.

Table 3. Distribution of NASA Funding by Center (in millions of dollars)

Center	FY1999 (Estimated)	FY2000 (Estimated)	FY2001 (Estimated)
Johnson Space Center	$4,279.3	$4,105.8	$4,112.3
Marshall Space Flight Center	2,312.5	2,303.4	2,346.1
Goddard Space Flight Center	2,286.2	2,269.8	2,278.0
Jet Propulsion Laboratory	1,143.2	1,263.8	1,424.8
Kennedy Space Center	941.7	919.0	1,043.8
Ames Research Center	617.5	612.5	693.4
Langley Research Center	631.9	566.7	645.8
Glenn Research Center	589.8	551.2	599.7
Headquarters	480.4	610.8	473.4
Dryden Flight Research Center	187.8	210.5	217.8
Stennis Space Center	163.2	167.4	168.0

Source: NASA. Prepared by CRS. Notes: For comparison purposes, NASA Headquarters funding is included in this list. NASA facilities affiliated with a center are funded within that center's budget.

---

[54] Federally Funded Research and Development Centers (FFRDs) were established during World War IIto meet defense research and development (R & D) needs that were not then readily able to be met by the military services or the private sector. Today, there are approximately 40 FFRDCs engaged in defense, energy, aviation, space, health and human services, and tax administration research and development (R&D) activities. All are organized as independent, not-for-profit entities with limitations and restrictions on their activities, such as being prohibited from manufacturing products or competing with industry. The centers are sponsored by government agencies, and privately administered by universities and other not-for-profit organizations. For more information, see: U.S. Congress, Office of Technology Assessment (OTA), *A History of the Department of Defense Federally Funded Research and Development Centers*, OTA-BP-ISS-157 (Washington: June 1995) [http://www.wws.Princeton.edu/cgi-bin/byteserv.prl/~ota/ disk1./1995/9501.PDF]; or U.S. National Science Foundation (NSF), Research and Development Statistics Program, Division of Science Resources Studies, "Annotated List of Federally Funded Research and Development Centers,: [http://www.nsf.gov/sbe/srs/anno96/start.htm].

Figure 2. Map of NASA Centers

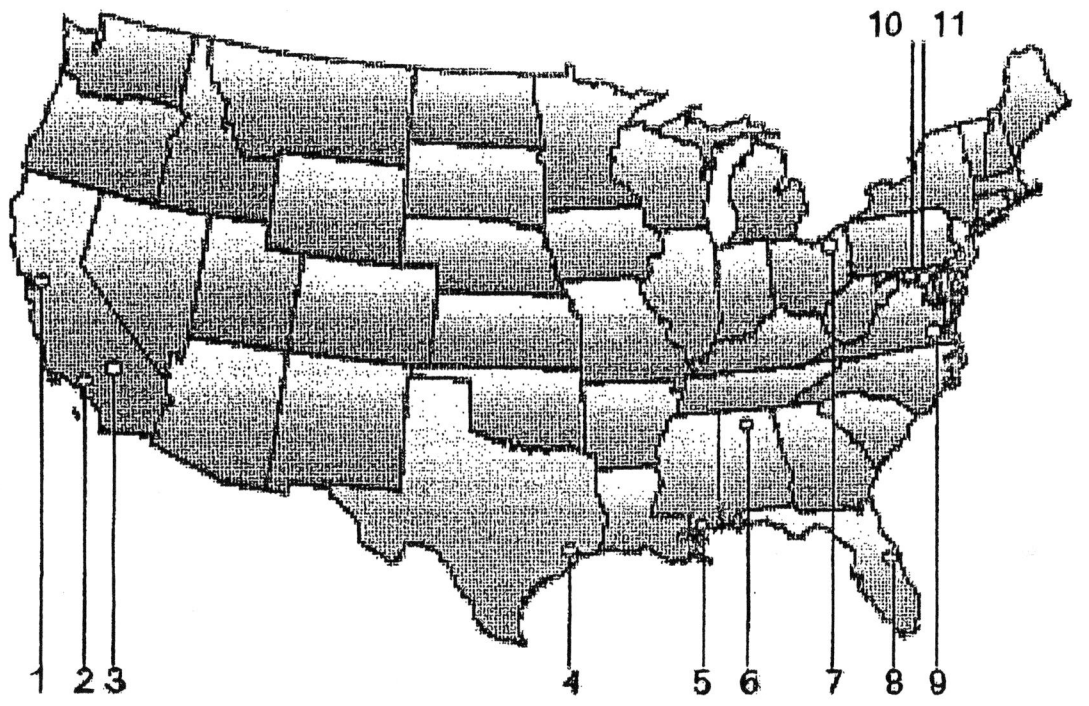

1. Ames Research Center
2. Jet Propulsion Laboratory
3. Dryden Flight Research Center
4. Johnson Space Center
5. Stennis Space Center
6. Marshall Space Flight Center
7. Glenn Research Center
8. Kennedy Space Center
9. Langley Research Center
10. NASA Headquarters
11. Goddard Space Flight Center

Source: NASA.

As part of activities in the 1990s to reduce overlap and streamline administrative and program functions across the agency, senior management established areas of excellence and specific missions for NASA Headquarters and each NASA center. Each "center of excellence" (Table 4) represents a focused agency-wide capability in an area of technical competence in which a center is in a position of preeminence within the agency. In step with transferring program management responsibilities from headquarters to the centers, each center has an identified primary mission for managing NASA programs.

Table 4. The Roles and Missions of the NASA Field Centers

Center/Facility	Center of Excellence	Primary Mission
Ames Research Center	Information Technology	Aviation Operation Systems, Astrobiology
Dryden Flight Research Center	Atmospheric Flight Operations	Astronautical Flight Research
Glenn Research Center at Lewis Field	Turbomachinery	Aeropropulsion
Goddard Space Flight Center	Scientific Research	Earth Science, Physics, Astronomy
Goddard Institute for Space Studies	------	Earth Science
Independent Validation and Verification Facility	Sophisticated Software Systems	----
Jet Propulsion Laboratory	Deep Space Systems	Planetary Science and Exploration
Johnson Space Center	Human Operations in Space	Human Exploration, Astromaterials
Kennedy Space Center	Launch and Cargo Processing Systems	Space Launch
Langley Research Center	Structures and Materials	Airframe Systems, Atmospheric Science
Marshall Space Flight Center	Space Propulsion	Transportation Systems Development, Microgravity Research
Moffett Federal Airfield	Shared Federal Facility	~
Stennis Space Center	Propulsion Testing Systems	Propulsion Testing
Wallops Flight Facility	Suborbital Research Programs	~
White Sands Test Facility	Testing and Evaluating Hazardous Materials, Components, and Rocket Propulsion Systems	~

Source: NASA. Prepared by CRS. Notes: ~ These facilities are affiliated with another center. As such, their work lies within that center's primary mission and "center of excellence" designations.

Ames Research Center, Moffett Field, CA.
Located in the heart of "Silicon Valley," Ames Research Center (ARC) [http://www.arc.nasa.gov/] serves as NASA's center of excellence for information technology. The center conducts research in the fields of supercomputing and networking, high-assurance software development, verification and validation, automated reasoning, planning and scheduling, and human factors. Ames also acts as the lead aeronautics center for aviation operations systems, concentrating its research on air traffic control and human factors. Also in aviation, ARC directs the agency's research efforts in rotocraft technology

and has major responsibility for the creation of design and development process tools, wind tunnel testing, and simulation. For space exploration, Ames manages the agency's astrobiology program. In this capacity, center officials develop science and technology requirements for current and future flight missions relevant to astrobiology, including identifying and developing astrobiology mission opportunities, life sciences experiments for space flight, and space science research components of astrobiology.

Dryden Flight Research Center, Edwards, CA.

Dryden Flight Research Center (DFRC) [http://www.dfrc.nasa.gov/Dryden.html] is NASA's primary installation for flight research, and serves as the center of excellence for atmospheric flight operations. DFRC conducts airborne science and flight operations missions and aids in the development of piloted and uninhabited aircraft testbeds for research and science missions. Dryden supports the development and operations of the Space Shuttle and future access-to-space vehicles. In this capacity, the center functions s the primary backup landing site for the space Shuttle and provides Shuttle servicing support, including mating the Shuttle with its carrier aircraft.

Glenn Research Center at Lewis Field, Cleveland, OH.

The primary mission of the Glenn Research Center (GRC) [http://www.grc.nasa.gov/] is to define and develop propulsion, space electrical power, and communications technologies for NASA's aeronautics and space missions. The center leads the agency's aeropropulsion research and serves as the center of excellence in turbomachinery. In aeropropulsion, Glenn conducts research and development activities in subsonic, supersonic, hypersonic, general aviation, and high-performance aircraft propulsion systems. In this capacity, the center performs materials, structures, internal fluid mechanics, instrumentation and controls, interdisciplinary technologies, and aircraft icing research. GRC also develops turbomachinery technology for future aerospace programs. Glenn leads NASA research in the microgravity science disciplines of fluid physics, combustion science, and some material science, and many Shuttle and space station science missions have an experiment managed by Glenn. The center also designs power and propulsion systems for space flight systems and leads the agency's space communications program.

Goddard Institute for Space Studies, New York, NY.

The NASA Goddard Institute for Space Studies (GISS) [http://www.giss.nasa.gov/] is affiliated with Goddard Space Flight Center. The institute conducts research emphasizing a broad study of global change, addressing natural and human-induced changes in the Earth's environment, occurring on various time scales, and affecting the habitability of the planet. Program areas include global climate modeling, Earth observations, climate impacts, planetary atmospheres, paleoclimate, radiation, atmospheric chemistry, and astrophysics and other disciplines. GISS works cooperatively with area universities and research organizations, including Columbia University.

Goddard Space Flight Center, Greenbelt, MD.

Goddard Space Flight Center (GDFC) [http://www.gsfc.nasa.gov/] is the lead center for NASA's Earth science programs and serves as the agency's center of excellence for scientific research. GSFC develops and operates Earth orbital flight experiments and spacecraft that conducts scientific investigations in our solar system and the universe, including directing tracking and data acquisition activities. In this capacity, Goddard operates the Hubble Space Telescope (HST),manages the Earth Observing System (EOS) (the centerpiece of the Earth science program), and its associated data system, the EOS Data Information System (EOSDIS). The center also operates an instrumented flight range for aeronautical and space research, and manages the procurement of expendable launch for small and medium payloads.

Independent Validation and Verification Facility, Fairmont, WV

The Independent Validation and Verification (IV & V) Facility [http://www.ivv.nasa.gov/] serves as the agency's center of excellence for sophisticated software systems, and is located in the heart of West Virginia's technology sector. The facility was established in 1993, following the *Challenger* accident, as part of an agency-wide strategy to increase the safety and cost-effectiveness of mission critical software. As indicated by its name, the facility provides independent verification and validation for NASA systems and software. The IV & V is affiliated with Ames Research Center, and conducts research in systems and software engineering, measurement, and independent assessment, and manages the agency's technology transfer programs.

Jet Propulsion Laboratory, Pasadena, CA.

The Jet Propulsion Laboratory (JPL) [http://www.jpl.nasa.gov/] is the only NASA center that is a government-owned, contractor-operated facility. As a federally-funded research and development center (FFRDC), JPL is operated under contract to NASA by the California Institute of Technology. The lab is NASA's lead U.S. center and center of excellence for robotic exploration of the solar system and deep space, space science research and analysis. In this capacity, the center manages the Deep Space Network (DSN) and spacecraft tracking and data acquisition for the agency. JPL also conducts research in the development of advanced spacecraft technologies, including propulsion, power, structures, guidance and control, thermal control, and electronics. In addition to its work for NASA, JPL conducts tasks for a variety of other federal agencies.

Johnson Space Center, Houston, TX.

Johnson Space Center's (JSC) [http://www.jsc.nasa.gov/] primary mission areas are human exploration and astromaterials. The center serves as the center of excellence for human operations in space. JSC is responsible for the selection and training of astronauts and is the lead center for the Space Shuttle program. Its main shuttle responsibilities also include flight operations, mission planning, operational procedures, and flight control (Mission Control Center is located at JSC). JSC also is the lead center for development of the International Space Station.

Kennedy Space Center, Cape Canaveral, FL.

Kennedy Space Center (KSC) [http://ww.ksc.nasa.gov/] is NASA's center of excellence for launch and cargo processing systems. The center serves as the starting point for all U.S. human space flights, including the checkout, launch, and landing of the Space Shuttle and its payloads. KSC personnel also provide management and technical expertise for the launch of NASA robotic missions from the adjacent Cape Canaveral Air Station and Vandenberg Air

Force Base in California. In addition, the center supports Space Station operational launch-readiness planning.

Langley Research Center, Hampton, VA.

Langley Research Center's (LaRC) [http://www.larc.nasa.gov/] primary mission assignments are airframe systems and atmospheric sciences. Langley also serves as the agency's center of excellence for structures and materials research, as well as the agency's focal point for wind tunnels and test facilities. Seventy percent of Langley's activities are in aeronautics research, including the development of technology for advanced space transportation systems and for small spacecraft and instruments. The center's research includes systems analysis, integration, and assessment; aerodynamics and aerothermodynamics; hypersonic propulsion; structures and materials; atmospheric sciences and remote sensing; and airborne systems, including crew station design and integration.

Marshall Space Flight Center, Huntsville, AL.

Marshall Space Flight Center's (MSFC) [http://www.msfc.nasa.gov/] primary mission areas are transportation systems development, microgravity research, and space product development. The center serves as the agency's center of excellence for space propulsion, and manages all Space Shuttle propulsion elements, including the Shuttle main engine, solid rocket booster, and external tank projects. The center also supports the development of advanced space transportation technologies, and manages the "X" series of launch vehicles that are aimed at testing new technologies in flight. Marshall conducts and develops microgravity experiments in materials processing. The center manages the Space Station's pressurized module development and permanently inhabited capability. The Destiny laboratory module was manufactured at Marshall facilities. The Space optic center at Marshall develops advanced optics manufacturing technologies for space observatories, such as the Chandra X-ray Observatory. The center's Earth science programs study hurricane and tornado formation, mitigation of the urban heat island (extreme heat zones), and the use of remote sensing technology to aid farm productivity and identify outbreaks of disease.

Michoud Assembly Facility, New Orleans, LA.

The Michoud Assembly Facility [http://mix.msfc.nasa.gov/MIXR/ABSTRACTS/MSFC-00403.html] has been managed by Marshall Space Flight Center since 1961. In the 1960's the facility was used for the design and assembly of Saturn launch vehicles. Since the end of the Apollo program, the facility's primary mission has been the manufacture and assembly of the Space Shuttle's External Tanks.

Moffett Federal Airfield, Mountain View, CA.

Moffett Federal Airfield [http://george.arc.nasa.gov/jf/mfa/] is a federal joint-use facility under the control of NASA Ames Research Center. As a former military base, it is a fully functional federal airport with all facilities necessary for aircraft operations including two parallel runways. Located at Moffett is the California Air & Space Center (CASC), a nonprofit educational and entertainment resource supported by Ames in partnership with the cities of Mountain View and Sunnyvale.

Stennis Space Center, Bay St. Louis, MS.

Stennis Space Center (SSC) [http://www.ssc.nasa.gov/] is NASA's primary center for testing and flight certifying rocket propulsion systems, and serves as the agency's center of excellence for rocket propulsion testing. In this capacity, center officials conduct or manage all NASA propulsion test programs, including testing the Shuttle main engines. Stennis Space Center is also the lead center for commercial remote sensing within the Earth Science Enterprise. As such, SSC works to assist companies involved in environmental consulting, land use planning and natural resource management. In addition Stennis scientists conduct research in biological, chemical, geological and physical processes, as well as humanity's influence on these processes.

Wallops Flight Facility, Wallops Island, VA.

Wallops Flight Facility (WFF) [http://www.wff.nasa.gov/] is an operational element and component installation of GSFC, and serves as the center of excellence for suborbital research programs. The facility launches sounding rockets and balloons, Small Shuttle Payload Projects, and University Class Explorers (UNEX). Wallops also manages and supports scientific aircraft that contain Earth-sensing instruments. The facility provides flight opportunities to aviation, space launch, and communications industries at its research airport and launch range. In addition, center officials track and communicate with the Space Shuttle and the Russian space station Mir and provide data to Mission Control using the facility's communications, telemetry, and radar instrumentation. In the future, Wallops will provide the same tracking and communications support to the Internal Space Station.

White Sands Test Facility, White Sands, NM.

The White Sands Test Facility (WSTF) [http://www.wstf.nasa.gov/] is affiliated with Johnson Space Center, and serves as the agency's center of excellent for testing and evaluating hazardous materials, components, and rocket propulsion systems. The facility's isolated location, expansive area (94 square miles), and favorable environmental conditions allow it to conduct large-scale and hazardous tests. In this capacity, White Sands develops and tests procedures, components, and systems used in the Shuttle program. WSTF also operates the White Sands Space Harbor, which provides approach and landing training for Space Shuttle astronauts and serves as an alternate Shuttle landing site.

PART II. ISSUES

Chapter 3

NASA'S EARTH SCIENCE ENTERPRISE

Erin C. Hatch

ABSTRACT

Since its creation, the National Aeronautics and Space Administration (NASA) has been studying the Earth and its changing environment by observing the atmosphere, oceans, land, ice and snow, an their influence on climate and weather. In 1991, NASA began a comprehensive program to study the Earth as an environmental system, called Mission to Planet Earth. Now known as the Earth Science Enterprise (ESE), it is the largest federal programs studying the Earth and its environment. Those goal is to increase understanding of Earth's natural processes and their interactions with humans. It seeks to provide scientific information to help policymakers and scientists formulate strategies to mitigate human impacts on Earth's environment, such as ozone depletion and deforestation, and impacts on the human environment of natural phenomena like volcanic eruptions and hurricanes. This chapter provides an overview of ESE programs, including the Earth Observation System and its data distribution system. It also discusses issues of interest to Congress, such as ESE program delays, and NASA's purchase of commercial remote sensing data.

EARTH OBSERVATION PROGRAMS

Earth Science Enterprise (ESE)[1] projects use a combination of space-based, airborne, and ground-based instruments to acquire long-term data on the global Earth system. These data primarily aid scientific understanding of environmental issues, and particularly global climate change. ESE programs also seek to enhance predictive capabilities about potential global environmental hazards, while also developing remote sensing[2] technologies that might reduce

---

[1] NASA consists of four enterprises: Earth Science, Space Science, Human Exploration and Development, and Aero-Space Technology. For more information, see CRS RL30577: *The National Aeronautics and Space Administration (NASA): History and Organization.*

[2] "Remote sensing" is the observation of the Earth from distant vantage points, usually by or from satellites or aircraft. Sensors, such as cameras, mounted on these platforms capture detailed pictures of the Earth that reveal features not always apparent to the naked eye.

the cost and increase the reliability of future missions. At the same time, ESE operates an information management system to capture, process, archive, and distribute data to scientists, policymakers, and the public. The agency is also a significant contributor to the U.S. Global Change Research Program (USGCRP), the International Geosphere-Biosphere Program, and the World Climate Research Program.[3]

Earth Observation System

The centerpiece of the enterprise's array of instruments is the Earth Observing System (EOS) spacecraft series. As originally conceived, EOS was intended to include three series of satellites collecting data over 15 years. However, in response to congressional direction, external reviews, and a recognition of the need for greater flexibility, NASA restructured the program. The current system is composed of large, multi-instrument satellites, some developed in cooperation with international partners, that will measure 24 parameters needed to understand global climate change. This first phase of EOS satellite launches began in 1997, and they will continue through 2003.[4] The ESE is in the process of developing a science implementation plan to drive the selection of follow-on missions. Current plans call for a suite of smaller spacecraft launched between 2003 and 2010. Agency officials have indicated that the National Polar-orbiting Operational Environmental Satellite System (NPOESS)- joint venture between NASA, the National Oceanic and Atmospheric Administration (NOAA), and the Department of Defense (DoD)-will likely be key among these follow-on missions.[5]

EOS Data Information System

To process data into useful information, NASA created the EOS Distributed information System (EOSDIS). This system manages data from past and current Earth science research satellites and field measurement programs, and provides data archiving, distribution, and information management services. With the launch of the Tropical Rainfall Measuring Mission (TRMM) in 1997, the system also began to command and control Earth science satellites and instruments. NASA plans to use EOSDIS to process and archive nearly all Earth

---

[3] For more information about issues surrounding global climate change, see the CRS *Global Climate Change Briefing Book:* [http://www.congress.gov/brbk/html/ebgcc1.html].
[4] Launched spacecraft in the first EOS series are: (1) Sea-viewing Wide Field-of-view Sensor (SeaWiFs), launched in August 1997); (2) TRMM (November 1997); (3) Landsat-7 (April 1999); (4) Quick Scatterometer (QuickScat, June 1999); (5) Terra (formerly EOS-AM-1, December 1999); and (6) Active Cavity Radiometer Irradiance Monitors (ACRIMSAT, December 1999). Yet-to-be launched planned for December 2000); (2) Jason-1 (February 2991); (3) Aqua (formerly EOS-PM, May 2001); (4) ICESat (formerly ALT-Laser, October 2001); (5) Advanced Earth Observing Satellite II (ADEOS-II, November 2001); (6) Solar Radiation and Climate Experiment (SORCE, July 2002); (7) SAGE III aboard the International Space Station (March 2003); and (8) Aura (formerly Chem-1, June 2003). For updates on missions and launches, see [http://eospsp.gfc.nasa.gov/eos_homepage/missions.html].
[5] In response to a 1994 presidential decision directive, NOAA, DoD, and NASA are merging NOAA's Polar-orbiting Operational Environmental Satellite (POES) program and the DoD's Defense Meterological Satellite Program (DMSP). The merged NPOESS program is managed by an Integrated Program Office (IPO), headed by NOAA. NASA is providing remote sensing and spacecraft technologies to improve NPOESS capabilities. See: [http://www.ipo.noaa.gov/].

science data, making it one of the largest civilian information systems ever developed.[6] The wide variety and utility of these data, and the enormous potential user community, have proven technically troublesome to the creation and distribution of EOSDIS data products. An initial version of EOSDIS (Version 0) became operational in 1994, and was implemented through cooperation with NOAA, the U.S. Geological Survey (USGS), and partnering international space agencies. Technical difficulties delayed the deployment of Versions 1 and 2, though the agency reports that both are now performing successfully. NASA plans to deploy the next phase (Version 3) in December 2000, and the last phase (Version 4) by April 2001.[7] The agency now characterizes EOSDIS as evolutionary, including the phased deployment of the EOS satellites and their enabling data transmission technology. Given upcoming EOS launches, NASA expects that the next few year will be very important for the information system, especially due to expected increases in the volume of archived climate data, and in user demand for timely delivery of data products.

Earth Probes

Complementing EOS is the Earth Probes program, which NASA defines as unique, specific, and highly-focused missions, including opportunities presented by international cooperative efforts, small satellites, and advanced technologies. Earth Probes investigate processes requiring special orbits or short development cycles of one to three years. Within this program are small and rapidly-developed Earth System Science Pathfinder (ESSP) missions. There are currently for ESSP missions, two of which are scheduled for launch in 2001 and 2002.[8] The Experiments of Opportunity projects include short duration flights of instruments on the Space Shuttle and other platforms. The University Earth System Science (UnESS) program is characterized by NASA as spaceborne investigations of modest science scope, led by U.S. university investigators with significant student involvement. Also contained within the Earth Probes program are various specialized Earth-observing projects. For example, the Total Ozone Mapping Spectrometer (TOMS) instruments measure long-term changes in total ozone, and verify chemical models of the stratosphere, so as to predict future trends.[9] Anther, the Triana spacecraft, will transmit a continuous Sun-lit full-disk image of the Earth.[10] The Shuttle Radar Topography Mission (SRTM) is an Earth-mapping

---

[6] EOSDIS archives approximately 368 terabytes of Earth science data from past and current NASA missions. According to NASA, these data would fill two columns of stacked CDs as high as the Empire State Building (566,000 CDs). The data yielded each day by the Terra satellite alone (190 gigabytes per day) is equal to the audio and visual data on 40 DVD-ROM movies.

[7] Version 0 is a working prototype with some operational elements, including providing direct access to existing and pre-EOS data. Version 1 is providing support for science data processing, archival, and management for the TRMM spacecraft. Version 2 is providing similar data support services plus flight operations support for Terra and Landsat-7. Version 3 will provide both science processing and flight operations support for Aqua and IceSat. Version 4 will provide the same for Aura. Also see the EOSDIS website:[http://spsosun.gsfc.nasa.gov/New_EOSDIS.html].

[8] The ESSP missons are: Vegetation Canopy LIDAR (VCL, planned launch in May 2002); Gravity Recovery and Climate Experiment (GRACE, November 2001); Pathfinder for Instruments for Cloud and Aerospace Spaceborne Observations-*Climatologie Etendue des Nuages et des Aersols* (PICASSO-CENA), March 2003); and ClodSat (March 2003). See the ESSP website for more information: [http://essp.gsfc.nasa.gov/].

[9] For more information, see the TOMS website: [http://jwocky.gsfc.nasa.gov/].

[10] Triana has been the subject of some congressional controversy. For more information, see CRS RS20252: *NASA's Triana Spacecraft: An Overview of Congressional Issues.*

mission that was conducted aboard the Space Shuttle in February 2000, supported jointly by NASA, the National Imagery and mapping Agency (NIMA), and international partners.

Scientific Research

The largest group of primary users of NASA's Earth science data is the scientific community. Using research and analysis (R & A) funding, the enterprise supports U.S research tasks in universities and the private sector, and cooperates with scientists from other nations. These projects include Earth system models, laboratory and field experiment, airborne observation projects, and instruments. Many R & A projects also help generate and test data application software for eventual use by universities, commercial firms, and state and local governments,

Technology Development

In order to test advanced technologies and mission concepts for future spacecraft, NASA's Earth Science and Space Science Enterprises developed the New Millennium Program (NMP). Through a series of deep space and Earth orbiting fights, NMP seeks to test promising but risky technologies that might be use din future space missions. The program is coordinated by the Space Science Enterprise, but incorporates technology for Earth-observing spacecraft. MNP's Earth Observer 1 (EO-1) will test land-imaging technologies that might reduce costs for a Landsat follow-on mission.[11] NASA currently plans to launch EO-1 in October 2000.

Education

The ESE supports programs that seek to increase public understanding of Earth system science through formal and informal educational opportunities. Teacher-training programs include in-service, workshops that incorporate Earth system science content into courses in science, mathematics, engineering, and technology. Programs for K-12 students included brief courses, summer workshops, and hands-on science experiences exposing students to Earth system science subjects and processes. Other activities range from a graduate student Earth System Science Fellowship program, to faculty member participation in the New Investigator Program, to the classroom-based Global Learning and Observations to Benefit the Environment (GLOBE)[12] program.

---

[11] The first Landsats were launched between 1972 and 1978. During that time, a second generation of Landsat satellite was developed. Landsat 4 launched in July 1982 and Landsat 5 in March 1984. Landsat 5 is still transmitting images, and is managed by a commercial firm. Landsat-6 was launched in October 1993, but the spacecraft failed to reach orbit. Landsat 7, operated and managed by USGS, was developed by NASA and launched in April 1999. NASA is in the process of investigation options for a Landsat follow-on mission. Also see the Landsat website: http://ltpwww.gsfc.nasa.gov/LNADSAT/CAMPAIGN_DOCS/<MAIN/Project.html].

[12] The GLOBE program is an interagency activity led by NOAA, in which NASA has a key role. For more information about GLOBE, see [http://www.globe.gov/]. For information about other ESE education programs, see [http://earth.nasa.gov/education/index.html].

Applications and Commercialization

NASA maintains that scientific knowledge of Earth processes, and space-based and airborne observing techniques and facilities will result in applications beneficial to the public. Such applications may include predicting environmental changes in fisheries, agriculture, and water resources; forecasting weather over an extended range of weeks; and predicting seasonal or longer-range climate, global air quality, and natural hazards variations. A significant part of this effort is the ESE's Commercial Remote Sensing Program (CRSP), which supports cooperative efforts with industry, universities, and state and local governments in developing the commercial remote sensing industry. Another significant effort is the Earth Science Applications Research Program (ESARP). The ESARP applies enterprise data and technologies to regional problems and assessments in partnership with the Department of Agriculture, the Department of Transportation, and the Environmental Protection Agency.

ISSUE FOR CONGRESS

Issues have arisen over the last several years primarily from continued delays and increased costs, the controversial nature of some of the subjects being studied (e.g. global climate change), and the relationship between NASA's remote sensing programs and those of the private sector. These questions and controversies has led to the redesign and restructuring of ESE programs on several occasions, sometimes resulting in additional delays and increased costs. The issues in combination have provoked some policymakers to question the value of the enterprise as a whole. The rest of this chapter discusses issues that are or are likely to be of current interest to Congress.

EOS Delays

In 1999, the first two EOS spacecraft, Landsat-7 and Terra, were launched. However, NASA recently announced launch delays for the next two satellites. Aqua, originally scheduled to launch in December 2000, was postponed to May 2001. Aura, originally scheduled for December 2002, is now scheduled to launch in June 2003. NASA attributes both delays to systems integration difficulties. Such launch delays have often occurred during the development of high-risk, complex spacecraft. However, some policymakers and scientists have expressed concern that such delays could cause difficulties in continuity of data, since several current orbiting Earth-observing spacecraft-that new spacecraft will replace-are already operating well beyond their design lifetime. NASA officials contend that data continuity will not be harmed.

EOSDIS Deployment

Some EOS program delays are attributes to difficulties in developing data management and satellite control software for the EOSDIS program. NASA has already spent more than $2.1 billion on the system, but has had significant difficulty in completing it, and has been forced to scale back the program more than once from its original design.[13] NASA officials now assert that the new EOSDIS timeline is both incremental and realistic. The agency plans for EOSDIS to be fully operational by the end of 2005. NASA officials project the remaining development costs for EOSDIS, from FY2001 through complete developing in 2005, to be $270,million.

Purchase of Commercial Remote Sensing Data

Until 1995, the only providers of remote sensing imagery were government-supported entities. Since then, a handful of commercial players has entered the market. For several years, however, forecasters have predicted a dramatic increase in the size of the international remote sensing industry. While these predicted growth rates have not materialized, another recent forecast has again predicted that the industry will grow dramatically in the next decade.[14]

In order to further growth in this emerging industry, Congress has directed NASA to purchase commercial remote sensing data. In 1996, Congress required the agency to begin an Earth science commercial data purchase program, and included $50 million in the agency's FY1997 appropriation for the purchases. The *Commercial Space Act of 1998* (P.L. 105-303) directed NASA to acquire remote sensing data, services, distribution, and applications from commercial providers to the maximum extent possible. In NASA's FY2001 spending bill, the House Appropriations Committee encouraged the agency to obtain global wind profile data through commercial sources (H.Rept. 106-674). In March 1999, NASA's Inspector General (IG) reported that the Earth science data buy initiatives have been successful in developing the remote sensing industry.[15] However, the IG also contended that NASA has been unable to fulfill its goal of reducing its costs of remote sensing science and technology programs through competition within the industry.

Integrated Science Plan

In 1999, the ESE was criticized by the National Research Council (NRC) for the lack of a "fully integrated science plan" for missions following the first EOS series.[16] Specifically, the NRC recommended that: "NASA develop the science plan with the participation of USGCRP agencies and the academic scientific community, and in consultation with international partners." As a result, the ESE developed a new targeted research program-including specific science questions0-for missions in 2003 and beyond, and submitted it to the NRC for review. THE NRC recently completed its review of the research strategy, and concluded that it is

---

[13] IN 1999, NASA announced another reduction in EOSDIS archiving abilities. See: Paula Shaki, "NASA Reduces Data Network's Archiving Ability," *Space News,* February 1, 1999, p.3.

[14] Teal Group Corporation," Teal Group Forecasts 43 New Commercial Imaging Satellites…To Be Built and Launched During 2001-2010," August 23, 2000, http://www.tealgroup.corm/.

[15] NASA IG, "Commercial Remote Sensing Program Office," March 1999, [http://www.hq.nasa.gov/office/oig/hq/fy99aud.html].

[16] NRC, Task Group on Assessment of NASA Plans for Post-2002 Earth Observing Missions, "NASA's Plans for Post-2002 Earth Observing Missions," April 1999, [http://www.nas.edu/ssb/post2000menu.htm].

useful, provides a firm foundation, and is worth pursuing.[17] However, the NRC questioned the ESE's prioritization methodology, and recommended either expanding the strategy to include other research areas, or being more clear about its prioritization criteria. NASA has not yet responded to those recommendations.

Global Climate Change

EOS has come under fire by some policymakers who have asserted that the program is based more on political than scientific goals.[18] Many Members of Congress support climate change research efforts, including ongoing work in several U.S. agencies. However, some have questioned whether EOS data could be used to verify violations of international environmental treaties. One of these treaties, the Kyoto Protocol, would legally obligate signatory nations to reduce greenhouse gas emissions. The United States has signed the treaty, but the Administration has not submitted it to the Senate for ratification. During the 104th Congress, the Senate declared (S.Res. 98) that it would not approve the treaty unless developing countries also participate in emissions limitations.

---

[17] NRC, Committee to Review NASA's ESE Science Plan, "Review of NASA's Earth Science Enterprise Research Strategy for 2000-2010," August 2000, [http://www.nas.edu/ssb/esemenu.htm].

[18] Joseph C. Anselmo, "Science Boon Or Data Dump?" *Aviation Week & Space Technology,* October 4, 1999, p. 59. Also see note 3.

Chapter 4

SPACE LAUNCH VEHICLES: GOVERNMENT ACTIVITIES, COMMERCIAL COMPETITION, AND SATELLITE EXPORTS

Marcia S. Smith

ABSTRACT

Launching satellites into orbit, once the exclusive domain of the U.S and Soviet governments, today is an industry in which companies in the United States, Europe, China, Russia, Ukraine, Japan, and India compete. In the United States, the National Aeronautics and Space Administration (NASA) continues to be responsible for launches of its space shuttle, and the Air Force has responsibility for launches associated with U.S. military and intelligence satellites, but all other launches are conducted by private sector companies. Since the early 1980s, Congress and successive Administrations have taken actions, including passage of several laws, to facilitate the U.S. commercial space launch services business. The Federal Aviation Administration (FAA0 regulates the industry

During the mid-1900s, demand for launching commercial communications satellites wad forecast to grow significantly through the early 21st Century. Those forecasts sparked plans to develop new launch vehicles here and abroad. In the United States, NASA and the department of Defense (DOD) created government-industry partnerships to develop reusable launch vehicles (RLVs) and "evolved" expendable launch vehicles (ELVS), respectively. The U.S. space shuttle is the only operational RLV today. All other operational launch vehicles are expendable (i.e., they can only be used once). Some U.S. private sector companies began direct government financial involvement, although some have sought government loan guarantees or tax incentives. Legislation to proved such assistance did not pass the 106th Congress, however.

In 1999 and 2000, projections for launch services demand decreased dramatically. At the same time, NASA's main RLV program, X-33, suffered delays and on March 1, 2001 NASA decided to end the program. Companies developing new launch vehicles are reassessing their plans, and NASA has initiated a new "Space Launch Initiative" to broaden the choices from which it can choose a new RLV design in 2005. Separately, DOD has restructured its EELV program in light of the lower market forecasts.

Until a replacement is developed, NASA will rely upon the space shuttle for launching humans into space, including to the international Space Station. Safe operation of the shuttle remains a top NASA concern.

In the commercial launch services market, U.S. companies are concerned about foreign competition, particularly with countries that have non-market economies such as China, Russia, and Ukraine. The U.S. signed bilateral trade agreements with each of those countries setting forth the conditions under which they can participate in the market, including quotas on how many launches they can conduct. The agreement with China expires in 2001; discussions have begun between the two countries on whether to renew it. The Clinton Administration ended quotas for both Ukraine and Russia in 2000. The U.S. has leverage because almost all satellites that require launch are made in the United States or have U.S. components and hence require U.S. export licenses. Export of U.S.-built satellites ahs become an issue in terms of whether U.S. satellite manufacturing companies provide military significant information to those countries in the course of the satellite launches.

MOST RECENT DEVELOPMENTS

The FY2002 NASA budget request includes $3.3 billion for the space shuttle program, an increase of $145 million over FY2001. For the "2nd generation RLV" program, the request is $475 million for FY2002, an increase over the $290 million it received in FY2001,byut less than the $610 million that has been anticipated. NASA has canceled two programs that were within the FY2001 SLI description (the X-34, and phase 2 of the Crew Return Vehicle) and reductions were made in "NASA Unique System Elements" and "Alternative Access."

Detailed FY2002 DOD budget figures are not yet available.

U.S. LAUNCH VEHICLE POLICY

The National Aeronautics and Space Administration (NASA) and the Department of Defense (DOD) have each developed expendable launch vehicles (ELVs) to satisfy their requirements. NASA also developed the partially reusable space shuttle. DOD developed the Atlas, Delta, and Titan families of ELVs (called expendable because they can only be used once0 from ballistic missile technology. NASA developed Scout and Saturn, both no longer produced. Atlas and Titan rockets today are built by Lockheed Martin. Delta is built by Boeing. Private companies also have developed ELVs: Pegasus and Taurus (orbital Sciences Corporation), and Athena (Lockheed Martin). Which launch vehicle is used for a particular spacecraft initially depends on the size, weight, and destination of the spacecraft.

From "Shuttle-Only" to "Mixed Fleet"

In 1972, President Nixon approved NASA's plan to create the first reusable launch vehicle, called the space shuttle, and directed that it become the nation's primary launch vehicle, replacing all the ELVs except Scout (later discontinued for unrelated reasons). This would have made NASA and DOD dependent on a single launch vehicle, but the resulting high launch rate was expected to reduce the cost per flight significantly. The shuttle was first launched in 1981, and was declared operational in 1982. The phase-out of the ELVs began, but in 1984 the Air Force successfully argued that it needed a "complementary" ELV as a backup t the shuttle for "assured access to space" and initiated what is now known as the Titan IV program. Production lines for the Delta and Atlas began to close down, and it was expected that only the shuttle, Scouts, and Titan IVs would be in use by the mid-1980s.

Everything changed on January 28, 1986, however, when the space shuttle *Challenger* exploded 73 seconds after launch. The apace shuttle program enjoyed 24 successful missions prior to *Challenger*. Apart from the human tragedy, the *Challenger* accident deeply affected U.S. space launch policy, demonstrating the vulnerability of relying too heavily on a single system. Many military and civilian satellites had been designed specifically to be launched on the shuttle, and could not have been transferred to ELVs even if the ELVs were not already being phased out. The few remaining ELVs had their own problems in 1986. A Titan exploded in April and a Delta failed in May, which also grounded Atlas because of design similarities. As a result of these failures, U.S. policy was significantly revised from primary depend dependence on the shuttle to a "mixed fleet" approach. The country once again has a wide variety of launch vehicles from which to choose. The shuttle is used principally for missions+ that require crew interaction, while ELVs are used for other spacecraft.

President Reagan also decided that commercial payloads could not be flown on the shuttle unless they were "shuttle-unique" (capable of being launched only by the shuttle or requiring crew interaction) or if there were special foreign policy considerations. That action

facilitated the emergence of a U.S. commercial space launch industry whose participants had long argued that they could not compete against government-subsidized shuttle launch prices. The White House and Congress had taken steps beginning in 1983 designation of the Department of Transportation as the agency responsible for facilitating and regulating the commercial space launch sector, and passage of the 1984 Commercial Space Launch Act (P.L. 98-575). But removing the shuttle as a competitor was the major factor in fostering the U.S. launch businesses. Passage of the Commercial Space Launch Act Amendments of 1988 (P.L. 100-657) and the Commercial Space Act of 1998 (P.L. 105-303) also have helped.

Clinton Administration Launch Vehicle Policy

On August 5, 1994, President Clinton released a National Space Transportation Policy that gave DOD lead responsibility for improving ELVs and NASA lead responsibility for upgrading the space shuttle and technology development and demonstration of new reusable launch vehicles. The policy set guidelines for the use of foreign launch systems and components, the use of excess ballistic missile assets for space launch, and encourages an expanded private sector role in space transportation R & D. Unless exempted by the President or his designee, U.S. government payloads must launched by U.S. manufactured launch vehicles. On September 19, 1996, the Clinton Administration released a comprehensive space policy, covering civil, military and commercial space activities.

U.S. LAUNCH VEHICLE PROGRAMS AND ISSUES

NASA's Space Shuttle Program

The space shuttle is a partially reusable launch vehicle (the large, cylindrical external tank is not reused) and is the sole U.S. means for launching humans into orbit. The 1986 *Challenger* accident and occasional shuttle launch delays have led to questions about the reliability of the shuttle system. *Challenger*, however, is the only failure so far in more than 100 launches since 1981. Nonetheless, concerns remain that cuts to the shuttle budget and associated personnel reductions, and NASA's decision to turn off much of the ground operations of the shuttle over to a "single prime contractor," could affect shuttle safety. NASA singed a $7 billion, 6-year Space Flight Operations Contract (SFOC) with United Space Alliance (USA)- a joint venture between Boeing and Lockheed Martin-to serve as single prime contractor on September 26, 1996 with the goal of reducing shuttle operational costs. The total shuttle budget for FY2001 is $3.1 billion; the request for FY2002 is $4.4 billion.

NASA is still deciding what the future holds for the shuttle. The 1994 Clinton policy directed NASA to pursue technology development and demonstration efforts to support a decision by the year 2000 on developing a new reusable launch system to replace a shuttle. This led to the X-33 program (see below), but delays in that program meant a delay in the decision time frame as well. Meanwhile, NASA outlines and began implementing a four-phase "shuttle upgrades" program to improve shuttle reliability, performance, and longevity. Phase I upgrades focused on improving shuttle safety and making the shuttle capable of supporting the space station program. Phase II upgrades also have begun and are designed to combat obsolescence. Phase III and IV upgrades would have been major modifications to the shuttle system costing hundreds of millions to billions of dollars. In its FY2001 budget, NASA slipped from 2000 to 2005 the date by which it will decide whether to continue relying on the space shuttle or on a new vehicle. NASA abandoned the 4-phase upgrade concept and instead added $1.3 billion to what it had planned for shuttle upgrades in FY2001-2005, for a total of $1.86 billion for "safety and supportability" upgrades. Shuttle advocates insist that the four space shuttle orbiters are less than 30% through their useful life and NASA says the shuttle will be one of the competitors in the 2005 decision on what vehicle NASA should use in the future.

The independent Aerospace Safety Advisory Panel (ASAP) oversees safety in NASA human spaceflight programs. For several years, ASAP ahs expressed concern about the loss of critical skills and experience in the shuttle workforce after years of downsizing and the transition to the single prime contractor. In August 2000, GAO released a report (GAO/NSIAD/GGD-00-186) that echoed the ASAP concerns. President Bush's "budget blueprint" calls for NASA to "aggressively pursue Space Shuttle privatization opportunities that improve the Shuttle's safety and operational efficiency" including continued implementation of planned and new privatization efforts through the prime contract (p. 157).

In its most recent report (February 2001), ASAP strongly recommended that NASA acknowledge that the space shuttle will be the primary U.S. launch vehicle to take crews to the International Space Station (ISS) through the station's lifetime and implement a program to ensure the shuttle's safe operation through that time frame.

Future Launch Vehicle Development Programs

Despite hopes that the space shuttle would reduce the cost of reaching orbit, U.S. launch systems remain expensive and less efficient and reliable than desired. Thus, efforts continue to reduce costs for both expendable and reusable U.S. launch systems. DOD and NASA initiated several efforts in the late 1980s and early 1990s to develop a new ELV system (See below), but each was terminated in turn because Congress or the agencies themselves were not convinced that the required investment had sufficient priority. In response to the 1994 Clinton policy, two programs were initiated: DOD's Evolved Expendable Launch Vehicle (EELV) program and NASA's Reusable Launch Vehicle (RLV) program.

DOD's Evolved Expendable Launch Vehicle (EELV) Program

The EELV program is the successor to several failed attempts to begin new ELV programs since 1985. DOD began what is now known as the EELV program in FY1995 (P.L. 103-335) with a $30 million appropriation. EELV was first formally identified in DOD's FY1996 budget. EELV's goal is to reduce launch costs by at least 25% and replace DOD usage of the existing Delta, Atlas, and Titan vehicles.

In 1996, the Air Force selected Lockheed Martin and McDonnell Douglas (later bought by Boeing) for pre-engineering and manufacturing development contracts worth $60 million. Originally, one of those companies would have been selected in 1998 to develop the EELV. Responding to indicators at the time that the commercial space launch market would be larger than expected, DOD announced in November 1997 that it would help fund development of both the Lockheed Martina and the Boeing vehicles (Atlas V and Delta IV, respectively). In October 1998, DOD awarded Boeing $1.88 billion for the Delta IV ($500 million for further development plus $1.38 billion for 19 launches). At the same time, it awarded Lockheed Martin $1.15 billion for the Atlas V ($500 million for further development plus $650 million for 9 launches). The companies are expected to pay the rest of the development costs themselves. The launches are scheduled to take place in 2002-2006. In2000, however, new market forecasts showed a reduction in expected demand, and DOD reevaluated its EELV strategy. It reportedly has renegotiated the contracts with both companies, relieving Lockheed Martin (reportedly at the company's request) of the requirements to build a launch pad at Vandenberg AFB, CA, and shifting two of the launches previously awarded to Lockheed Martin to Boeing instead. Thus Boeing now has 21 launches while Lockheed Martin has seven. DOD also reportedly has agreed to seek federal funds to pay for a test launch of Boeing's Delta IV. The FY2001, DOD authorization and DOD appropriations acts (P.L. 106-398, P.L. 106-259) funded EELV at $333 million for R & D and $283 million for procurement (a reduction of $5 million from the procurement request).

Government-Led Reusable Launch Vehicle (RLV) Programs

The 1994 Clinton policy gave NASA lead responsibility for technology development for a next-generation reusable space transportation system, such as the single-stage-to-orbit (SSTO) concept. NASA initiated the Reusable Launch Vehicle (RLV) program to develop and flight test experimental RLVs to form the basis for next-generation vehicles to replace the space shuttle and replace or augment ELVs. As already discussed, in 2000 NASA significantly changed its approach to developing new RLV technology, delaying from 2000 to 2005 the decision point at which to choose whether to continue to rely on the space shuttle for the indefinite future or expect a new "2nd generation" RLV early in the early 2000s. NASA had focused on developing single stage to orbit (SSTO) technology through the X-33 program (see below) that involves a rocket that can attain orbit with only one stage (instead of two or more as is common today) carrying people or cargo. Now it is reassessing its plans. The goal is to develop a vehicle capable of being launched, returning to Earth, being serviced quickly, and flying again within a very short time. Proponents believe that RLV technology could dramatically lower the cost of accessing space.

"X" Programs

From 1995 to 2000, NASA's approach to developing new RLVs was based on establishing new forms of cooperation with industry by sharing the costs of developing technology with the intent that industry take over development, operation, and financing of the operational vehicle. Two "X" (for "experimental") flight test programs were begun under this philosophy: X-33, a large RLV based on SSTO technology to demonstrate technologies in the Mach 13-15 range (13-15 times the speed of sound) and X-34, a small RLV "testbed" to demonstrate reusable technologies at Mach 8.

In March 2001, NASA announced the termination of X-33 and X-34. X-33 was a cooperative program between NASA and Lockheed Martin. According to the contract signed in 1996, NASA's costs were fixed at $912 million (not including civil service costs, which take NASA's cost to about $1.2 billion). Lockheed Martin says that by the end of the program it had spent $356 million of its own funding on the program. X-33 was a suborbital prototype of a vehicle which, if it has been built, would been called VentureStar. Technical problems with the X-33, particularly with its new "aerospike" engines and construction of it composite hydrogen fuel tanks, led to delays in test flights from 2000 to 2003. NASA concluded that the cost it would to complete the program was to high compared to the benefits and terminated its participation in the program. Although Lockheed Martin could continue the program without NASA, it apparently does not intend to do so. X-34 was a "technology testbed" being built under contract to NASA by Orbital Sciences Corporation. The program had begun as a cooperative program like X-33, but the companies (Orbital and Rockwell International) that partnered with NASA decided not to continue it under those terms. NASA later modified the program and signed a traditional contract with Orbital. As with X-33, NASA concluded that the cost to complete the program was too high relative to the value of the technologies to be demonstrated. NASA spent $205 million on X-34. Lockheed Martin and Orbital reportedly are each trying to interest the Air Force in continuing their respective programs.

NASA also has a cooperative agreement with Boeing to build the X-37, a "technology testbed" designed to demonstrate technologies in orbit and during reentry. The government (NASA and the Air Force) and Boeing agreed to split the cost of the $173 million project. Plans are eventually to launch the X-37 into orbit on the space shuttle, release it, and after a period of time on orbit, reenter and land. It is an automated vehicle. Advocates say that the technologies could be used for future reusable launch vehicles. Drop tests of the Air Force's X-40A (part of a program to develop a "military spaceplane," or "Space Maneuver Vehicle"), a related vehicle also being built by Boeing, from helicopters or B-52 aircraft are being used to reduce the risk of developing the X-37. Before being launches into orbit on the shuttle, drop tests of the X-37 from B-52s also are planned.

Space Launch Initiative (SLI)

NASA restructured its program to develop new space launch vehicles in its FY2001 budget. The time frame for choosing between the shuttle and a 2nd generation RLV slipped to 2005. In the meantime, NASA announced it would spend $1.86 billion (FY2001-2005) in "safety and supportability" upgrades to the space shuttle and $4.4 billion (FY2001-2005) in a new Space Launch Initiative (SLI). SLI includes the 2nd generation RLV program, designed to encourage development of a variety of RLV technologies rather than focusing on a single concept such as SSTO as was done with X-33. Thus NASA hopes to have more than one competitor for a new RLV system in 2005. SLI also funds development of "NASA Unique" launch systems, i.e., for taking humans to or from space, which at this time is a uniquely NASA activity; and of "alternative access to the space station" to help private companies compete to service the space station. For Fy2001, NASA requested and received $290 million for SLI. (X-33 and X-34 were submitted as candidates for SLI funding, but NASA did not choose either program and both have been terminated as discussed already.) NASA has not announced the proposals that won SLI funding, pending completion of negotiations with the companies involved. *Space News* reported on March 12, 2001 (p.8), however, that the following companies would receive SLI funding, although details of what work would be performed was not provided: Kistler Aerospace, Andrews Space and Technology, Universal Space Lines, Boeing, Lockheed Martin, Northrop Grumman, and Orbital Sciences Corp.

Private Sector RLV Development Efforts

In addition to the government-led programs, several U.S. Companies have been attempting to develop RLVs through private financing. A January 1998 DOT study summarizes these efforts as well as those of the U.S. government and foreign countries (*Reusable Launch Vehicle Programs and Concepts*, at [http://ast.faa.gov]). The House Science Committee's Space and Aeronautics Subcommittee held a hearing on these efforts on October 13, 1999. Many of the companies have described the difficulties in obtaining financing from the financial markets, and have been seeking government loan guarantees or tax credits. NASA's Space Launch Initiative may also be a source of funding for them.

U.S. COMMERCIAL LAUNCH SERVICES INDUSTRY

Congressional Interest

The 106th Congress debated several issues involving satellite export issues (discussed below) and the domestic launch services industry. Many of these issues are expected to be of interest to the 107th Congress as well. One issue is what the government should do to stimulate development of new launch vehicles by the private sector. In the 106th Congress, S. 469 (Breaux) would have provided at least $550 million in loan guarantees to U.S. companies trying to build low-cost space launch vehicles; the source of the funds is not specified. A hearing was held on the bill May 20, 1999 by a Senate Commerce subcommittee. While two companies supported the bill, others opposed loan guarantees because they allow the government ot choose winners and losers. Tax incentives have been suggested as an alternative, though they also are controversial since they reduce revenues to the government. H.R. 4676 (Cook) would have provided tax incentives to investors in launch companies building low cost launch vehicles. A House Science subcommittee held a hearing on H.R. 4676, S. 469, and other commercial space legislation on July 18, 2000. There was no legislative action on any of these bills.

The House Science Committee held hearings on related issues, including barriers to the commercial space launch industry (June 10, 1999) and modernization and privatization of the nation's space launch sites or "ranges" (March 24 and June 29, 1999, the latter jointly with the House Armed Services Committee; and September 28, 2000). Commercial space launch companies use Air Force launch ranges in Florida and California, raising issues about who should pay for range improvements now that commercial launches outnumber those for the government. The White House released a range management and use plan in February 2000. Several states also have or plan to establish "spaceports" for commercial space launches; spaceports are licensed by the FAA. Bills to make spaceports, like airports, eligible for tax exempt bonds was introduced in the 106th Congress (H.R. 2289 and S. 1239), but there was no action on them. S. 1236 (Graham) would have allowed private sector companies to lease NASA property for the purpose of expanding commercial launch capacity. It also was not acted upon.

Foreign Competition (including Satellite Export Issues)

Europe, China, Russia, Ukraine, India, and Japan offer commercial launch services in competition with U.S. companies. The Department of Transportation estimated that for the period January 1995-June 2000, the United States captured 39% of internationally competed launches (excluding small launch vehicles), Europe 33%, Russia 17%, China 8%, "international" (Sea Launch) 3%, and Ukraine less than 1% (*Quarterly Launch Report, 1st Quarter 2000*, p. 15). Most satellites are manufactured by U.S. companies or include U.S. components and hence require export licenses, giving the United States considerable influence over how other countries participate in the commercial launch services market. The United States negotiated bilateral trade agreements, with China, Russian, and Ukraine on "rules of the road" for participating in the market to ensure they did not offer unfair competition because of their non-market economies. Launch quotas were set in each of the agreements, but President Clinton terminated the quotas for Russia and Ukraine in 2000. They remain in place for China.

Europe
The European Space Agency (ESA) developed the Ariane family of launch vehicles. Ariane was first test-launched in 1979, and began operational launches in 1982. ESA continues to develop new variants of Ariane. Operational launches of Ariane 4 and Ariane 5 are conducted by the French company Arianspace, which is owned by the French space agency (CNES) and European aerospace companies and banks. Arianespace also markets Russia's Soyuz launch vehicle as part of a French-Russian joint venture, Starsem.

In 1985, a U.S. company (Transpace Carriers Inc.) filed an unfair trade practices complaint against Arianespace, asserting that European governments were unfairly subsidizing Adriane. The Office of the U.S. Trade Representative (USTR) investigated and found that Europe was not behaving differently from the United States in pricing commercial launch services (then offered primarily on the government-owned space shuttle). The incident raised questions about what "rules of the road" to follow in pricing launch services. In the fall of 1990, USTR and Europe began talks to establish such rules of the road and asses show to respond to the entry of non-market economies into the launch services business. The only formal negotiating session was held in February 1991.

Each side has been concerned about how much the respective governments subsidize commercial launch operations, but another controversial topic (not formally part of the talks) is whether Arianespace should be able to bid for launches of U.S. government satellites, which not must be launched on U.S. launch vehicles as a matter of U.S. policy. Arianespace wants that restriction lifted. France and other European governments do not have written policies requiring the use of Ariane for their government satellites. However, the member governments of ESA agreed to pay a surcharge of as much as 15-20% if they chose Ariane. The surcharge led some cost-conscious European governments to buy launch services from other (notably U.S.) suppliers. In the fall of 1995, ESA's member governments reached agreement with Arianespace to reduce the surcharge to encourage use of Ariane. (ESA itself does give preference to using Ariane.)

China

The People's Republic of China offers several versions of its Long March launch vehicles commercially. China poses special issues not only because of its non-market economy, but because of technology transfer and political concerns. Launch services are offered through China Great Wall Industry Corp. (CGWIC).

U.S. China Bilateral Trade Agreements for Launch Services

In 1989, China and the United States signed a 6-year bilateral trade agreement restricting the number of Chinese commercial space launches to ensure China, with its nonmarket economy, did not unfairly compete with U.S. companies. A new agreement that was reached in 1995 and amended in 1997 will expire on December 31, 2001. Under the existing agreement, China is allowed to launch up to 20 foreign satellites to geostationary orbit (GEO). There are no numerical limits on the number of launches to low Earth orbit (LEO), but if China, Russia, and Ukraine combined launch more than 50% of any individual LEO system, the United States would consider that cause for concern. GEO launches must be priced on a par with Eastern prices. If the price is within 15%, it will normally be considered consistent with this obligation. Prices more than 15% below will be examined in detail. LEO launches must be priced on a par with Western LEO launch prices. Discussions about whether to renew the agreement began in March 2001.

U.S. Satellite Exports to China: 1998-1997

In September 1988, the U.S. government agreed to grant three export licenses for satellites manufactured by Hughes to be launched by CGWIC. Two were Opetus communications satellites (formerly called AUSSAT) built for Australia and the third was AsiaSat 1, owned by the Hong Kong-based Asiasat Co. (of which China's International Trust and Investment Corp. is one-third owner). The Reagan Administration granted the export licenses on the conditions that China sign three international treaties related to liability for satellite launches and other subjects; agree to price its launch services "on a par" with Western companies and establish a government-to-government level regime for protecting technology from possible misuse or diversion. China met the conditions and the two countries signed a 6-year agreement in January 1989. The now-defunct Coordinating Committee on Multilateral Export Controls (COCOM) approved the licenses that March.

On June 5, 1989, after the Tiananmen Square uprising, President George H. Bush suspended all military exports to China. At the time, exports of communications satellites were governed by the State Department's Munitions List. The Satellites counted as military exports and the licenses were suspended. Then Congress passed language in the FY1990 Commerce, Justice, State and Judiciary appropriations (P.L. 101-162) and the 1990-91 Foreign Relations Authorization Act (P.L. 101-246), Section 902) prohibiting the export of U.S.-built satellites to China unless the President reported to Congress that (1) China has achieved certain political and human rights reforms, or (2) it was in the national interest of the United States. In December 1989, President Bush notified Congress that export of the satellites was in the national interest and the licenses were reinstated. AsiaSat-1 became China's first commercial launch of a U.S.-built satellite in April 19909. Final export approval for Optus 1 and 2 was granted in April 1991. They were launched in 1992.

A different issue arose in 1990. China signed a contract to launch an Arabsat Consortium satellite for $25 million, much less than what many consider "on par" with Western companies. The main competitor was Arianspace, which turned to both the French and U.S. governments to prohibit export of the satellite (the prime contractor was French and it included American components). No formal action was taken by the United States. In 1991, the Arabsat Consortium terminated the contract with the Chinese and signed an agreement with Arianespace, so the case became moot, but the issue of what constituted "on a par" remained. China argued that because its costs are so low, it could offer lower prices and still adhere to international norms as to what costs are included in setting the price. Yet another issues arose in 1991-linkage of satellite export licenses with U.S. concern over China's ballistic missile proliferation policies. On April 30, 1991, the bush Administration approved final export licenses for Optus 1 and 2, and for U.S. components of a Swedish satellite called Freja (launched by China in October 1992). To emphasize its concern about Chinese missile proliferation, however, the White House disapproved export of U.S. components for a satellite China itself was building (Dong Fang Hone 3). Then, on June 16, the White House announced that it would be "inappropriate for the United States to approve any further export licenses for commercial satellite launches at this time." On July 17, the State Department identified CGWIC as one of two Chinese entities engaged in missile technology proliferation activities that require the imposition of trade sanctions in accordance with the Arms Export Control Act, including denial of license applications for export items covered by the Missile Technology Control Regime (MTCR). Although the MTCR does not cover satellites (only satellite launch vehicles, which are close cousins of ballistic missiles), the identification of CGWIC as a cause of concern complicated China's marketing plans. China agreed to adhere to the MTCR, and the sanctions were lifted on February 21, 1992.

China's fortunes improved. In May 1992, the International Telecommunications Satellite Organization (Intelsat) agreed to launch at least one of its satellites on a Chinese launch vehicle. On September 11, 1992, the State Department notified Congress that it was waiving legislative restrictions on U.S. exports for six satellite projects with China: APSAT, AsiaSat-2, Intelsat 7A, STARSAT, AfriStar, and Dong Fang Hong 3. The first five were satellites China wanted to launch; the sixth was for satellite components for which export was disapproved in April 1991. (The satellite was launched in 1994, but failed once it was in orbit). Many observers saw the move as a conciliatory gesture in the wake of the U.S. decision to sell F-16s to Taiwan.

On August 25, 1993, however, the U.S. government again imposed sanctions against China for ballistic missile proliferation activities, and the State Department said that satellite exports would not be permitted. The State Department announced October 4, 1994 it would lift the sanctions after China pledged to abide by the MTCR. During this period, U.S. tensions were acute between those who view the sanctions as harmful to U.S business interests (notably satellite manufacturers Hughes and Lockheed Martin), and those who want to prevent sensitive technology from reaching China and/or to punish China for MTCR infractions. The debate centered on whether the satellites should continue to be governed by export guidelines of the State Department (Munitions List) or the Commerce Department (Commerce Control List). Some responsibility for export of commercial communications satellites was transferred from the State Department to the Commerce Department in 1992, and in October 1996 primary responsibility was transferred to Commerce.

In January 1995, the launch of the Hughes-built APStar-2 satellite failed in-flight. Falling debris killed 6 and injured 23 on the ground. On February 6, 1996, President Clinton approved the export of four satellites to China for launch (2 COSAT satellites, Chinasat 7, and Mabuhay) despite concerns about China exporting nuclear weapons-related equipment to Pakistan. The CoSAT (now Chinastar) satellites are built by Lockheed Martin and the first was successfully launched on May 30, 1998. Chinasat 7 was built by Hughes, and Mabuhay (now Agila 2) by Loral. The 6-year trade agreement was signed on March 13, 1995. On February 14, 1996, a Long March 3B rocket carrying the Intelsat 708 communications satellite built by Loral malfunctioned seconds after lift off impacting the ground and spreading debris and toxic fumes over the launch site and a nearby village. The Chinese reported 6 dead and 57 injured, but other reports suggested a higher figure. After this second Chinese launch failure involving fatalities, some customers, including Intelsat, canceled contracts.

In May 1997, USTR stated that it believed China violated the pricing provisions of the bilateral agreement for the launching of Agila 2 (formerly called Mabuhay) for the Philippines. Chinese officials disagreed. On September 10, 1997, the Washington Times published a story that Chinese and Russian entities (including CGWIC) were selling missile technology to Iran. China denies the allegations.

Satellite Exports to China: 1998-Present (including the "Loral/Hughes" Issue and the Cox Committee Report)

On February 19, 1998, the President notified Congress that it was in the national interest to export Loral's Chinasat 8 to China. On April 4, 1998, the New York Times reported that a 1997 classified DOD report alleged that Space Systems/Loral (part of Loral Space & Communications) and Hughes Electronics (a subsidiary of General Motors) provided technical information to China that improved the reliability of Chinese nuclear missiles. The assistance was provided in the wake of the February 1996 Intelsat 708 launch failure (see above). The Intelsat satellite was built by Loral, which participated and inquiry into the accident at the request of insurance companies seeking assurances that the Chinese had correctly diagnosed and solved the cause of the failure. Loral formed a review committee that included representative of other satellite companies, including Hughes. According to Loral, the review committee did not itself investigate the accident, but listed to Chinese officials explain their investigation and then wrote a report. Loral conceded that a copy of the report was given to the Chinese before it was provided to the State Department, in violation of Loral's internal policies. Loral says it notified the State Department when it learned that the Chinese had been given a cop. According to media sources, DOD's 1997 report says that the companies provided technical information in violation of the export license that allowed the export of the satellite to China for launch. The companies insist they did nothing that violated the export license. The Justice Department investigated the allegations and reportedly expanded the probe to include Hughes' response to the 1995 APStar-2 failure. A grand jury reportedly was empaneled in 1999. On December 11, 2000, the Los Angeles Times reported that the grand jury had adjourned and the government was exploring a plea agreement with the two companies that "would probably include monetary penalties. Loral and Hughes continue to deny any wrongdoing."

Many hearings on the "Loral/Hughes" issue were held by various House and Senate committees. In addition, on June 18, the House established the Select Committee on U.S> National Security and Military/Commercial Concerns with the People's Republic of China chaired by Representative Cox t investigate the issues. The Cox committee unanimously adopted its report on December 30, 1998, but public release was delayed until May 25, 1999 pending preparation of an unclassififed version. The Cox committee concluded that Hughes and Loral deliberately transferred technical information and know-how to China during the course of accident investigations. The committee also investigated other cases of China acquiring technical information from the United States and made 38 recommendations, including that the United States should increase its space launch capacity.

The FY2000 DOD authority Act (P.L. 106-65) included language implementing many of the Cox committee recommendations. In brief, the Department of Justice must notify appropriate congressional committees when it is investigating alleged export violation in connection with commercial satellites or items on the munitions list if the violation is likely to cause significant harm or damage to national security with exceptions to protect national security or ongoing criminal investigations; companies must be provided with timely notice of the status of their export applications; enhanced participation by the intelligence community in export decisions is required; adequate resources must be provided for the offices at DOD and the State Department that approve export licenses; individuals providing security at overseas launch sites do not have to be DOD employees, but must report to a DOD launch monitor; and DOD must promulgate regulations concerning the qualifications and training for DOD space launch monitors and take other actions regarding those monitors and the records they maintain.

In February 1999, the Clinton Administration denied Hughes permission to export two satellites for the Asia Pacific Mobile Telecommunication (APMT) system to China for launch. Export permission for APMT had been granted in 1997 (the President notified Congress on June 25, 1997), but Hughes changed the spacecraft design, necessitating new export approval. That application was denied. On May 10, 2000, the White House made its first certification to Congress under the new process detailed in the FY1999 DOD authorization bill, approving the export to China of satellite fuels and separations stems for the iridium program. On August 18, 2000, the State Department stated it would continue the suspension of a technical assistance agreement for Loral regarding launch of Chinasat 8 because the concerns that initiated the suspension in December 1998 have not been rectified. In January 2001, *Space News* reported that the Chinasat 8 export application was returned to

Loral without action. Loral President Bernard Schwartz was subsequently quoted by *Reuters* (March 29, 2001) as remaining optimistic about ultimately getting approval to export Chinasat 8, however.

Lockheed Martin

In April 2000, it became known that Lockheed Martin also was under investigation, in this case for performing a technical assessment, without an export license, of a Chinese "kick motor" used to place a satellite into its final orbit. On June 14, 2000, the State Department announced it has reached agreement with Lockheed Martin involving $13 million in penalties-$8 million that the company will pay over a 4-year period and $5 million that was suspended and that the company can draw upon to fund a series of remedial compliance measures specified in the consent agreement.

Agency Jurisdiction over Satellite Export Licenses

Between 1992 and 1996, the Bush and Clinton Administrations transferred responsibility for decisions regarding export of commercial satellites from the State Department to the Commerce Department. A January 1997 GAO report (GAO/NSIAD-97-24) examines that decision. In response to concerns about the Loral/Hughes issue, Congress directed in the FY1999 DOD authorization bill (P.L. 105-261) that export control responsibility be returned to the State Department effective March 15, 1999. Which agency should control these exports remains controversial, and on May 10, 2000 Representative Gejdenson introduced a bill (H.R. 4417) to return control the Commerce Department. A hearing on the issue was held by the Senate Foreign Relations Committee on June 7, 2000. The Department of Commerce witness called for Commerce to regain jurisdiction over these exports. The State Department's witness said State neither sought nor welcomed the decision to return jurisdiction to them, but the department is committed to administering those responsibilities. The Security Assistance Act (P.L. 106-280) called for a reexamination of the jurisdiction question.

Some of the controversy reflects concerns of the aerospace and space insurance industries in the Untied States abroad that the new regulations are being implemented too broadly and vigorously and exports for launches non-Chinese launch vehicles (such as Europe's Ariane) also are being affected. DOD officials and others have cited potential harm to the U.S. defense industrial base if U.S. exports are stifled, too. One of the concerns is the length of time needed to obtain a State Department approval, one factor being whether State has sufficient export license examiners. Section 309 of the FY2000 State Department authorization act (incorporated into the FY2000 Consolidated Appropriations Act, P.L. 106-113) directed the Secretary of State to establish tan export regime that includes expedited approval for exports to NATO allies and major non-NATO allies. The State Department announced those new rules in May 2000; they took effect July 1. Also in may, the State Department reportedly notified France that it would not apply strict technology export control on satellites to be launched by Ariane (*Space News*, May 29, 2000, p. 1). Other reforms to broader U.S. export controls for NATO allies also were announced the same month. The Security Assistance Act (P.L. 106-280) reduces from 30 days to 15 days the time Congress has to reviews decisions on exporting commercial communications satellites to Russia, Ukraine, and Kazakhstan, making the time period the same as for NATO allies.

Russia

Following the collapse of the Soviet Union, interest developed in loosening U.S. policy to permit export of U.S. –made satellites to Russia for launch. In June 1992, President George H. Bush said he would not oppose Russia launching an Inmarsat (International Maritime Satellite Organization) satellite and the United States would negotiate with Russia over "rules of the road" for the future commercial launches. Discussions were held in the fall of 1992, agreement in principle was reached in May 1993, and the agreement was signed on September 2, 1993, after Russia agreed to abide by the terms of the MTCR. On January 30, 1996, the countries amended the agreement (see table). Prior to Russia's first launch of a U.S.-built satellite, an agreement to protect American technology was reached. For subsequent launches, an exchange of diplomatic letters extended that agreement, but the State Department decided in 1998 that a formal Technology Safeguard Agreement was needed. The agreement, among the United States, Russia, and Kazakstan (where the launch site is located) was signed in January 1999. A similar agreement for launches from Russia's Plesetsk, Svobodny, and Kapustin Yar launch sites was signed in January 2000.

The 1993 agreement was signed only after Russia agreed to comply with the MTCR in a case involving a Russian company, Glavkosmos, that planned to sell rocket engine technology to the Indian Space Research Organization (ISRO). The Untied States declared it violated the MTCR and imposed 2-year sanctions against Glavkosmos and ISRO. In June 1993, the Untied States threatened to impose sanctions against Russian companies that did business with Glavkosmos. The two countries finally agreed that Russia would cease transferring rocket engine technology (the engines themselves were not at issue) to India.

As noted, on September 10, 1997, the *Washington Times* published a story that Russian and Chinese entities, including the Russian Space Agency, were selling missile technology to Iran. In July 1998, Russia announced that it had identified nine entities, not including the Russian Space Agency, that might be engaged in illegal export activities. The United States imposed sanctions against seven of them on July 28 and three more on January 12, 1999. The State Department said the Untied States would not increase the quota of geostationary launches that Russia can conduct under the 1996 agreement unless Russian entities cease cooperation with Iran's ballistic missile program. The launches are conducted primarily by a U.S.-Russian joint venture composed of Lockheed Martin and Russia's Khrunichev and Energia, companies that have not been sanctioned. Lockheed Martin was anxious to have the quota raised to 20 and eventually eliminated. On July 13, the White House agreed to raise the quota to 2. A Senate Governmental Affairs Committee hearing was held on July 21, 1999. The agreement that set the quotas was due to expire on December 31, 2000. The *Wall Street Journal* reported on December 1, 2000 (page A4) that the White House decided to eliminate the quota. That action was taken even though Russia had informed the United States, as of December 1, 2000, it would withdraw from a 1995 agreement to stop selling conventional arms to Iran.

Ukraine

Ukraine also offers commercial launch services, chiefly as part of the Sea Launch joint venture amount Boeing, Ukraine's Yuzhnoye, Russia's Energomash, and Norway's Kvaerner. The Sea Launch vehicle consists of a Ukranian two-stage Zenit rocket with a Russian third stage. The vehicle is launched from a mobile ocean oil rig built by Kvaerner. The rig is stationed in long Beach, CA, where the launch vehicle and spacecraft are mated, and then towed into the ocean where the launch takes place. The United States and Ukraine signed a bilateral trade agreement in February 1996 (see table), that would have expired in 2001, but President Clinton terminated it on June 6, 2000 in recognition of "Ukraine's steadfast commitment to international nonproliferation norms." The first successful commercial launch was in October 1999. In 1998, Boeing agreed to pay $10 million in civil penalties for not abiding by export regulations in its dealings with Russia and Ukraine. An investigation into whether criminal laws were broken is ongoing.

Separately, Ukraine signed an agreement with the U.S. company Globalstar to launch its satellites on Zenit from Baikonur. The first attempt failed in September 1998, destroying 12 Globalstar satellites. Globalstar switched to Russian Soyuz launch vehicles (marketed through Starsem) for subsequent launches.

India and Japan

India conducted its first successful orbital space launch in 1980. Its ASLV and PSLV launch vehicles can place relatively small satellites in low Earth orbit. India conducted its first commercial launch (of German and South Korean satellites) using the ASLV to low Earth orbit in May 1999. India is developing a larger vehicle (GSLV) capable of reaching geostationary orbit. The first GSLV test launch was completed in April 2001. The GSLV uses Russian cyrogenic engines that were the subject of a dispute between the United States and Russia (discussed elsewhere in this report).

Japan successfully conducted the first launch of its H-2 launch vehicle in 1994, the first all-Japanese rocket capable of putting satellites in geostationary orbit. Previous rockets used for this purpose were based on U.S. technology and a 1969 U.S.-Japan agreement prohibited Japan from launching for third parties without U.S. consent. With the H-2, Japan was freed from that constraint. Japan's Rocket Systems Corp. (RSC). Created in 1990, offers commercial launch services, but H-2 is not cost effective and encountered technical problems that led to the Japanese government abandoning the program in 1999. A new version, is in development. RSC had contracts with two U.S. satellite builders, Hughes and Loral, for 10 launches each between 2000 and 2005, but Hughes canceled its contract in May 2000. In June 1997, the Japanese government reached agreement with the fishing industry to allow more launches from Tanegashima. Fishermen must evacuate the area near the launch site during launches. The agreement extends from 90 to 190 the number of days per year that launches may be conducted, and permits up to eight launches a year instead of two.

106TH CONGRESS: LAWS PASSED IN THE SECOND SESSION

P.L. 106-259, H.R. 4576

FY2001 DOD appropriations bill. Reported from House Appropriations Committee June 1 (H.Rept. 106-644); passed House June 7. S. 2593 reported from Senate Appropriations, Committee May 18 (S.Rept. 106-298); incorporated into H.R. 4576 and passed Senate, amended June 13. Conference report (H.Rept. 106-754) filed July 17; passed House July 19, Senate, July 27. Signed into law August 9, 2000.

P.L. 106-280, H.R. 4919

Security Assistance Act. Introduced and passed House July 24 (no formal committee action). Passed Senate September 7 in lieu of S. 2901. Conference report (H.Rept. 106-868) filed September 19; passed Hosue September 21; Senate September 22. Signed into law October 10.

P.L. 106-377, H.R. 4635

FY2001 VA-HUD-IA appropriations act, including NASA. Reported from House Appropriations Committee June 12, 2000 (H.Rept. 106-674); passed House June 21. Reported by Senate Appropriations Committee September 13 (S.Rept. 106-410); passed Senate, amended, October 12. Conference report (H.Rept. 106-988) passed House and Senate October 19. Signed into law October 27.

P.L. 106-391, H.R. 1654

FY2000-2002 NASA Authorization Act. H.R. 1654 reported from Committee on Science May 18, 1999 (H.Rept.106-77), as amended. Conference report filed September 12, 2000 (H.R. Rept. 106-843), passed House September 14, 2001, Senate October 13. Singed into law October 30.

P.L. 106-398, H.R. 4205

FY2001 DOD authorization act. H.R. 4205 reported from House Armed Services Committee May 12, 2000 (H.Rept. 106-616); passed House May 18, passed Senate July 13 in lieu of S. Rept. 106-292). Conference report filed October 6 (H.Rept. 106-945); passed House October 11, Senate October 12. Signed into law October 30.

P.L. 106-405, H.R. 2607 (Rohrabacher)

Commercial Space Transportation Competitiveness Act. Extends government indemnification of third party liability for commercial space launches until Dec.31, 2004. House Science Space and Aeronautics subcommittee markup completed July 29, 1999. No further committee action. Passed House, amended, October 4, 1999. Passed Senate, amended, October 13, 2000. Senate version passed House October 17, 2000. (The Senate Commerce Committee had reported a related measure, S. 832, on August 4, 1999, S. Rept. 106-135). Singed into law November 1.

Chapter 5

SPACE STATIONS

Marcia S. Smith

ABSTRACT

Congress continues to debate NASA's program to build a permanently occupied space station in Earth orbit where astronauts live and conduct research. NASA expects that research performed in the near-zero gravity environment of the space station will result in new discoveries in life sciences, biomedicine, and materials sciences. The program is currently called the International Space Station (ISS); the facility itself does not have a name although individual modules do.

The space station is being assembled in Earth orbit. Almost 90 launches are needed to take the various segments, crews, and cargo into orbit. The first launches have taken place. The original date to complete assembly, June 2002, has slipped to April 2006, with at least 10 years of operations expected to follow. Crews are expected to rotate on 4-6 month shifts. The first crew rotation has taken place and the "Expedition 2" crew is now aboard. Congress appropriated about $27.6 billion for the program from FY1985-2001. For FY2002, $2.1 billion is being requested.

Canada, Japan, and several European countries became partners with NASA in building the space station in 1988; Russia joined in 1993. Brazil also is participation, but not as a partner. Except for money paid to Russia, there is no exchange of funds among the partners. Europe, Canada, and Japan collectively expect to spend about $9 billion of their own money.

President Clinton's 1993 decision to bring Russia into the program was a dramatic change. Under the 1993 agreement, Phase I of U.S./Russian space station cooperation involved flights of Russians on the U.S. space shuttle and Americans on Russia's *Mir* space station. Phases II and III involve the construction of ISS as a multinational facility.

In 1993, NASA said the space station would cost $17.4 billion for construction; no more than $2.1 billion per year. The estimate did not include launch or other costs. NASA exceeded the $2.1 billion figure in FY1998, and the $17.4 billion estimate grew to $24.1-$26.4 billion. Congress legislated spending caps on part of the program in 2000. Cost growth of another $4 billion has been announced since. NASA has proposed indefinitely deferring construction of some hardware to reduce costs to stay within the cap.

Controversial since the program began in1984, the space station has been repeatedly designed and rescheduled, often for cost-growth reasons. Congress has been concerned about the pace station for that and other reasons. Twenty-two attempts to terminate the program in NASA funding bills, however, were defeated (3 in the 106th Congress, 4 in the 105th Congress, 4 in the 104th, 5 in the 103rd, and 5 in the 102nd). Three other attempts in broader legislation in the 103rd Congress also failed.

Current congressional debate focuses on whether Russia can fulfill its commitments to ISS and the cost and schedule implications if it cannot; whether NASA is correctly interpreting the Iran Nonproliferation Act in terms what items may be purchased from Russia whether or not Russian entities are not proliferating certain technologies to Iran; and the recently revealed $4 billion in cost growth in NASA's part of the program, and the resulting possibility that portions of the space station may not be built.

MOST RECENT DEVELOPMENTS

The "Expedition 2" crew-Yuri Usachev, Commander; James Voss and Susan Helms, flight engineers-continues its work aboard the International Space Station (ISS). The Expedition 2 crew has been on the station since March and was scheduled to be replace by a new crew in July. However, difficulties have been encountered with software that controls a remote manipulator system delivered to ISS in April. The system, Canadarm2, is designed to assist in assembling the station and is expected to be used to attach the next segment, an airlock. According to press reports, NASA has decided to delay the shuttle launch that will deliver the airlock from June until July to provide time to resolve the Canadarm2 problems. The subsequent shuttle flight, that will take a new crew to ISS and return the Expedition 2 crew to Earth, therefore would be postponed from July until August.

In February, NASA revealed $4 billion in cost growth over the next 5 years last year's $24.1-26.4 billion estimate. That level already was an increase over the original 1993 estimate of $17.4 billion. The new estimate would be $28-30 billion, higher than the cap of $25 billion set by Congress last year. To reduce costs, NASA has proposed termination ISS construction after the "U.S. Core" is complete and European and Japanese laboratory modules are launched. The proposal is to indefinitely defer the Habitation Module and Crew Return Vehicle (CRV), and reduce the ISS scientific research budget by 36%. NASA already has cancelled the Propulsion Module. One alternative to the CRV would be to purchase additional Soyuz lifeboats from Russia. Each can accommodate three people, so if two were simultaneously docked at the station, six people could be returned in an emergency allowing crew size to increase to six. The current agreement is for only one at a time, however. ISS will remain dependent on Russia for Soyuz lifeboats and crew size will be limited to three, until and unless NASA or another partner builds an alternative. ISS will be dependent on Russia reboost to keep the space station at the proper altitude (except for limited capabilities provided by the space shuttle) at least until 2004 when a European alternative is expected to be available.

INTRODUCTION

NASA launched its first space station, Skylab in 1973. Three successive crews were sent to live and work there in 19730-74. It then was unoccupied until it reentered Earth's atmosphere in July 1979, disintegrating over Australia and the Indian Ocean. Skylab was never intended to be permanently occupied space station with crews rotating on a regular basis was high on NASA's list for the post-Apollo years. In 1969, Vice President Agnew's Space Task Group recommended a permanent space station and a reusable space transportation system (the space shuttle) to service it as the core of NASA's program in the 1970-s and 1980s. Budget constraints forced NASA to choose to build the space shuttle first. When the shuttle was declared operational in 1982, NASA was ready to initiate the space station program.

In his January 25, 1984 State of the Union address, President Reagan directed NASA to develop a permanently occupied space station within a decade and to invite other countries to participate in the project. On July 20, 1989, the 20th anniversary of the first Apollo landing on the Moon, President George H. Bush gave a major space policy address in which he voiced his support for the space station as the cornerstone of a long-range civilian space program eventually leading to base s on the Moon and Mars.

President Clinton was strongly supportive of the space station program, and dramatically changed its character in 1993 by adding Russia as a partner to this already international endeavor. Adding Russia made the space station part of the U.S foreign policy agenda to encourage Russia to abide by agreements to stoop the proliferation of ballistic missile technology, and to support Russia economically and politically.

President George W. Bush has not yet made a statement about his position on the space station program, although his FY2002 budget calls for indefinitely deferring some space station capabilities in order to curb costs, while still meeting U.S. commitments to Europe and Japan to launch their laboratory modules.

THE INTERNATIONAL SPACE STATION (ISS) PROGRAM

NASA began the current program to build a space station in 1984 (FY1985). In 1988, the space station was named *Freedom*. Following a major redesign in 1993, NASA announced that the *Freedom* program had ended and a new program begun, through NASA asserts that 75% of the design of the "new" station is from *Freedom*. The new program does not have a formal name and is simply referred to as the International Space Station (ISS). Individual ISS modules have various names. ISS is a laboratory in space for conducting experiments in near-gravity ("microgravity"). Life sciences research on how humans adapt to long durations in space, biomedical research, and materials processing research on new materials or processes are underway or contemplated. From FY1985 through FY2001, Congress appropriated approximately $26.7 billion for the space station program (a year-by-year table is included below).

Table 1. U.S. Space Station Funding (in $ millions)

Fiscal Year	Request	Appropriated
1985	150	150
1086	230	205
1987	410	410
1988	767	425
1989	967	900
1990	2,050	1,750
1991	2,430	1,900
1992	2,029	2,029
1993	2,250	2,100
1994	2,106	2,106
1995	2,113	2,113
1996	2,115	2,114
1997	2,149	2,149
1998	2,121	2,441*
1999	2,270	2,270
2000	2,483	2,323
2001	2,115	2,115
2001	2,087	

The numbers here reflect NASA's figures for "the space station program." Over the years, what is included in that definition has changed.

*NASA's FY1999 Budget documents show $2.501 billion for FY1998 based on the expectation that Congress would approve additional transfer requests, but it did not.

Space Station *Freedom*: 1984-1993

When NASA began the space station program in 1984, it said the program would cost $8 billion (FY1984 dollars) for research and development (R & D-essentially the cost for building the station without launch costs) through completion of assembly. From FY1985-1993, NASA was appropriated $11.4 billion for the *Freedom* program. Most of the funding went for designing and redesigning the station over those years. Little hardware was built and none was launched. Major redesigns were made in 1986, first in response to cost issues and user requirements and later in the wake of the space shuttle *Challenger* tragedy; in 1987 due to rising costs, which led NASA to split the program into two phases, the second of which ultimately disappeared; in 1989 due to rising costs and expected future budget constraints; and in 1990-1991 due to congressional concerns over rising costs, higher than expected requirements for astronauts to perform extravehicular activity (EVA, or spacewalks) to assemble and maintain the station, and other issues.

The 1991 restructuring evoked concerns about the amount of science that could be conducted on the scaled-down space station. Both the White House Office of Science and Technology Policy (OSTP) and the Space Studies Board (SSB) of the National Research Council (part of the National Academy of Sciences) concluded that materials science research could not justify building the space station, and questioned how much life sciences research could be supported. Criticizing the lack of firm plans for flying a centrifuge, considered essential to this research. NASA subsequently agreed to launch a centrifuge.

In 1988, after 3 years of negotiations, Japan, Canada and 9 European countries under the aegis of the European Space Agency (ESA) agreed to be partners in the space station program (two more since have joined). An Intergovernmental Agreement (IGA) on a government-go-go government level was signed in September, and Memoranda of Understanding (MOUs) between NASA and the other relevant space agencies were signed then or in 1989. The partners agreed to provide hardware for the space station at their own expense, a total o f$8 billion at the time.

Cost estimates for *Freedom* varied widely depending on when they were made and what was included. *Freedom* was designed to be operated for 30 years. As the program ended in 1993, NASA's estimate was $90 billion (current dollars): $30 billion through the end of construction, plus $60 billion to operate it for 30 years. The General Accounting Office (GAO) estimated the total cost at $118 billion, including 30 years of operations.

1993 Redesign

In early 1993, NASA revealed $1 billion in cost growth on the *Freedom* program. President Clinton gave NASA 90 days to develop a new, less costly, design with a reduced operational period of 10 years. A new design, *Alpha*, emerged on September 7, 1993, which NASA estimated would cost $19.4 billion. It would have used some hardware bough from Russia, but Russia was not envisioned as a partner. Five days earlier, however, the White House announced it has reached preliminary agreement with Russia to build a joint Russian/American space station. Now called the International Space Station (ISS), it superseded the September 7 *Alpha* design. NASA asserted it would be more capable space station and ready sooner at less cost to the United States. Compared with the September 7 *Alpha* design, ISS was to be completed 1 year earlier, have 25% more usable volume, 42.5 kilowatts more electrical power, and accommodate 6 instead of 4 crew members. ISS is being built in an orbit inclined at $51.6°$, the same as that used by Russian space stations, rather than the $28.8°$ orbit NASA planned to use, so Russian as well as U.S. launch vehicles can service the station.

In 1993, President Clinton pledged to request $10.5 billion ($2.1 billion a year) for FY1994-1998. NASA said the new station would cost $17.4 billion to build, not including money already expended on the *Freedom* program. That estimate was derived from the $19.4 billion estimate for the September 7 Alpha design minus $2 billion that NASA said would be saved by having Russia in the program. The $2.1 billion and $17.4 billion figures became known as "caps," though they were not set in law. By 2000, the program's cost had risen to $24.1-26.4 billion, and the $2.1 billion "cap" was exceeded in FY1998. Congress now legislated caps on parts of the program (see ISS Costs, Caps…)

The *Freedom* program was criticized for its complex management structure. As part of the 1993 redesign, NASA named Boeing as the prime contractor with McDonnell Douglas and Rockwell as subcontractors. Boeing later bought Rockwell's space and defense business, and then merged with McDonnell Douglas. NASA's Johnson Space Cneter (JSC) near Houston, TX was given all program implementation responsibilities. (In February 2001, program management reporting responsibilities were at least temporarily shifted to NASA Headquarters instead of JSC because of the $4 billion in cost growth discussed below.)

Current Program: The International Space Station (ISS)

The current International Space Station program began in 1993 with Russia added as a partner, joining the Untied States, Europe, Japan, and Canada. The 1993 and subsequent agreements with Russia on space station cooperation established three phases of cooperation (see below) and the payment to Russia, originally, of $400 million to ($100 million per year for FY1994-1997). In 1996, NASA increased from $400 million to $473 million the amount of money it would pay Russia fo rhtese purposes. Of the $473 million, approximately $323 million was for Phase I and $150 million is for Phase II. (Through the end of 2000, NASA had sent a total of approximately $800 million to Russia for space station cooperation through this and other contracts. A request to transfer another $24 million is pending.)

Phase I: The Shuttle-Mir Program

Phase I has been completed. During that phase (1999-1998), seven U.S. astronauts remained on Russia's space station *Mir* for long duration (several month) missions with Russian cosmonauts. Russian cosmonauts flew on the U.S. space shuttle seven times, and nine space shuttle missions docked with *Mir* to exchange crews and deliver supplies. Repeated system failures and two life-threatening emergencies on *Mir* in 1997 raised questions about whether NASA should leave more astronauts on *Mir*, but NASA decided *Mir* was sufficiently safe to continue the program.

Phase II and III: ISS Design, Schedule, and Lifetime

NASA identifies Phases II and III of space station cooperation separately, but they blend into each other and ultimately would result in a multinational space station built by the United States, Russia, 10 or 11 European countries, Japan, Canada, and Brazil (which is not a partner on the program, but has a bilateral agreement with NASA to participate). The White House and NASA have recently proposed dramatic changes to the ISS program in response to $4 billion of cost growth. Since that is still a proposal, the following paragraphs describe the program as it existed in January 2001. The proposed changes are discussed in the next section.

Boeing is the U.S. prime contractor for ISS. Construction is expected to take more than 7 years and be completed around April 2006 (originally it was to have been completed in June 2002). NASA originally stated that ISS would be operated for 10 years, with a possibility for 5 additional years if the research was considered worthwhile. Using the original completion date of 2002, that would have meant guaranteed operations through 2012. As the time frame for building the station slipped beyond 2002. NASA stated that it would operate the station until 2012 regardless of when construction is completed, with subsequent peer review determining whether continued operation was warranted. That would mean a shorter guaranteed lifetime. By 200, NASA had returned to stating that it would operate the station for at least 10 years. Whether NASA, a non-governmental organization, or the private sector will operate the space station is currently being discussed) see Issues below).

ISS is being taken into orbit in segments and assembled there. Four major modules are now in orbit: the first two, Zarya (Sunrise) and Unity, were launched in 1998; followed by

Zvezda (star) in 2000 and Destiny in 2001. NASA paid for, built, and launched Unity and Destiny. NASA paid for Zarya; it was built and launched by Russia. Russia paid for, built, and launched Avezda, which was launched more than 2 years late. NASA continues to revise the schedule for launching the remaining segments (the "assembly sequence"). The most recent public edition, "Rev F" (August 2000), shows completion of assembly in April 2006.

There are 50 launches in the "Rev F" assembly sequence of which nine have been accomplished (Zarya, Unity, Avezda, Destiny, and five other shuttle missions). Of the 50 launches, 40 are American, 9 are Russian, and one is listed as unassigned (of the European Automated Transfer Vehicle) although Europe plans to launch ATV on its Ariane launch vehicle. In addition, Russia is expected to provide each year about two flights of its Soyuz spacecraft tot take crews to the station and three to six Progress spacecraft to "reboost" the station's orbit periodically. NASA is concerned that Russia may not provide all of the Progress reboost flights and modified the space shuttle orbiters so they can provide a limited amount of reboost in case sufficient Progresses are not available. NASA established two other contingency plans: an Interim Control Module (ICM) built by the U.S. Naval Research Lab (NRL), and a Propulsion Module NASA itself planned to build. Following the successful launch of Avezda, NASA directed NRL to put the ICM in "cold storage" such that if it were needed in the future, it would take 24-30 months to get ready. Delays and costs estimate increases led NASA o reevaluate its Propulsion Module concept, and led Congress to request a GAO study of the program. NASA then announced a new plan to procure a Propulsion "System" using a "structural test article" constructed as part of building the Unity module. Because of the cost growth announced in February 2001, NASA has canceled the Propulsion System.

The number of astronauts who can live on the space station is limited by how many can be returned to Earth in an emergency by "lifeboats" docked to the station. NASA has been planning to build a U.S. Crew Return Vehicle (CRV), but its availability date slipped to 2005. Until then, only Russian Soyuz spacecraft will be available as lifeboats. Such Soyuz can hold three people, limiting the space station crew size to three if only one Soyuz is attached. Each Soyuz must be replaced every 6 months. The U.S. CRV is expected to accommodate six or seven people and would have a lifetime of 3 years, reducing operational costs. Because of the cost growth revealed in February 2001, NASA is considering canceling the CRV.

Astronauts performing spacewalks (or "extravehicular activity"—EVA) need to help assemble the segments in space. The number of spacewalks has grown from 434 hours (with 70-80 hours that NASA said it would eliminate) when the design was approved in 1993, to 888 hours in February 1995, to 1104 hours in May 1996, to 1519.5 hours in April 1997, to 1729 hours in December 1997. The number of spacewalks is important in terms of risk to the astronauts and potential program schedule slippage if they cannot be competed on time.

ISS Costs, Caps, Overruns, and Additional Money to Russia

In 1993, NASA said it would cost $17.4 billion to build the space station (variously called its "development cost," "construction cost," or "R & D cost") from FY1994 through completion of assembly in June 2002 ($206 million was carried over from the *Freedom* program, for a total program cost of $17.6 billion). NASA also estimated the space station program's life-cycle cost from FY1985-FY2012 (including funding spent prior to 1993, construction costs, associated shuttle launch costs, civil service salaries, and 10 years of operations) at $72.3 billion. A more recent NASA life-cycle estimate is not available. In

1998, GAO said that the life-cycle cost would be $95.6 billion (GAO/NSIAD-98-147). As of early 2000, NASA's estimate for construction alone (FY1994 through completion of assembly, slipped to 2006) was $24.1-$26.4 billion.

Caps

In 1993, NASA agreed that it would spend no more than $2.1 billion per year on ISS. The $17.4 billion and $2.1 billion figures became knows as "caps," although they were not set in law. They did not include the cost of space shuttle launches needed to place ISS in orbit. Both were exceeded in 1997-1998. In 2000, Congress legislated caps on certain parts of the ISS program in the F2000-2002 NASA authorization act (P.L. 106-391). The act also authorizes an additional $5 billion for development and $3.5 billion for associated shuttle launches in case of specified contingencies.

$4 Billion Cost Growth

In February 2001, NASA revealed cost growth in the program that would exceed the legislated cap. NASA is still in the process of determining the extent of the cost growth, but its current estimate (March 2010) is $4 billion over the next 5 years (Fy2002-2006). That would increase the cost to $28-30 billion, a 61-72% increase over the 1993 estimate. Some NASA officials expect that number to decrease, perhaps to $2-3 billion, once the agency has fully "scrubbed" the figures. Conversely, the $4 billion figure could grow if certain "threats" to the program that NASA has identified materialize. NASA characterizes the $4 billion a "50-50" estimate in terms of the probability that it is correct. The cost growth was announced coincident with the release of President Bush's FY2002 budget blueprint, which proposed major changes to the program to bring it back under the cap. However, NASA and the Office of Management of Budget (OMB) still appear to be evaluating how to proceed. NASA emphasizes that what is in the budget is a *proposal.* Some NASA officials apparently hope that even though some segments may be deferred into the indefinite future, in time, the space station ultimately will have all its originally planned capabilities. It is not clear where the funding would come from, however.

If the proposal in the FY2002 budget is adopted, space station construction will end after the "U.S. core" is completed in late 2003 and modules built by Europe and Japan are launched. NASA estimates that these steps would reduce the space station construction cost to $22-23 billion. The proposal calls for indefinite deferral of the Habitation Module (the "Hab"), the Crew Return Vehicle (CRV), and the Propulsion Module, and could mean that several other segments-the Centrifuge Accommodation Module (CAM), "Node 3," a cupola, and the fourth solar array—scheduled for launch after completion of the U.S. core also may not be launched. However, the fate of those latter segments is quite unclear. CAM, Node 3, and the cupola are being built for NASA in exchange for NASA launching Japanese and European hardware (Japan is building CAM; Europe is building Node 3 and the cupola). Thus the cost to NASA would be only to integrate those segments into the space station, not to build them. NASA hopes the integration costs are small enough to be accommodated within the new budget envelope not only because it wants those capabilities, but because otherwise it would have to renegotiate the barter agreements. NASA appears to hope that remaining costs for the fourth solar array might also be accommodated to ensure adequate electrical power levels on the station.

Regarding the Hab, CRV, and Propulsion Module, NASA already has canceled the Propulsion Module. As discussed, plans to build that module developed because of concern that Russia could not fulfill its commitment to provide Progress spacecraft for reboost. NASA indicated in mid-2000 that it wanted a U.S. propulsion capability to reduce reliance on Russia and to ensure that ISS was not dependent on *any* other country for reboost. Hence, it did not at that time consider Europe's ATV as an alternative. NASA now expresses optimism that Europe will proceed with its plans to build ATV and apparently feels comfortable relying on Europe for this capability. The first of nine planned ATV flights is scheduled for 2004, meaning that ISS will remain dependent on Russia for reboost at least until then. Questions remain about Russia's financial ability to provide sufficient Progresses.

The Hab module and CRV are important for allowing crew size on ISS to grow from three to six or seven people. The additional crew members would conduct the scientific research that was foreseen as the major purpose of building a space station. If crew size is limited to three, research will be severely constrained because NASA estimates that "2 ½" people are required to operate ISS, leaving only half of one person's time for research. NASA also has proposed reducing the research budget by 36%. Instead of building the Hab module, as a barter arrangement, NASA and Italy reached a "framework" agreement in April 2001 for Italy to provide a module that would offer many "Hab" capabilities. Details are pending. If NASA proceeded with construction after "Core Complete" and Italy provided "Hab" functions, the question would still remain as to how to return a larger crew in an emergency. One Soyuz can satisfy the lifeboat function for only three people. IF CRV is not built, additional Soyuzes (which must be replaced every 6 months) would have to be provided by NASA or another partner buying them from Russia (it seems unlikely that Russia would provide them at no cost). Another partner might choose to develop a lifeboat capability, although this would take considerable time and money. NASA is discussing the possibility of a joint CRV development program with Europe and with Japan. NASA plans to continue with the X-38 program that is expected to lead to a flight demonstration of a prototype CRV; Europe is participating in the X-38 program. NASA says that the X-38 program will reduce the technical risk that would be associated with a CRV if a decision is made to build one.

Evolution of Cost Growth and Funding Transfers within NASA

Program cost growth concerns had first emerged publicly in early 1996. In March 1996, NASA Administrator Daniel Goldin gave the space station program manager control of money allocated fro (and previously overseen by) the science offices at NASA for space station research. Congress gave NASA approval to transfer $177 million from those science accounts to space station construction in the FY1997 VA-HUD-IA appropriations act (P.L. 104-204). A similar transfer was approved for FY996 ($50 million). NASA changed its accounting methods so future transfers would not require congressional action, and transferred $235 million from space station science into construction in FY 1998. ("Space station science" funding is for scientific activities aboard the space station. It is separate from NASA's other "space science" funding, such as Mars exploration, astrophysics, or earth sciences.)

One factor in the cost growth was concern that Russia would not launch its Zvezda module on time. As insurance against Zvezda delays or a launch or docking failure, NASA decided to build the ICM (discussed earlier), which could substitute for Zvezda's guidance, navigation, and control functions. To cover cost growth associated with Avezda's delay and the need to procure at least one ICM, NASA requested permission to move $200 million in FY 1997 from the space shuttle and payload utilization and operations accounts to the space station program, and to transfer $100 million in Fy1998 from unidentified NASA programs to the space station program, The latter request was not approved, but he appropriations committees did approve transferring the $200 million in FY1997.

In September 1997, NASA and Boeing's prime contract would have at least a $600 million overrun at completion, and that NASA needed $430 million more than expected in FY1998. Boeing's estimate of its contract overrun grew to $968 million in 1999, where it has remained; NASA estimates that overrun at $1.14 billion. Boeing's contract is currently valued at $9.6 billion and runs through December 31, 2003.

In March 1998, NASA announced that the construction cost estimate had grown from $17.4 billion to $21.3 billion. In April 1998, NASA released a review of space station cost conducted by an independent "cost assessment and validation" task force reporting to the NASA Advisory Council. Headed by Jay Chabrow, the report concluded that the space station's cost through assembly complete could be $24.7 billion and assembly could take 10-38 months longer. NASA agreed its schedule was optimistic and there would be about $1.4 billion in additional costs, but Administrator Goldin refused to endorse the $24.7 billion estimate. NASA's estimate as of early 2000 was $24.1-$26.4 billion. As noted, another $4 billion of cost growth now has been identified.

Additional Money to Russia

Meanwhile, NASA decided it needed to provide funding to Russia to ensure completion of Zvezda. This was in addition to more that $700 million NASA has transferred to Russia in exchange for goods and services since 1994. NASA formally notified Congress on September 29, 1998 of its plan to sent another $60 million to Russia (from its FY1998 budget). The agency said the money would buy "up to" all the research time (4,000 hours) allocated to Russia during the assembly period, together with access to stowage space on Russian modules. The House and Senate appropriations subcommittees that oversee NASA (VA-HUD-IA) approved the transfer of the $60million, but said they wouod view with "grave concern" a request for an additional $40 million NASA also had suggested it would need. NASA's authorizing committees (House Science and Senate Commerce) did not approve the $60 million transfer.

NASA also said it expected to transfer $100 million of its FY1999 space station funds to Russia. Of that, $65 million was expected to be for a Russian Soyuz spacecraft and the remainder for other Russian hardware and services. However, NASA learned that Russia was selling one Soyuz, two Progresses, and 45 days of research time on the *Mir* space station to an American venture capitalist for $20 million. NASA thereupon withdrew its plans to spend $65 million for a Soyuz and proceeded, on February 14, 2000, to request permission to spend $35 million of the FY1999 funding to buy Russian hardware and services. OF that amount, $14 million was for a "pressure dome" that would enable the U.S. ICM to dock with the Russian-built Azrya module in case of problems with Zvezda launch. The request for the $14 million was approved, but after Zvezda was launched, the pressure dome was no longer needed. NASA had spent $11 million at that point. This means that the request to spend $24 million of the $35 million is still pending. Whether NASA will obtain approval to spend that money in Russia, and what it would buy, is undecided and could be affected by the Iran

Nonproliferation Act. No funds are included in NASA's FY2000 or FY2001 budget for transfer to Russia.

Risks and Benefits of Russian Participation, Including Proliferation Issues

In the report to a company the FY1994 VA-HUD-IA appropriations bill (P.L. 103-124), Congress stated that Russian participation "should enhance and not enable" the space station. The current design, however, can only be viewed as being "enabled" by Russian participation. Today it is dependent on Russian Progress vehicles for reboost, to keep the station from reentering Earth's atmosphere; on Russian Soyuz spacecraft for emergency crew return; and on Russia's Zvezda module for crew quarters, which allow ISS to be permanently occupied.

The extent to which the program is dependent on Russia is important in terms of program risk. While there will be technical challenges, Russia's financial ability to meet its commitments has been a major issue for several years. Congressional and Clinton Administration concerns that Russia was not providing adequate funding to the companies (Khrunichev and Energia) building early space station hardware led to exchanges of letters, conversations, and meetings between then-Vice President Gore and Russian Prime Minister Chernomyrdin and his successors since the spring of 1996. Many assurances were not upheld, leading to skepticism on the part of U.S. officials as to Russia's financial ability and political resolve to meet its commitments. Mr. Koptev estimated in 1997 that Russia would spend $3.5 billion on its portion of the ISS (later he said $6.2 billion if launch costs wee included), but it is not clear at this point how much money Russia will put into the program.

Political issues are also crucial. The overall relationship between the Untied States and Russia is one major factor. Another is the linkage between the space station adherence to the Missile Technology Control Regime (MTCR) designed to stem proliferation of ballistic missile technology. Getting Russia to adhere to the MTCR appears to have been a primary motivation behind the White House's decision to add Russia as a partner. The United States wanted Russia to restructure a contract with India that would have given India advanced rocket engines and associated technology and know-how. The Untied States did not object to giving India the engines, but to the technology and know-how. Russia claimed that restructuring the contract would cost $400 million. The 1993 agreement to bring Russia into the space station program included the Untied States paying Russia $400 million for space station cooperation. At the same, time Russia agreed to adhere to the MTCR. The question is what the United States will do if Russia violates the MTCR. Some Members of Congress believe Russia already has done so. This was quite controversial during the Clinton Administration, which sanctioned 10 Russian entities for providing technology to Iran. Neither the Russian Aviation and Space Agency (RAKA, or Rosaviakosmos) nor any major Russian ISS contractors or subcontractors were among those sanctioned.

On March 14, 2000, President Clinton signed into law (P.L. 106-178) the Iran Nonproliferation Act (INA). The law, *inter alia*, prohibits NASA from making payments after January 1, 1999 in cash or in kind to Russia for ISS unless Russia takes the necessary steps to prevent the transfer of weapons of mass destruction and missile systems to Iran and the President certifies that neither the Russian space agency nor any entity reporting to it has made such transfers for at least one year prior to such determination. Exceptions are made for payments needed to prevent imminent loss of life by or grievous injury to individuals aboard ISS (the "crew safety" exception); for payments to construct, test, prepare, deliver, launch, or maintain Avezda as long as the funds do not go to an entity that may have proliferated to Iran and the Untied States receives goods or services of commensurate value; and the $14 million for hardware needed to dock the U.S. ICM (see above). President Clinton provided Congress with the required certification with regard to the $14 million on June 29, 200, but no certification was forthcoming for the remaining $24 million. Without such a certification, NASA would only be able to spend more money in Russia for ISS by meeting the one of the remaining exceptions-maintenance of Avezda (which is further defined in the law) and crew safety.

At a House International Relations Committee hearing on October 12, 2000, Members sharply criticized NASA's legal interpretation of the crew safety exception, which is worded as "imminent loss of life." NASA argued that "imminent" did not mean "immediate" and therefore many more situations were covered. Representative Rohrabacher, who was instrumental in placing the exception into the law, asserted that he meant for it to apply only in emergency situations. On September 11, 2000, President Clinton delegated responsibility for functions and authorities in the ISS-related sections of the Act to the NASA Administrator so it will be Mr. Goldin's decision as to what items to attempt to purchase from Russia under this or any other exception.

Russian adherence to MTCR was cited by the Clinton Administration as one of the benefits of involving Russia. That benefit is now in question along with another—financial savings. Clinton Administration and NASA officials asserted repeatedly in 1993 that a joining space station would accelerate the schedule by 2 years and reduce U.S. costs by $4 billion. This was later modified to one year and $2 billion, and in an April 1, 1994 letter to Congress from NASA said 15 months and $1.5 billion. NASA officials continued to use the $2 billion figure thereafter, however. A July 1994 GAO report (GAO/NSIAD 94-248) concluded that Russian participation would cost NASA $1.8 billion, essentially negating the $2 billion in expected savings. In 1998, NASA's Associate Administrator for Human Spaceflight conceded that having Russia as a partner added $1 billion to the cost. Other benefits cited by the Clinton Administration were providing U.S. financial assistance to Russia as it moves to market economy, keeping Russian aerospace workers employed in non-threatening activities, and the emotional impact and historic symbolism of the two former Cold War adversaries working together in space.

CONGRESSIONAL ACTION

FY2001

For FY2001, NASA requested and Congress appropriated (P.L. 106-377) $2.11 billion for the ISS. A Roemer amendment to terminate ISS was defeated (98-325) during consideration of the FY2001 VA-HUD-IA appropriations bill (H.R. 4635).

The FY2000-2002 NASA authorization bill (H.R. 1654) was signed into law October 30, 2000 (P.L. 106-391). During floor debate in the House on May 18, 1999, three Roemer amendments were rejected: one to cap funding at the same levels as in the Senate version (defeated 114-315); one to remove Russia as a partner in the program while still allowing NASA to purchase items from Russia (defeated 117-313); and one to terminate the program (defeated 92-337). As signed into law in 2000, the Act fully funds ISS in FY2001; sets a cap of $25 billion for development costs for ISS plus $17.7 billion for related launch costs, excluding operations, research, and crew return activities after "substantial completion" of ISS; establishes an additional contingency fund of $5 billion for ISS development plus $3.5 billion for related launches; directs NASA to establish a non-governmental organization to manage research and commercial activities on ISS after assembly; prohibits NASA from funding Transhab, a proposed inflatable module to replace a traditionally designed module NASA plans to build for crew habitation, but permits NASA to lease such a module if the private sector builds it as long as the cost is the same or less than the traditional design and does not delay the schedule or increase risks; encourages NASA to seek reductions in ISS utilization rights for international partners that "willfully violate" their commitments' requires a study by the National Research Council and the national Academy of Public Administration of ISS life and microgravity research; and reduces from 5 to 3 years an ISS commercial demonstration program that was enacted in the FY2000 NASA appropriations act (P.L. 106-74).

FY2002

For FY2002, NASA is requesting $2.097 for ISS. This is $229 million above what NASA said last year would be needed for FY2002. In total, the 5-year budget runout shown in the FY2002 budget request includes about $1 billion more for FY2002-2006 than had been planned last year. According to NASA, the increase is offset by redirecting the funding that had been planned for the Crew Return Vehicle, which previously had been carried in a different part of NASA's budget (under Aerospace Technology).

INTERNATIONAL PARTNERS

The Original Partners: Europe, Canada, and Japan

Canada, Japan, and most of the 15 members of the European Space Agency (ESA) have been participating in the space station program since it began. Formal agreements were signed in 1998, but had to be revised following Russia's entry into the program, and two more countries also joined in the interim. The revised agreements were signed on January 29, 1998, among the partners in the ISS program: United States, Russia, Japan, Canada, and 11 European countries—Belgium, Denmark, France, Germany, Italy, the Netherlands, Norway, Spain, Sweden, Switzerland, and the Untied Kingdom. Representatives of the various governments signed the government-to-government level Intergovernmental Agreement (IGA) that governs the program. (The United Kingdom signed the IGA, but is not financially participating in the program so the number of European countries participating in the program is variously listed as 10 or 11. NASA also signed Memoranda of Understanding for implementing the program with its counterpart agencies: the European Space Agency, the Canadian Space Agency, and the Russian space agency, RAKA. The IGA is considered a treaty in all the countries except the Untied States and must be ratified by those governments (in the United States it is considered an Executive Agreement). NASA has a bilateral agreement under which Italy is providing three "mini-pressurized logistics modules" (MPLMs). The first of these, Leonardo, which launched in March 2001. They are designed to be attached to ISS while cargo is transferred to the station, then filled with trash and returned to Earth. Another bilateral agreement was signed with Brazil in October 1997 for Brazil to provide payload and logistics hardware. According to NASA data provided to CRS in June 2000, these countries jointly have spent $4.5 billion of their own funding in the space station so far and expect to spend a total of $8.6 billion.

Canada is contributing the Mobile Servicing System (MSS) for assembling and maintaining the space station. In February 1994, the new prime minister of Canada had decided to terminate Canada's role in the program, but later agreed to reformulate Canada's participation instead. The first part of the MSS (the "arm") was launched in April 2001; the remainder (the Special Purpose Dextrous Manipulator, or the "fingers") is scheduled for late 2003. ESA is building a laboratory module called Columbus and an Automated Transfer Vehicle (ATV). The major contributors are Germany (41%), France (27%) and Italy (17%). Budgetary difficulties over the years led ESA to cancel other hardware it was planning, but the agency remains committed to ATV and Columbus. ESA also is paying for Italy to build two of the three "nodes" (Node 2 and Node 3) needed for the ISS in exchange for free shuttle flights to launch its ISS hardware. Japan is building a laboratory module, the Japanese Experiment Module (JEM), named Kibo (Hope). Part of it will be pressurized and another part will be exposed to space (for experiments requiring those conditions). Japan is also planning to provide NASA with a large centrifuge and a module ("CAM") for accommodating it in exchange for free shuttle flights to launch JEM and its equipment. The fates of CAM and Node 3 are in question as described earlier.

Russia

Issues associated with Russia's participation in ISS are discussed elsewhere. This section explains Russian space station activities from 1971 to the present. The Soviet Union launched the world's first space station, Salyut 1, in 1971 followed by five more *Salyuts* and then *Mir*. At least two other *Salyuts* failed before they could be occupied. The Soviets accumulated a great deal of data from the many missions flown to these stations on human adaptation to weightlessness. The data were often shared with NASA. They also performed microgravity materials processing research, and astronomical and Earth remote sensing observations. Importantly, they gained considerable experience in operating space stations.

Russia's most recent space station was *Mir*, a modular space station that was build and operated between 1986 and 2001. Crews were ferried back and forth to *Mir* using Soyuz spacecraft (reminiscent of Apollo capsules). A Soyuz spacecraft was always attached to *Mir* when a crew was aboard in case of an emergency, and Soyuz capsules now are used as Crew Return Vehicles, or lifeboats, for ISS.

Crews occupied *Mir* from 1986-2000. For almost ten of those years (1989-1999), *Mir* was continuously occupied by crews on a rotating basis. Although occasionally crews stayed for very long periods of time to study human reaction to long duration spaceflight, typically crews remained for 5-6 months and then were replaced by a new set of cosmonauts. The longest continuous amount of time spent by a single individual on *Mir* was 14 months. From 1995-1998, seven Americans participated in long duration (up to six months) missions aboard *Mir*, and nine space shuttle missions docked with the space station. Individuals from Japan,

Britain, Austria, Germany, France, and the Slovak Republic also paid for visits to *Mir*. Russia deorbited *Mir* into the Pacific Ocean on March 23, 2001.

ISSUES FOR CONGRESSIONAL CONSIDERATION

Rationale

When NASA, the Reagan Administration, and Congress considered the rationale for building a space station in the early 1980s, NASA summed it up by calling a space station "the next logical step" in the space program. In many respects, that is the fundamental rationale for the space station program. Human exploration of space appeals to what many believe is an innate desire to push the frontiers of human experience. They view the space station as the next step in America's—and humanity's inexorable desire to explore new worlds. Life sciences research on the effects of long durations in weightlessness on human physiology is considered by some as a prerequisite to sending people to Mars, research for which a space station is required. Other supporters believe materials research conducted on a space station will lead to new profitable industries, although this rationale was dismissed by the White House science office and the National Academy of Sciences in 1991.

Human spaceflight is felt by many in the space community to be the heart and sol of the space program. For them, the debate over the space station is a debate over America's future in space and NASA's purpose. A rejection of the program would be viewed as an abandonment of the vision they perceive as inherent in a strong national program of civilian space activities. As a visible symbol of America's technological prowess, human spaceflight is often perceived as a centerpiece of an image of American preeminence.

This somewhat romantic view is in stark contrast to those who view human exploration of space as, at best, a waste of money, and at worst, an unnecessary exposure of humans to the hazards of space travel. These observers argue that there is much yet to explore her on Earth, and robotic spacecraft should be used to explore the heavens for safety and cost-effectiveness reasons. They see the Apollo, space shuttle, and space station programs as successive drains on resources that could be better used for robotic space activities, or non-space related activities.

Cost and Cost Effectiveness

Cost effectiveness involves what can be accomplished with the facility that is ultimately built versus its cost. In 1993, NASA said it would cost $17.4 billion to build the U.S. portion of the space station. That rose to $24.1-$26.4 billion by early 2000, with $4 billion more in cost growth announced in 2001. Cost estimates for the earlier *Freedom* design had risen significantly as the years passed, and with each *Freedom* redesign, the amount of science diminished. Many wondered whether the same fate awaited ISS. In FY1996, FY1997, and FY1998 NASA transferred a total of $462 million from the space station science accounts into space station construction. In response to the $4 billion in cost growth, NASA has proposed reducing the research budget by 36% and to indefinitely defer building hardware that would enable six or seven crew members to live aboard the station. Without it, crew size could be limited to three. Since NASA states that 2 ½ crew members are needed to operate the station, only half of one person's time would be available for research. The fate of the centrifuge and its accommodation module is uncertain. Many worry that as costs rise further, other NASA activities may suffer, despite assurances that cost growth will have to be accommodated within NASA's human spaceflight budget.

Congress has several options in response to the announced $4 billion in cost growth. It could choose to terminate the ISS program, although 22 attempts to do so since 1991 have failed. Congress could decide to stop construction, temporarily or permanently, now or at some other point along the assembly sequence, with reduced science capability and continued dependence on Russia for certain functions. Or Congress could decide to provide additional resources and relax the cap to ensure the station is completed as originally planned, or to build some, but not all, of the capabilities NASA is proposing to defer.

Operations and Commercialization Issues, Including Transhab

As NASA continues to struggle with building ISS, attention is also turning to who should operate the facility and how to encourage commercial use of it. At congressional urging, NASA has embraced the concept of space station commercialization, both in terms of station operations and getting the private sector to use research facilities on ISS on a commercial basis. In 1998, NASA proposed creation of a non-governmental organization (NGO) to oversee research on the space station, similar to the Space Telescope Science Institute at Johns Hopkins University that operates the Hubble Space Telescope.

The NGO would report to NASA. Others wants the private sector, not the government, to manage and operate the space station at some point in the future. Still others think there is a role for the private sector in building, not just operation the space station. In December 1999, the U.S. company Spacehab announced agreement with the Russian company Energia to build a commercial module to be attached to the Russian part of ISS. The companies planned to provide space-originated news, information, education, entertainment, and business advertising and promotion, broadcasting from the module for viewing on television and the Internet. In March 2001, however, they announced that they no longer expected substantial revenue from those activities, and would wait until one of the space station partners other than Russia committed to leasing the module (perhaps as crew quarters) before they construct it. On June 2, 2000, NASA announced a deal with DREAMTiME, a company that said it would, among other things, broadcast multimedia images from ISS and make documentaries about its construction.

NASA also has been exploring whether the private sector would build a module called "Transhab" for ISS. Intheory, Transhab would replace the Habitation Module as the long term crew quarters. Transhab would be an inflatable module that its supporters argue could be a prototype for a craft to take crews to Mars. Inflatable modules are an innovative concept, making reliable cost estimating difficult. The idea was first broached and studies within NASA, but congressional concerns that it might add costs to the already overrun ISS program led to language in the conference report on the FY2000-2002 NASA authorization bill (P.L. 106-391) prohibiting NASA from spending funds on Transhab, but allowing NASA to lease such a module if the private sector builds it, with conditions.

More broadly, language in the FY2000 VA-HUD-IA approporiations act (P.L.106-74) permits NASA to conduct a demonstration commercialization program for 5 years. Receipts collected from commercial use of ISS would be used first to offset costs incurred by NASA in support of commercialization with any remainder retained by NASA for promoting further ISS commercialization activities. NASA was directed to establish a pricing policy for use of ISS by commercial entities and it was released in February 2000. The chairs of the House and Senate Committees that authorize NASA activities (House Science and Senate Commerce) both objected to including the language because of concern that it would allow NASA to pick and choose winners. The FY2000-2002 NASA authorization act (P.L.106-391) limits the project to 3 years.

Russian Non-Performance and Proliferation Issues

The risks and benefits of Russia's participation in the program already have been discussed. Currently, the main issue is how to cope with the fact that the Russian government may not provide the funding needed to fulfill its commitments to the program. Assuming that U.S. policy remains to build the space station and include Russia, the keys are: how to limit theamount of money that is transferred to Russia and help ensure that it is used for the space station program nd not for other purposes, and how to manage the construction of ISS admist the uncertainty of when or if Russian hardware and services will be available. NASA's decision to put the ICM in "cold storage" and cancel Propulsion System ensures ISS dependence on Russia to reboost (except for the very limited reboost capabilities of the U.S. space shuttle) at least through 2004. At that time, an alternative may be Europe's ATV. If the Crew Return Vehicle is canceled, ISS would remain dependent on Russia for "lifeboat" spacecraft indefinitely. As discussed earlier, the Iran Nonproliferation Act (INA) prohibits U.S. payments to Russia for ISS, with some exceptions, unless the government of Russia prevents Russian nuclear and missile technology from reaching Iran. NASA's interpretaion of that law has stirred contoversy and remains unresolved. The key question is what will happen if Russia insists it cannot fund reboost or lifeboat missions yet NASA is not permitted to transfer money to Russia for sucj missions because Russia is not in compliance with INA.

Chapter 6

MILITARY SPACE ACTIVITIES: HIGHLIGHTS OF THE RUMSFELD COMMISSION REPORT AND KEY ORGANIZATION AND MANAGEMENT ISSUES

Marcia S. Smith

ABSTRACT

Congress created three commissions in 1999 to assess certain aspects of space activities conducted by the Department of Defense (DOD) and the Intelligence Community (IC). One of these, the Commission To Assess U.S. National Security Space Management and Organization, was chaired by Donald Rumsfeld and issued its report in January 2001. Now that Mr. Rumsfeld is Secretary of Defense, the conclusions and recommendations of the "Rumsfeld Commission" are expected to receive increased attention. This chapter provides an overview of the Rumsfeld Commission's report and identifies key issues about the organization and management of national security space activities on which Congress is expected to focus.

Concerned about how the Department of Defense (DOD) and the Intelligence Community (IC) are managing and executing the nation's national security space program, Congress created the Commission to Assess United States National Security Space Management and Organization in the FY2000 DOD authorization act (P.L. 106-65). Chaired by Donald Rumsfeld, it is referred to as the Rumsfeld Commission. Mr. Rumsfeld served as Secretary of Defense (SecDef) under President Ford and was sworn in again as SecDef on January 26, 2001. He resigned as chairman of the Commission on December 28, 2000 when he was nominated for Defense Secretary. Other Commissioners were: Hon. Duane P. Andrews; Mr. Robert V. Davis; Gen Howell M. Estes III, USAF (Ret.); Gen. Ronald R. Fogelman, USAF (Ret.); LTG Jay M. Garner, USA (Ret.); Hon. William R. Graham; Gen. Charles A. Horner, USAF (Ret.); ADM David E. Jeremiah, USN (Ret.); Gen. Thomas S. Moorman, Jr., USAF (Ret.); Mr. Douglas H. Necessary; Gen. Glenn K. Oits, USA (Ret.); and Sen. Malcom Wallop (Ret.). The report was released on January 11, 2001. The text is available at [http://www.space.gov] or [http://www.house.gov/hasc/reports/misc materials.html].

RUMSFELD COMMISSION CONCLUSIONS AND RECOMMENDATIONS

The Executive Summary of the Rumsfeld Commission report states (p. vii, p. xv) that---
...it is in the U.S. national interest to:

- Promote the peaceful use of space.
- Use of nation's potential in space to support its domestic, economic, diplomatic and national security objectives.
- Develop and deploy the means to deter and defend against hostile acts directed at U.S. space assets and against the uses of space hostile to U.S. interests

The Commission's report presented five conclusions (pp. ix-x of the Executive Summary; pp. 99-100 of the full report). They are paraphrased here.

1. The extent of U.S. dependence on space, the rapid pace at which that dependence is increasing, and the vulnerabilities it creates, demand that U.S. national security space activities be recognized as a top national security priority. Specific guidance and direction from the very highest governmental levels, including the President, is needed. Only Presidential leadership can ensure the cooperation needed from all space sectors—commercial, civil, defense and intelligence.
2. The U.S. government, especially DOD and the IC is not yet arranged or focused to meet the national security space needs of the 21st century. A number of disparate space activities should be merged promptly, chains of command adjusted, lines of communication opened and policies modified to achieve greater responsibility and accountability. Only then can necessary trade-offs be made and priorities established to realize opportunities for improving U.S. military and intelligence capabilities. Only with senior-level leadership, when properly managed and with the right priorities, will U.S. space programs both deserve and attract required funding.
3. U.S. national security space programs are vital to peace and stability. The two officials primarily responsible and accountable are the Secretary of Defense and the Director of Central Intelligence. Their relationship is critical to the development and deployment of space capabilities needed to support the President in war, in crisis, and in peace. They must work together in partnership.
4. Every medium—air, land and sea—has seen conflict. Reality indicates that space will be no different. Therefore, the Untied States must develop the means to deter and defend against hostile acts in and from space. The United States has not yet taken the necessary steps.
5. Investment in science and technology resources—facilities and people—is essential for the United States to remain the world's leading space-faring nation. The U.S. government needs to play a role in expanding and deepening the pool of military and civilian talent in science, engineering and systems operations. It also needs to sustain its investment in enabling and breakthrough technologies.

The Commission made 10 recommendations that appear on pp. xxxi-xxxv of the Executive Summary. An 11th recommendation, and elaboration on all of them, are in the full report, pp. 82-98. They are paraphrased here.

1. **Presidential Leadership.** The President should consider establishing space as a national security priority.
2. **Presidential Space Advisory Group.** The President should consider the appointment of a Presidential Space Advisory Group to provide independent advice on developing and employing new space capabilities.
3. **Senior Interagency Group for Space.** The President should direct that a Senior Interagency Group for Space be established and staffed within the National Security Council structure.
4. **SecDef/DCI Relationship.** The Secretary of Defense and the Director of Central Intelligence should meet regularly to address national security space policy, objectives and issues.
5. **Under Secretary of Defense for Space, Intelligence and Information.** Such a position should be established to oversee DOD's research and development, acquisition, launch and operation of its space, intelligence and information assets; coordinate the military intelligence activities within DOD; and work with the IC on long-range intelligence requirements for national security.
6. **Commander in Chief of U.S. Space Command and NORAD and Commander, Air Force Space Command.** The Secretary of the Air Force should assign responsibility for the command of Air Force Command to a four-star officer other than INCSPACE/CINCNORAD. The Secretary of Defense should end the practice of assigning only Air Force flight-rated officers to the positions of CINCSACE and CINCNORAD to ensure that an officer from any Service with an understanding of combat and space should be assigned to this position.
7. **Military Services.** The Air Force should realign headquarters and field commands to more effectively organize, train, and equip for prompt and sustained space operations. Air Force Space Command should have responsibility for providing the resources to execute space research, development, acquisition and operations. The Army and Navy would still establish requirements and develop and deploy space systems unique to each Service. Title 10 U.S.C. should be amended to assign the Air Force responsibility to organize, train and equip for prompt and sustained offensive and defensive air and space operations. Also, the SecDef should designate the Air Force as Executive Agent for Space within DOD.
8. **Aligning Air Force and NRO Space Programs.** The Under Secretary of the Air Force should be assigned as Director of the National Reconnaissance Office and as Acquisition Executive for Space.
9. **Innovative Research and Development.** The SecDef and DCI should direct the creation of an organization to focus on the requirement for innovative research and development. The SecDef should direct the Defense Advanced Research Projects Agency and the Service's laboratories to undertake development and demonstration of innovative space technologies and systems for dedicated military missions.

10. **Budgeting for Space**. The SecDef should establish a Major Force Program for Space to provide better visibility into the level and distribution of fiscal and personnel resources, improving management and oversight of space programs.
11. **Congress**. Congress will play a key role in reviewing and coordination many of the recommendations in this report and helping promote greater public understanding of the importance of national security space.

The report also calls for an early review of nationals pace policy and a review of the approach the United States takes to intelligence collection from space. Two other themes are emphasized. One is that U.S. government policy should ensure that conditions exist so that the U.S. commercial space industry can field systems one generation ahead of its foreign competitors, and that the U.S. government can field systems one generation ahead of the commercial sector. The other is that the United States needs to accelerate space control efforts to prevent a "Space Pearl Harbor," including making better assessments of the threat environment to space systems (including satellites in orbit, their launch sites, and the ground stations needed to communicate with the satellites.)

KEY ISSUES REGARDING ORGANIZATION AND MANAGEMENT

The Commission took a broad look at DOD and IC space activities. Following are three recommendations from the report regarding organization and management that may be an initial focus of congressional attention.

The Concept of a "Space Force."

One of the factors that led Congress to create the Rumsfeld Commission was concern that the Air Force was not devoting sufficient attention to space policy and programs. According to the Commission's report (p. xxii), 85% of DOD's space-related budget activity is within the Air Force. An Air Force General serves as Commander in Chief of U.S. Space Command (CINCSPACE). U.S. Space Command is one of the nine U.S. unified commands, with component commands from the Army, Navy, and Air Force. CINCSPACE also serves as Commander in Chief of the North American Aerospace Defense Command (CINCNORAD), a joint U.S. –Canadian organization that monitors objects in Earth orbit and detects, validates, and warns of attacks against North America by aircraft, missiles, or space vehicles. He also serves as Commander of Air Force Space Command. Still, according to the report, "Many believe the Air Force treats space solely as a supporting capability [to]…air operations. Despite official doctrine that calls for the integration of space and air capabilities, the Air Force does not treat the two equally." (p. xxii-xxiii)

Some argue that it is now time to create a Space Force separate from the Air Force, just as the Air Force was separated from the Army in 1947, to increase attention to and resources for national security space activities. The Commission did not recommend establishing a Space Force today, but stated that it almost certainly would happen sometime in the future. For now, it made sweeping recommendations about reorganizing management of national security space programs. In the mid-term, it suggested a "Space Corps" might be created within the Air Force, similar to the Marine Corps within the Department of the Navy, someday leading to a separate Space Force.

Among the advocates of a Space Force is Senator Bob Smith, who is widely credited with spearheading creation of the Rumsfeld Commission to address that issue in particular. In a January 11, 2001 press release, Senator Smith stated "The Commission's recommendations lay the foundations for what I have often maintained—that we should evolve to the eventual creation of a separate Space Force. These near-term management and organization reforms will begin to put in place the leadership and advocacy for space programs that have long been lacking." Separately, Air Force Chief of Staff Michael Ryan was quoted (*Aerospace Daily*, February 9, 2001, p. 217) as saying that neither a Space Force nor a Space Corps will be needed for at least 50 years.

Also, the Commission recommended a change in the practice of assigning only flight-rated Air Force Generals as CINCSPACE/CINCNORAD so that officers from any of the services with knowledge of combat and space could be eligible. It also recommended that two different four-star officers serve as CINSCPACE/CINCNORAD and Commander of Air Force Space Command instead of the same person.

Organization within DOD and the IC

For the near-term, the Commission made many recommendations tor reorganize DOD and IC to manage space activities more effectively. The report includes a pull-out chart showing the dozens of organizations within DOD and IC involved in national security space activities. DOD and the IC have tried a number of organizational models already. The Commission recommended another rearrangement. Two of its proposals that are garnering attention are to create a new Under Secretary of Defense for Space, Intelligence, and Information, and to expand the duties of the Under Secretary of the Air Force to include serving as Director of the National Reconnaissance Office (NRO, which builds and operates the nation's reconnaissance satellites) and as Air Force Acquisition Executive for Space.

Today, the Assistant Secretary of Defense for Command, Communications, Control, and Intelligence, or ASD (C31), serves as the focal point within DOD for space and space-related activities. The ASD (C32) coordinates space policy and acquisition and has responsibility for certain aspects of DOD intelligence agencies. In addition, the National Security Space Architect (NSSA) develops mid- and long-term space architectures for DOD and intelligence space mission areas, reporting both to the ASD (C31) and to the Community Intelligence Staff under the Director of Central Intelligence. The NSSA has no authority over budgets or acquisition, however. The Director of the NRO also serves as an Assistant Secretary of the Air Force, but has no responsibility for non-NRO Air Force space activities, which are under the Air Force Acquisition Executive.

The Commission would create a new position of Under Secretary of Defense for Space, Intelligence, and Information, or USD (SII) to provide policy, guidance, and oversight for space within a single organization in the Office of the Secretary of Defense (OSD). The new position would absorb the duties of the ASD (C31) and serve as the senior OSD advocate for space. Within the Air Force, the Commission recommended that the Director of the NRO also serve as an Under Secretary of the Air Force (a higher level than an Assistant Secretary), and as the Air Force Acquisition Executive for Space, integrating NRO and other Air Force space activities. Furthermore, the Commission called for emendation of Title 10 of the U.S. Code to give the Air Force responsibility to organize, train and equip for prompt and sustained offensive and defensive air *and* space operations, instead of only air operations as currently stated.

According to *Space News* (January 22, 2001, p. 14), as he ended his term as Secretary of the Air Force, Whitten Peters criticized the Commission's recommendations, saying that giving the undersecretary of the Air Force additional responsibilities for NRO would overburden that individual. (Others note, however, that the Director of the NRO used to serve also as an Under Secretary of the Air Force.) Mr. Peters added that creating new positions and reassigning duties could create difficult situations because there would be "two potentially divergent defense acquisition executives...and two service acquisition executives." Conversely, a February 8, 2001 *Reuters* story quoted Air Force Maj. Gen. Brian Arnold as saying the Air Force "strongly supports the...report and is already moving to implement many of (its) recommendations."

White House Organization: SIG/Space versus a Space Council

Some space advocates had hoped that the Rumsfeld Commission would recommend reactivation of the National Space Council within the Executive Office of the President to coordinate military, civilian, and commercial space policy. The 1958 Act (P.L. 85-568) that created the National Aeronautics and Space Council within the White House to coordinate between the two agencies. President Nixon abolished that Council in 1973. Several mechanisms were tired in the ensuing years to coordinate space policy. The Reagan Administration used a Senior Interagency Group/Space (SIG/Space) within the National Security Council (NSC) to serve that role. A number of criticisms were levied against SIG/Space. In particular, many were dismayed by the length of time needed to make space policy decisions in the wake of the space shuttle Challenger tragedy in1986. Congress subsequently passed a bill creating a National Space Council in the Executive Office of the President. President Reagan vetoed that bill, but two years later, at the end of his second term, signed into law (P.L. 100-685) a less prescriptive version of the language. President George H. Bush formally created the National Space Council by Executive Order 12675 in April 1999. By law, the Space Council is chaired by the President.

The Rumsfeld Commission did not recommended reactivation of the Space Council, however. Instead, it first called for the President to create a "President's Space advisory Group" of high-level outside advisors as a counterpart to the President's Foreign Intelligence Advisor Board (PFIAB). Second, it recommended a return to the SIG/Space model used in the Reagan Administration to coordinate space policy across the defense, intelligence, civil, and commercial sectors. It noted that the current NSC official responsible for space has too many areas to cover and insufficient resources, resulting in a case-by-case approach to space policy that "has not allowed the development of a coherent, persistent and deliberate national process..." (page 50). The Commission's report does not discuss the Space Council option and why it chose DIG/Space instead. In response to a question following a February 1, 2001 speech, however, one of the commissioners commented that some members of the Commission thought the Space Council had been overly bureaucratic. Other observers note that a similar complaint about SIG/Space in the Reagan White House led to the 1989 creation of the Space Council.

Chapter 7

COMMONLY USED ACRONYMS AND PROGRAM NAMES IN THE SPACE PROGRAM

Marcia S. Smith and David P. Radzanowski

ACE	**Advanced Composition Explorer.** One of the Explorer series of spacecraft (see Explorer). ACE is scheduled for launch in 1997. It will be placed at the L1 Lagrange coordinate in the Sun-Earth system (where the gravitational forces of the Sun and Earth are in equilibrium) in order to track particles from the solar corona and intergalactic and interplanetary space before they reach Earth. The Sun-Earth L1 point is 1.5 million kilometers (930,000 miles) from Earth, and 148.5 million kilometers (92 million miles) from the Sun.
ACRIM	**Active Cavity Radiometer Irradiance Monitor.** This instrument measures the Sun's total output of optical energy form the ultraviolet to infrared wavelengths—called the total solar irradiance. This measurement is an important factor in the study of the Earth's climate. The instrument ahs flown on the Spacelab 1 space shuttle mission and three ATLAS shuttle missions (see below). It is currently operating on-orbit on UARS (see below) which was launched in 1991. As part of EOS (see below), the instrumet is again scheduled in 1998.
ACTS	**Advanced Communications Technology Satellite.** A NASA satellite launched in 1993 to demonstrate advanced communications satellite technologies.
ADEOS	**Advanced Earth Observing Satellite.** Japanese satellite launched in 1996 carrying Japanese and U.S. sensors for studying Earth's environment. The U.S. instruments NSCAT and TOMS are aboard ADEOS (see below).
ALH84001	Designation of a meteorite studies by a NASA science team that, they say, shows evidence of life on Mars 3.6 billion years ago. ALH refers to Allan Hills, Antarctica, where it was found in 1984.

ALI	**Advanced Land Imager** (see EO1)
APU	**Auxiliary Power Unit.** Hydrazine-fueled power unit that supplies power to devices (called control surfaces, such as ailerons) that steer the shuttle orbiter. The shuttle orbiter has three APUs.
ARC	**Ames Research Center.** NASA field center located in Moffett Field, CA. ARC is primarily involved in aeronautics, atmospheric sciences, and life sciences. ARC also is NASA's Center of Excellence in information systems technologies.
ARIANE	Family of European expendable launch vehicles.
ARIANSPACE	French company that operates commercial launches of Ariane.
ASLV	**Advanced Space Launch Vehicle.** One of India's expendable launch vehicles
AST	**Advanced Smallsat Technology** (See SSTI)
ATLAS	Family of U.S. expendable launch vehicles built by Lockheed Martin.
ATLAS	**Atmospheric Laboratory for Applications and Science.** Part of NASA's Mission to Planet Earth program, ATLAS is a space shuttle Spacelab mission which studies the complex interrelationships of the Sun, Earth's atmosphere and the near-Earth environment. Three ATLAS missions have been flown, in 1992, 1993, and 1994.
AXAF	**Advanced S-Ray Astronomy Facility.** One of NASA's three "great observatories," AXAF is designed to study the universe at x-ray wavelengths. Information from AXAF would reveal information about the birth and death of stars, and the nature of quasars, pulsars, neutron stars and black holes. After being split into two, and one portion later canceled, the remaining part of the AXAF program is scheduled for launch in 1998. (See Great Observatories.)
BION	A Russian program to study the effects of space conditions on living organisms, including monkeys. NASA and other countries cooperate with Russia on these missions. The most recent was Bion-11 which generated controversy in the United States as to whether NASA should be participating in missions that use primates as research specimens. Russia plans the next flight, Bion-1, in 1998.
CATSAT	**Cooperative Astrophysics and Technology Satellite.** First of the University Explorer (UNEX, see Explorer), missions, CATSAT is a collaborative effort by the University of New Hampshire, Weber State University (Ogden, Utah) and the University of Leiceister (United Kingdom). The satellite is designed to determine the origin of gamma ray bursts.
CASSINI	U.S. spacecraft under development for the robotic exploration of Saturn and its moons. Planned for launch in 1997.
CCAFS	**Cape Canaveral Air Force Station.** Air Force-managed launch site adjacent to NASA's Kennedy Space Center. CCAFS is used for launches of expendable launch vehicles.
CCDS	**Commercial Center for the Development of Space.** A NASA/industry/university nonprofit consortia focused on

commercial space research and technology. The CCDS program was established by NASA in 1985 to increase private sector interest and involvement in commercial space-related activities in areas such as materials processing, life sciences, remote sensing, robotics, space structures, power, and propulsion. There are currently 11 CCDSs.

CENTAUR   An upper stage (see below) used with Titan and Atlas launch vehicles to take heavy payloads to very high orbits or in to interplanetary trajectories. Its use with the space shuttle was canceled for safety reasons.

CGRO   **Compton Gamma Ray Observatory.** The second of NASA's three "great observatories," CGRO is studying the universe at gamma ray wavelengths. Launched in 1991, it is studying quasars, pulsars, black holes and others objects of interest in the universe. Once called simply GRO. (See Great Observatories.)

CLARK   A small Earth observing satellite that is being developed by CTA Inc. as part of NASA's SSTI program (see below). Clark will have a high-resolution panchromatic (black and white) imager with 3-meter resolution and a multispectral imager with 15-meter resolution. The satellite is scheduled for launch in 1997.

CLEMENTINE   A DOD spacecraft whose primary purpose was to test sensors for detecting and tracking cold objects in space, in this case, the asteroid Geographos. As a further sensor test, Clementine was first sent to orbit the Moon, taking detailed images of its surface. A software error prevented the spacecraft from accomplishing its primary objective, but data were returned about the Moon. The program is often cited as a success in demonstrating the concept of "smaller, faster, cheaper, better" satellites because it was built, launched and separated at low cost ($80 million total), in a short period of time (22 months), and accomplished some of its objectives. "Clementine 2" is now being developed by DOD.

CRV   **Crew Return Vehicle.** Proposed NASA spacecraft to ensure that crews aboard the international space station can quickly return to Earth in an emergency. Initially, the Russian Soyuyz spacecraft will perform this role. NASA is planning to build the CRV because it could carry more people and remain in orbit for a longer period of time. NASA's design program is called the X-38. Unlike the Crew Transfer Vehicle (CTV) discussed below, it would be designed for return trips only, not for taking astronauts into space. (Formerly called the Crew Emergency Rescue Vehicle, Assured Crew Return Vehicle, X-35, or X-CRV).

COLUMBUS   A pressurized laboratory module being built by the European Space Agency as part of its contribution to the international space station.

COSMOS   Russian expendable launch vehicle. Many Russian spacecraft are also called "Cosmos," the Russian word for space. Historically, the Russians (formerly the Soviets) designated spacecraft for which they did not wish to release specific information(including military

	spacecraft) simply as Cosmos. More than 2300 satellites designated "Cosmos" have been launched to date.
COSPAS/ SARSAT	A search and rescue satellite system launched and operated jointly by Russia (COSPAS) and the United States, France and Canada (SARSAT). Other countries also participate in the program.
COSTAR	**Corrective Optics Space Telescope Axial Replacement.** Corrective mirror package placed on the Hubble Space Telescope to correct for the spherical aberration of Hubble's primary mirror.
CTV	**Crew Transfer Vehicle.** A European proposal to build a spacecraft for taking astronauts to and from the international space station, launched on an Ariane. NASA and the European Space Agency are discussing cooperation on this concept, which may build on U.S. plans for a Crew Return Vehicle (CRV).
CYCLONE (or TSIKLON)	Russian expendable launch vehicle.
DAAC	**Distributive Active Archive Center.** Part of NASA's EOSDIS (see below), DAACs process raw data from Mission to planet Earth satellites into useful products, handle all user product searches, requests, and orders, and distribute data and information directly to the user community via the Internet and through other means. The DAACs also permanently archive all Mission to Planet Earth data and information for future use. There are currently nine DAACs, each focussing on the data needs of a specific segment of the user community.
DELTA	Family of U.S. expendable launch vehicles built by McDonnell Douglas. (McDonnell Douglas is currently merging with Boeing.)
DFRC	**Dryden Flight Research Center.** NASA field center located at Edwards Air Force Base in California (see below). It is NASA's main installation for flight research and also supports the space shuttle program as a backup landing site. Dryden is NASA's Center of Excellence for atmospheric flight operations.
DISCOVERY	A NASA program to develop small, low-cost spacecraft for planetary exploration. Spacecraft in this series are not supposed to exceed $150 million (FY199 dollars) for spacecraft design and development (launch and mission operations and data analysis costs are not included). The first four Discovery-series spacecraft are Mars Pathfinder, NEAR, Lunar Prospector, and Stardust (see those entries).
DOD	**Department of Defense.** Federal agency charged with, among other things, conduct of the U.S. military program.
DS1	**DEEP SPACE 1.** The first of NASA's New Millennium program's deep spae missions (see below), Deep Space 1 is scheduled for launch in 1998. An asteroid-comet flyby mission to determine how the vehicle performs with a rapidly moving target, it will also test solar electric propulsion.

DS2	**DEEP SPACE 2.** The second New Millennium program deep space mission (see below), Deep Space 2 will be launched "piggyback" with the Mars' 98 Lander (see Mars Surveyor, and Mars '98 Lander/Mars '98 Orbiter). DS2 is a pair of penetrators to bore into the Martian surface to collect data on subsurface chemistry.
ECLSS	**Environmental Control Life Support System.** System on the shuttle that supplies the shuttle astronauts with a comfortable and safe environment, maintaining the proper atmospheric pressure, humidity, carbon dioxide level and temperatures, and removing odors from the cabin area.
EDWARDS AIR FORCE BASE	Air Force base in the Mojave Desert in California where aircraft are flight tested. It is also a backup landing site for the space shuttle.
EELV	**Evolved Expendable Launch Vehicle.** Air Force space lift modernization program to replace the current launch vehicle fleet (Delta, Atlas, and Titan) with a more affordable family of launch vehicles evolved from current expendable systems. The program's goal is a 25 to 50 percent reduction in cost over current systems between 2002 and 2020.
ELV	**Expendable Launch Vehicle.** The type of space launch vehicle typically described as a rocket. ELVs can only be used once. The United States has four operational families of ELVs today; Pegasus, Delta, Atlas, and Titan. ELVs themselves are automated (people are not required on-board to operate them), but can launch spacecraft that carry crews. The Mercury, Gemini, and Apollo programs all used ELVs (Redstone, Atlas, Titan and Saturn, respectively).
ENERGIA (or ENERGIYA)	Russian expendable launch vehicular similar in capability to the U.S. Saturn V that was used to send the Apollo missions to the Moon. Energiya was launched only twice (n 1987 and 1988), after which the program was mothballed. (Energia is also the name of the Russian company that manufactured the rocket and many other space systems, including the Mir space station.)
ENVISAT	ESA Earth observing satellite carrying instruments to analyze several of Earth's environmental processes. Envisat is scheduled for launch 1999
EO1	**EARTH ORBITER 1.** The first of the New Millennium program's Earth orbiting spacecraft, EO1 is an advanced land imagery mission (and is also know as the Advanced Land Imager) that is to demonstrate new instruments and spacecraft systems that my enable future land-imaging satellites to be much smaller than they are currently. Launch is planned for 1999.
EOS	**Earth Observing System.** A series of small-to intermediate-sized spacecraft that is the centerpiece of the Mission to Planet Earth program (see EOS-AM, EOS-CHEM, EOS-PM entries). Each of the EOS spacecraft will carry instruments designed to study the Earth system on a regional and global scale focusing mainly on global climate change.

EOS-AM	EOS "MORNING". Series of EOS polar orbiting satellites that cross the Earth's equator during morning daylight hours. EOS-AM! Is scheduled for launch in 1998.
EOS-CHEM	EOS CHEMISTRY. Series of EOS polar orbiting satellites that are to study the Earth's atmospheric chemical species and their transformations. EOS-CHEM1 is currently scheduled for launch in 2002, however, the launch date and satellite design are being reviewed.
EOSDIS	Earth Observing System Data Information System. Information system under development to manage the data from NASA's Mission to Planet Earth program.
EOS-PM	EOS "AFTERNOON". Series of EOS polar orbiting satellites that cross the Earth's equator during afternoon daylight hours. EOS-PM1 is scheduled for launch in the year 2000.
EP	Earth Probe. Series of U.S. spacecraft that form part of the Mission to Planet Earth program. Earth Probes complement EOS (see above) by providing the ability to investigate Earth processes with sensors that require special orbits or have unique requirements. Earth probes also offer the opportunity to take advantage of international cooperative efforts or technical innovation.
ERS	European Remote Sensing Satellite. European Space Agency satellite for remote sensing of the Earth using a synthetic aperture radar (see SAR below).
ESA	European Space Agency. A group of 14 European countries jointly working on space programs (Austria, Belgium, Denmark, Finland, France, Germany, Ireland, Italy, Netherlands, Norway, Spain, Sweden, Switzerland, and the United Kingdom). Canada has a technical agreement for cooperation with ESA, too.
ESSP	Earth System Science Pathfinder. A category of Earth Probes (see above), ESSP is a science-driven program intended to identify and develop small missions to accomplish scientific objectives in response to national and international research priorities not addressed by current programs. The total life cycle cost (including launch vehicle) of an ESSP mission is limited to $120 million or less with a development time of 24 to 36 months.
ET	External Tank. Expendable propellant storage tank that contains fuel (liquid hydrogen and liquid oxygen0 for the space shuttle's main engines. The space shuttle orbiter is attached tot he ET during launch and the ET is jettisoned when the orbiter is just short of attaining orbit. The ET then breaks up in the atmosphere and impacts in a remote ocean area.
EVA	Extravehicular Activity. A "spacewalk" by astronauts in spacesuits outside the space shuttle or space station. **EXPENDABLE LAUNCH VEHICLE** (See ELV)
EXPLORER	A series of comparatively small earth-orbiting satellites launched by the Untied States for specific objectives infields such as astronomy,

atmospheric research and studies of solar-terrestrial relationships. The first U.S. satellite, launched in 1958, was named Explorer 1. More than 70 have been launched since. In 1994, NASA restructured the Explorer program to permit more frequent, low-cost missions, with no more than three years from design and development to launch. Three categories of Explorer missions were created: Medium-class Explorer (MIDEX), Small Explorer (SMEX), and University Explorer (UNEX, formerly the Student Explorer Demonstration Initiative, STEDI). (See MIDEX, SMEX, UNEX, and STEDI).

FUSE — **Far Ultraviolet Spectorscopic Explorer.** One of the Explorer series of spacecraft, designed to study the universe in ultraviolet wavelengths. Scheduled for launch in 1998.

FGB — **Functional Cargo Block.** (Russian acronym) The first module for the international space station. Built by Russia with U.S. funds ($215 million). Scheduled for launch in November 1997.

GAILELO — A NASA spacecraft that is in orbit around Jupiter, studying the planet and its moons. A probe was released into the Jovian atmosphere when the spacecraft reached the planet in late 1995, sending back data for 57 minutes before it was destroyed by Jupiter's environment.

GEO — **Geostationary Orbit.** An orbit located 35,800 kilometers (22,300 miles) above the Earth. At this altitude, a satellite makes one orbit of the Earth at the same rate as the Earth rotates around its axis—once every 24 hours. These are "geosynchronous" orbits (synchronized with the Earth). If the satellite is located directly above the equator at this altitude, it will maintain a fixed position relative to a point on Earth and is called "geostationary." Satellites in any other geosynchronous orbit will trace a "figure 8" pattern relative to a fixed point on Earth. Also abbreviated GSO.

GGS — **Global Geospace Science.** GGS is the U.S. contribution to the International Solar-Terrestrial Physics Program (ISTP, see below). NASA launched two satellites, Wind and Polar, as part of the GGS program.

GLOBE — **Global Observations to benefit the Environment.** Part of NASA's Mission to Planet Earth, the GLOBE program links the education process with scientific study of the Earth as an integrated system. The objective is to bring school children teachers, and scientists together to enhance environmental awareness and help students achieve higher levels of achievement in science and mathematics.

GOES — **Geostationary Operational Environmental Satellite.** One of the two series of civilian weather satellites used by the United States. These satellites are in geostationary orbit; the other series (NOAA, see below) is in polar orbit.

GP-B — **GRAVITY PROBE B.** A spacecraft designed to test a key prediction of Einstein's general theory of relativity, it is scheduled

	for launch in the year 2000. It also is know as the "Relativity Misison."
GPC	**General Purpose Computer**. Computer hardware (IBM AP-101S machines) on the shuttle orbiter that, with the associated software, manages the following functions: guidance and navigation, flight control, main engine interfaces, intercomputer data transfer, memory storage, display system, payload operations, and ground interface. There are five GPCs on the shuttle.
GPS	**Global Positioning System**. DOD navigation satellite system (also called NAVSTAR) that provides users with three-dimensional information (latitude, longitude and altitude) on their exact location. GPS is available to civilian as well as military users.
GREAT OBSERVATORIES	In the 1980s, NASA planned to launch four "great observatories" to study the universe at different wavelengths: Hubble Space Telescope (HST) for visible and ultraviolet wavelengths, Compton Gamma Ray Observatory (CGRO) for gamma rays, Advanced X-Ray Facility (AXAF) for x-rays, and Space Infrared Telescope Facility (SIRTF) for infrared. HST and CGRO are in orbit today (see those entries). AXAF was split into two programs, and one was subsequently canceled by Congress (See AXAF). NASA scaled back the SIRTH mission and it is no longer categorized as a "great observatory."
GRO	(See CGRO)
GSFC	**Goddard Space Flight Center**. NASA field center located in Greenbelt, MD. GSC is primarily involved in Earth sciences, astronomy and space physics mission, robotics, and spacecraft command and control. GSFC also manages NASA Wallops Flight Center, VA from which sounding rocket, and occasionally orbital, launches are conducted. GSFC is NASA's Center of Excellence for scientific research.
GSO	**Geostationary Orbit or Geosynchronous Orbit** (see GEO)
GTO	**Geostationary Transfer Orbit**. AN elliptical intermediate orbit between low Earth orbit (LEO) and GEO into which satellites are initially placed; either an upper stage ro an onboard engine is then used to boost the satellite to GEO.
H-2	Japanese expendable launch vehicle.
HOPE	Japanese spaceplane under development (see Spaceplane).
HST	**Hubble Space Telescope**. The first of NASA's three "great observatories," HST was launched in 1990 to study the universe in visible and ultraviolet wavelengths. NASA soon announced that the Hubble's mirrors were flawed, but a repair mission in 1993 rectified the problem. A second Hubble servicing mission by the shuttle is scheduled for February 1997. (See Great Observatories).
ICSU	**International Council of Scientific Unions**. International organization of national academics of science around the world. Headquartered in Paris, ICSU coordinates many international

	science programs, including the International Geosphere-Biosphere Program (see below).
IGBP	**International Geosphere-Biosphere Program.** IGBP is conducted under the auspices of the International Council of Scientific Unions; the U.S. contribution is called the U. Global Change Research Program (it once was called Global Habitability). IGBP's goal is to provide a scientific understanding of the entire Earth system on a global scale by describing how its component parts and their interactions have evolved, how they function, and how they may be expected to continue to evolve.
IKI	**Russian Institute for Space Research.** (Russian acronym) Part of the Russian Academy of Sciences responsible for the Russian space science program.
IMAGE	**Imager for Magnetopause-to-Aurora Exploration.** One of the first two of the MIDEX series of Exploer spacecraft (see MIDEX), IMAGE will use three dimensional imaging techniques to study the global response of the Earth's magnetosphere to variations in the solar wind. Launch is planned for the year 2000.
IMU	**Inertial Measurement Unit.** Instrument on the shuttle orbiter that supplies data on the vehicle's attitude (orientation) and acceleration. The orbiter has three IMUs.
ISPM	(see Ulysses)
ISS	**International Space Station.** Designation currently used for the space station being built by the United States, Russia, Japan, Canada, and 10 of the 14 members of the European Space Agency (ESA). The space station does not have a formal name. From 1993-1995, NASA called it International Space Station Alpha (ISSA), and some of the partners in the program still refer to it as Alpha. Prior to 1993, NASA was designing a space station called Freedom, but that name was dropped when the program transitioned into its new designed in 1993.
ISTP	**International Solar-Terrestrial Physics Program.** ISTP is an international program to better understand solar-terrestrial physics. The U.S. contribution to ISTP is called the GGS program (see above). There are two U.S. spacecraft (**Wind** and **Polar**), two Japanese (**Geotail** and **CRRES**), and two European (**Cluster** and **SOHO**). (All are in orbit except Cluster, which was lost in the 1996 Ariane 5 launch failure.)
ISU	**Inertial Upper Stage.** A type of upper stage (see below) that can be used with Titan and the space shuttle)
J-1	Japanese expendable launch vehicle.
JEA	**Joint Endeavor Agreement.** Cooperative agreement between NASA and industry where NASA provides a launch opportunity on the space shuttle for experiments by private companies, who in return usually share resulting data with NASA.

JEM	**Japanese Experiment Module.** Japanese contribution to the international space station program. JEM is being designed as a laboratory module for materials processing experiments. One section will be pressurized where astronauts can work, while another is for experiments that must be exposed to the space environment.
JERS	**Japan Earth Resources Satellite.** A Japanese satellite for remote sensing of the Earth using a synthetic aperture radar (SAR see below).
JPL	**JET Propulsion Laboratory.** NASA field center located in Pasadena, CA. JPL is a government-owned contractor-operated (GOCO) facility that is operated and managed by the California Institute of Technology. The center is primarily responsible for NASA's robotic planetary exploration missions and associated spacecraft command and control. JPL is NASA's Center of Excellence for deep space missions.
JSC	**Johnson Space Center.** NASA field center located in Houston, TX. JSC is primarily involved in astronaut training, developing the space station, and managing the space shuttle program. JSC takes over responsibility for each shuttle mission after the shuttle has "cleared the tower" (left the launch pad) at KSC and is the site of the Mission Control Center. JSC also manages NASA operations at the Army's White Sands Missile Range in New Mexico where sounding rockets are launched. JSC is NASA's Center of Excellence for human operations in space.
KSC	**Kennedy Space Center.** NASA filed center located at Cape Canaveral, Fl. KSC is the launch site for the space shuttle (and was the site of the Apollo launches). ELV launches are conducted at the adjacent Cape Canaveral Air Force Station (see above). KSC is NASA's Center of Excellence for launch and cargo processing systems.
LANDSAT	A series of satellites (five have been successfully launched since 1972) for studying the surface of the Earth from space. Data are used for crop forecasts, pollution monitoring, land use studies, minerals surveys, etc. The next satellite, Landsat 7, is scheduled for launch in 1998.
LaRC	**Langley Research Center.** NASA field center located in Hampton, VA. LaRC is primarily involved in aeronautics. LaRC in NASA's Cneter of Excellence for structures and materials.
LCC	**Launch Control Center.** Facility at Kennedy Space Center form which space shuttle launches are controlled. KSC has responsibility for shuttle launches until the shuttle "clears the tower" (leaves the launch pad) at which time control is transferred to Johnson Space Center and its Mission Control Center.
LEO	**Low Earth Orbit.** Satellite orbit closest to the Earth. O formal definition of LEO exists, although it is generally considered to be any orbit approximately up to 8000 kilometers (5000 miles).

LeRC	**Lewis Research Center.** NASA field center located in Cleveland, OH. LeRC is primarily involved in aeronautical and space propulsion, energy systems, and advanced communications satellite technology. LeRC is NASA's Center of Excellence for turbomachinery.
LEWIS	A small Earth observing satellite that is being developed by RW as part of NASA's SSTI program (see below). Lewis will have a hypserspectral imager with 30-meter resolution and 384 channels. The satellite is scheduled for launch in 1997.
LIGHTSTAR	A low-cost SAR (see below) mission currently being developed by NASA's JPL. The mission would use advanced technologies to reduce the cost and size and enhance the quality of SARs. In January 1997, NASA began soliciting proposals for design and definition studies of the LightSAR satellite. The goal of the mission is to demonstrate that advanced technologies can reduce the cost of SAR missions and provide valuable scientific and commercial remote sensing data.
LONG MARCH	Family of Chinese expendable launch vehicles.
LUNAR PROSPECTOR	Scheduled for launch in 1997, Lunar Prospector is the third of the Discovery series of spacecraft (see Discovery). It is designed to orbit the Moon, making maps of the chemical composition of the surface (including searching for ice at the poles) and of the Moon's global magnetic and gravity fields.
MAGELLAN	A NASA spacecraft that provided detailed radar mapping of the surface of Venus.
MAP	**Microwave Anisotropy Probe.** One of the first two of the MIDEX series of Explorer spacecraft (see MIDEX), MAP will investigate the cosmic microwave background to help understand the large scale structure of the universe. Planned for launch in the year 2000.
MARINER	A series of 10 spacecraft launched sine 1962 to study Mercury, Venus and Mars. The most well known are Mariner 9, which provide detailed maps of the surface of Mars, and Mariner 10, which flew past Venus and Mercury.
MARS'98 LANDER/MARS '98 ORBITER	A pair of spacecraft in the Mars Surveyor series (see Mars Surveyor). NASA plans to launch one of the backup Mars Observer instruments on the Mars '98 orbiter. Experiments for the Mars '98 lander include a robotic arm to dig into the Martian soil, and the Deep Space 2 (DS2, see above) penetrators. Launch dates for the orbiter and lander are December 1998 and January 1999, respectively.
MARS OBSERVER	A NASA Spacecraft designed to study Mars, NASA lost contact with it shortly before it was to have entered Martian orbit in 1993. Backup versions of the 7 instruments that were aboard Mars Observer are planned for reflight on Mars Survey on spacecraft (see below).

MARS PATHFINDER	The first of the Discovery series (see Discovery), Mars Pathfinder was launched in December 1996 and is scheduled to land on Mars in July 1997. It carries a small robotic rover, named Sojourner. The primary goal is to demonstrate new technologies for Mars exploration, but it also is expected to provide scientific data about Mars.
MARS SSURVEYOR	A series of U.S. spacecraft planned for launch between 1996 and 2005 at 26-month intervals (when the planetary alignment between Earth and Mars permits such launches). The first, launched in November 1996, was Mars Global Survey on (MGS, see that entry). NASA currently plans to follow with launches of orbiter/lander pairs in 2001 and 2003, and a sample return mission in 2005. Backup versions of the instruments lost on the Mars Observer mission are being reflown on several of the Mars Surveyor orbiters (5 are on MGS< one will be on the Mars '98 orbiter, and NASA is considering flying the 7th on the Mars '01 orbiter). Details of what investigations later probes in the series will perform are not finalized.
MED-LITE	NASA's medium-light ELV services program to fill the gap between the payload capability of current small launch vehicles and medium-sized launch vehicles. In 1996, McDonnell Douglas (which is currently merging with Boeing) was awarded the Med-Lite contract to provide 5 firm launches with options for 9 more launches through 2004. The ELV used for this program is a scaled down version of the Delta II rocket.
MEO	**Medium Earth Orbit.** No formal definition of MEO exists, although it is generally considered to be any orbit between LEO and Geo (See those entries).
METOP	**European Meterological Satellite.** A polar-orbiting weather satellite being developed by ESA and EUMETSAT. The satellite would provide the "morning meterological satellite" in the NOAA and/or NPOESS system (see below).
METSAT	**Meterological Satellite.** Another term for weather satellites.
MIDEX	**Medium-class Explorer.** One of the three categories of Explorer programs (see Explorer. Spacecraft design and development are not expected to cost more than $70 million (FY 1994 dollars) or take more than three years. Launch, mission operations and data analysis costs are excluded. One launch per year is planned. The first two MIDEX missions are MAP and IMAGE (see those entries).
MGS	**Mars Global Surveyor.** The first of the Mars Surveyor series (see above). AN orbiter, it was launched in November 1996 and carries backup version of 5 of the 7 instruments that were aboard the ill-fated Mars Observer spacecraft.
MIR	The Russian space station complex now in operation (Mir means Peace). The Mir core module was launched in 1986. Five more modules have since been added: Kvant (1987), an astrophysical

	observatory; Kvant 2 (1989), a logistics module; Kristall, (1990) a laboratory for materials processing experiments; Spektr (1995) for atmospheric studies; and Priroda (1996) for remote sensing of the Earth.
MOLNIYA	Russian expendable launch vehicle (a variant of the Soyuz) for placing satellites into highly elliptical orbits. It is also the name of Russian communications satellites that utilize that type of orbit.
MPS	**Material Processing in Space**. Use of the microgravity environment for producing purer substances or new materials that are difficult or impossible to produce in the one-gravity environment at Earth's surface.
MSFC	**Marshall Space Flight Center**. NASA field center located in Huntsville, AL. MSFC is primarily involved in space propulsion, space-based astronomy, management of Spacelab flights, and the development of the space station. MSFC is NASA's Center of Excellence for space propulsion.
MSS	**Mobile Servicing System**. Canada's contribution to the international space station. The MSS is a robotic arm that will assist in assembly and maintenance of the space station and servicing payloads attached to the outside of the station.
MTPE	**Mission to Planet Earth**. NASA's contribution to the U.S> Global Change Research Program (see IGBP). MTPE includes the EOS satellites, Earth probes, EOSDIS (see those entries), and a cadre of scientists and research programs to analyze collected data.
MU	Family of Japanese expendable launch vehicles.
NAVSTAR	(see GPS)
NASA	**National Aeronautics and Space Administration**. Federal agency charged with conducting most governmental civilian space activities (the Department of Commerce and the Department of Transportation also have some civilian space program responsibilities, and other agencies are involved in space policy formulation and/or utilization of space data).
NSCAT	**NASA Scatterometer**. A NASA Earth Probe (see above) launched on the Japanese ADEOS (see above) spacecraft. NSCAT measures near-surface wind speeds and directions over the global oceans every two days.
NEAR	**Near Earth Asteroid Rendezvous**. A NASA spacecraft, launched in 1996, for studying asteroids whose orbits bring them close to Earth. Part of the Discovery program (see Discovery).
NEW MILLENNIUM	A term used by NASA and the Office of Management and Budget (OMB) to refer to a category of science and technology programs they seek to protect from funding cuts. It includes the Explorer, Mars Surveyor, and Discovery programs in their entirety, plus the Smallsat program, the Earth System Science Pathfinder (ESSP), and the New Millennium programs under the Office of Space Science and Office of Mission to Planet Earth.

NEW MILLENIUM INITIATIVE	A set of science programs for which NASA requested $342 million in out-year budget authority for FY 1997. It included five specific spacecraft—three in the Explorer series (MIDEX 1, SMEX 6, and UNEX 1); the fourth Discovery mission (Stardust); and the Earth System Science Pathfinder. No FY 1997 NASA authorization bill cleared Congress, and the final appropriations bill that includes NASA (VA-HUD_IA, P.L 104-204) did not include this funding.
NEW MILLENIUM PROGRAM	A set of NASA programs under the Office of Space Science and the Office of Mission to Planet Earth to launch technology demonstration experiments. NASA plans 6 New Millennium missions in 5 years to test "revolutionary" technologies. The first New Millennium experiments are DS1 (Earth Space 1) and DS2 (Deep Space 2) for solar system exploration, and EO1 (Earth Orbiter 1, also known as Advanced Land Imager) for remote sensing (see those entries).
NGST	**Next Generation Space Telescope**. A component of NASA's Origins program. NGST would be a space-based telescope of aperture greater than 4 extremely deep exposures at near-in infrared wavelengths. NGST would study the first stars and galaxies of early universe formulation. NASA is currently studying the concept's feasibility and technology requirements.
NICMOS	**Near Infrared Camera and Multi-Object Spectrometer**. Instruments to be placed on the Hubble Space Telescope during the second Hubble servicing mission scheduled fro February 1997. NICMOS will be used to obtain high spatial resolution images and low resolution spectra of astronomical objects at near infrared wavelengths.
NOAA	**National Oceanic and Atmospheric Administration**. Part of the Department of Commerce (DOC)), NOAA is responsible for U.S. civilian weather satellites. The civilian weather satellites in polar orbit are also designated "NOAA." They are complemented by the GOES satellites (see above). NOAA also executes DOC's responsibility for licensing commercial remote sensing systems.
NPOESS	**National Polar Operational Environmental Satellite System**. On May 5, 1994, the White House announced the Clinton Administration's decision to merge the nation's military and civilian operational meterological polar satellite systems into a single national system. The joint program formed as a result of this direction is know as NPOESS. DOD, NOAA, and NASA have formed an integrated program office to manage the program. Launch of the first dedicated NPOESS satellite is planned for 2006. In the meantime, the program will use DOD and NOAA polar metsats that are already in orbit or being developed under current contracts.
NPS	**Nuclear Power in Space**. The use of nuclear sources to produce electricity on spacecraft, either RTGs (see below) or nuclear reactors. Russia used nuclear reactors in space for many years for

one if its military reconnaissance programs, but no longer does today. The United States had a program to develop space nuclear reactors (the SP-100 program), but it was terminated, although DOD has been conducting ground-based research using Russia's Topaz II space nuclear reactors.

NSTL	(see SSC—Stenis Space Center)
NSTS	**National Space Transportation System** (see STS)
OA	**Office of Aeronautics**. NASA program office responsible for developing aeronautics and space transportation technologies and for facilitating the transfer an commercialization of these technologies. Also known as "Code R."
OLMSA	**Office of Life and Microgravity Sciences and Applications**. NASA program office responsible for life and microgravity sciences and applications. Also known as "Code U."
OMPTE	**Office of Mission to Planet Earth**. NASA program office responsible for the mission to Planet Earth environmental satellite and research program. Also known as "Code Y."
ORIGINS	A NASA program designed to answer fundamental questions about the universe: where did galaxies, stars and planets come from; are there other planets in the universe, are they habitable, and does life exist there; and what is the origin of the universe? Several existing or planned NASA telescopes already are being used to find the answers (HST, WIRE, FUSE, SOFIA and SIRTF—see those entries).
OSMA	**Office of Safety and Mission Assurance**. NASA program office charged with ensuring design safety, operations logistics, and program management through the implementation of procedures and standards that reduce risks in programs like the space shuttle and space station. Formerly called Safety, Reliability, Maintainability and Quality Assurance (SRM & QA). Also know as "Code Q."
OSS	**Office of Space Sciences**. NASA program office responsible for astrophysics, planetary exploration, and solar-terrestrial physics. Also referred to as "Code S."
OSF	**Office of Space Flight**. NASA program office responsible for the space shuttle and space station programs. Also referred to as "Code M."
PAM	**Payload Assist Module**. A type of upper stage (see below) used with the Delta or space shuttle. Built by McDonnell Douglas (which is currently merging with Boeing).
PEGASUS	Air-launched expendable launch vehicle built by Orbital Sciences Corporation.
PF	**Planet Finder**. A component of NASA's Origins program, PF is currently conceived to be a suite of 4 telescopes each .5 meters across, precisely located on a 73 meter long truss. The infrared light collected by each telescope would be combined with that from the other telescopes in such a way that the light from a star is rejected,

	but the light from a Earth-size planet around the star is collected and analyzed. To limit interference form the Sun, PF would be located at a distance from the Sun beyond Jupiter's orbit.
PIONEER	Series of numbered spacecraft (0-11) plus Pioneer-Venus 1 and 2 launched since 1958 to study the Moon, other planets, and interplanetary space. Pioneer 10 and 11 were the first spacecraft to fly past Jupiter and Saturn.
PLUTO EXPRESS	Proposed mission to conduct the first reconnaissance of Pluto and its large moon Charon with two small flyby spacecraft. The mission also is to serve as a pathfinder for lower cost exploration of the outer solar system and may involve cooperation with Russia. Launch of Pluto Express could occur in 2001 with arrival at Pluto in 2013.
POLAR	Part of NASA's GGS initiative (see above), Polar was launched in February 1996 and is studying the interaction between the Earth and Sun in the Earth's polar region.
POLAR ORBIT	An orbit that circles the Earth's poles.
PROTON	Russian expendable launch vehicle used to place satellites in geostationary orbit, to place the Russian space stations (Salyut and Mir) in low Earth orbit, and to launch Russian spacecraft. Today, Proton is marketed commercially by a Russian-U.S. joint venture called International Launch Service s(ILS). Lockheed Marin is the U.S partner.
PSLV	**Polar Space Launch Vehicle.** AN Indian expendable launch vehicle designed to place satellites in polar orbits.
RADARSAT	Canadian SAR satellite launched by the United States in November 1995.
RELATIVITY MISSION	(see GP-B)
RLV	**Reusable Launch Vehicle.** Type of space transportation system that can be used more than once. The U.S. space shuttle is the only reusable system in existence today. The shuttle actually is only partially reusable (the External Tank cannot be reused). NASA is now involved in a program to develop fully reusable launch vehicle technologies through the X-33 and X-34 programs (see below).
RMS	**Remote Manipulator System.** Remotely controlled manipulator arm of the shuttle orbiter that is used for deployment and/or retrieval of payloads from the shuttle cargo bay. The RMS was funded, designed, developed, and manufactured by Canada.
RTG	**Radioisotope Thermal Generator.** A plutonium-fueled device to generate electricity for spacecraft. The United States uses RTGs for spacecraft whose missions preclude the use of solar energy (because they travel too far from the Sun or are placed on the lunar or planetary surfaces where "nights" are long). Russia also occasionally uses RTGs for planetary missions.
RXTE	**Rossi X-Ray Timing Explorer.** One of the Explorer series of satellite (see Explorer) launched in 1995 to study neutron stars and

black holes form the fluctuations in the intensity of x-rays they produce.

SALYUT A series of Soviet space stations launched from 1971 to 1982. Six were successfully operated (Salyut 1,3,4,5,6 and 7). Russia's seventh successful space station, still in operation today, is named Mir (see above).

SAR **Synthetic Aperture Radar.** An imaging radar that synthesizes a very long antenna by combining signals received by the radar as it moves along its flight track. Aperture means the opening used to collect the reflected energy that is used to form an image. For radar, the aperture is the antenna size. The length of a radar antenna determines the resolution in the direction of the image. The longer the antenna, the finer the resolution in the tracking direction. The SAR takes advantage of its movement, thus creating a "synthetic" aperture. The aperture appears longer than the actual length of the antenna. Since SARs provide their own illumination (the radar pulses). They can image at any time of day or night. Since the radar wavelengths are much longer than those of visible or infrared light, SARs can also "see" through cloudy and dusty conditions that visible and infrared instruments cannot.

SARSAT (see COSPAS/SARSAT)

SATURN Formerly a U.S. expendable launch vehicle used to send Apollo missions to the Moon, and to launch the Skylab space station. It is no longer produced.

SCA **Shuttle Carrier Aircraft.** NASA-owned Boeing-747 airplane that is used to transport the shuttle orbiter form one place to another (often from Edwards Air Force Base where it sometimes lands to Kennedy Space Center from which it is launched again). NASA has two SCAs.

SCOUT Formerly a U.S expendable launch vehicle built by LTV Aerospace. It is no longer produced.

SEASTAR Spacecraft developed by Orbital Sciences Corporation ((OSC) that will carry the SEA-WIFS (see below) instrument to low Earth orbit. Launch is currently planned on a Pegaus launch vehicle in April 1997. As part of the MTPE program, NASA has agreed to purchase data from OSC that will be collected by SEASTAR.

SEA-WIFS **Sea-Viewing, Wide-Field-of-View Sensor.** Instrument designed to provide ocean color and temperature information. The instrument will be launched on Orbital Sciences Corporation's SEASTAR spacecraft (see above).

SERVICE MODULE A Russian module to be launched as part of the international space station. It will provide crew quarters, and guidance, navigation and control. Delays in its construction due to Russian funding constraints have raised schedule concerns at NASA for the ISS program as a whole. Launch was scheduled for April 1998, but Russia has said it may be 8 months late.

SETI	**Search for Extraterrestrial Intelligence.** Scientific effort to use ground-based astronomical facilities to search for messages that might come from an intelligent civilization elsewhere in the universe. Once funded by NASA, Congress ended government funding in FY1994. Private funds are now used for a smaller-scale program.
SHAVIT	Israel's expendable launch vehicle.
SHUTTLE-C	Concept for an automated version of the space shuttle that would be used for carrying cargo into space without a crew.
SIM	**Space Interferometry Mission.** A component of NASA's Origins program, SIM is an effort (capped at $700 million in FY 1996 dollars) to detect extra-solar planets and other phenomena by making extremely precise measurements of the positions of stars. SIM will use two telescopes separated from each other at the ends of a long boom to survey the motion of the 100 or so closest stars with precise accuracy. SIM also will be able to generate images with a resolution that exceeds that of Hubble by about a factor of 4.
SIR	**Shuttle Imaging Radar.** Also known as the Spaceborne Imaging Radar, the DIR program uses synthetic aperture radars (SARS, see above) on the shuttle to perform land and ocean studies. NASA has flown three SIR instruments on the shuttle, SIR-A, SIR-B, and SIR-C.
SIRTF	**Space Infrared Telescope Facility.** Once planned to be the fourth of the "great observatories," SIRTF is to study the universe in infrared wavelengths. Funding constraints caused NASA to scale back the original concept for this spacecraft and it is no longer considered a "great observatory." NASA hopes to launch the spacecraft in 2002, but the schedule is not finalized. Although not yet a "new start," the project is currently undergoing Phase-B definition studies.
SKYLAB	The first space station. Launched in 1973, it hosted three crews from 1973-1974 staying 28, 59 and 84 days respectively. Skylab was not designed for permanent occupancy, and reentered the atmosphere in 1979, spreading debris over Australia and the Indian Ocean.
SLF	**Shuttle Landing Facility.** Runway at Kennedy Space Center for space shuttle landings.
SLR	**Space Radar Laboratory.** Space shuttle mission that carried the SIR-C (see above) instrument on board.
SLS	**Spacelab Life Sciences.** A designation for space shuttle missions using the Spacelab module and devoted to life sciences disciplines.
SLV	**Space Launch Vehicle.** Generically used to refer to launch vehicles. Also, the name of one of India's expendable launch vehicles.
SMEX	**Small Explorer.** One of the three categories of Explorer missions (see Explorer). Spacecraft in the SMEX series are expected to cost no more than $35 million (FY 1994 dollars) for design and development, and take less than three years to develop. Launch,

mission operations and data analysis are not included. One mission per year is planned for launch. NASA now has a "SMEX-lite" initiative in progress to lower the cost of SMEX missions by half. Two SMEX spacecraft already have been launched (Sampex-Solar Anomalous Magnetospheric Particle Explorer, in 1992; and FAST—Fast Auroral Satellite Explorer, in 1996). Three more have been approved so far: SWAS, TRACE and WIRE (see those entries).

SNOE **Student Nitric Oxide Explorer.** One of the STEDI Explorer missions (see STEDI). SNOE will be designed, developed and operated by the University of Colorado, and will study nitric oxide density in the Earth's lower thermosphere. Launch is scheduled for 1997.

SOFIA **Stratospheric Observatory for Infrared Astronomy.** A 2.5 meter optical/infrared/submillimeter telescope scheduled to be flown on a 747 aircraft for studies of the universe. In December 1996, NASA selected the University Space Research Association as the prime contractor for SOFIA. Science flights are scheduled to begin in 2001.

SOJOURNER Robotic rover that is part of Mars Pathfinder (see above).

SOUNDING ROCKERS Small rockets that launch payloads into sub-orbital flight paths. Many are used to create a few minutes of microgravity conditions for the conduct of materials processing experiments. Others are used for atmospheric studies.

SOYUZ Russian spacecraft used to transport crews into space. The original Soyuz debuted in 1967. It has been upgraded twice, and the current version is referred to as Soyuz TM. (The name Soyuz is also used to denote the launch vehicle used to launch the Soyuz spacecraft and other Russian spacecraft).

SPACELAB A module built by the European Space Agency to fly in the cargo bay of the U.S. space shuttle to provide a shirt-sleeve environment for conducting scientific experiments. Spacelab cannot fly independently of the shuttle and therefore is not considered a space station.

SPACEPLANE Reusable spacecraft launched by a rocket. The Japanese are developing the JOPE spaceplane. French plans for the HERMES spaceplane have been reduced to studies due to budget constraints. HERMES was to carry a crew; HOPE is designed only for cargo. Not to be confused with "aerospace planes," a concept for a craft which would be launched like an airplane and operate in both air and space. U.S., British, and German aerospace plane programs all have been reduced to studies. Those programs were named National Aero-Space Plane (NASP), Horizontal Take-Off and Landing (HOTOL), and Saenger, respectively.

SPACEPORT	A commercially run launch and/or re-entry facility used for ELVs, RLVs, sounding rockets, or re-entry vehicles. In the United States, Alaska, California, Florida, New Mexico, and Virginia and in various stages of developing commercial spaceports.
SPACE SHUTTLE	Partially reusable space transportation system for taking people and cargo into low Earth orbit. The United States has the only space shuttle in existence today, It consists of an orbiter (the airplane-like part), two solid rocket boosters (on either side of the external tank), and a large external tank which holds the fuel for the engines on the orbiter. (Russia developed a space shuttle called Buran and launched it once, in 1988, without a crew. The program since has been terminated.)
SPACE STATION	A space facility intended for repeated or permanent occupancy by humans. The United States had one space station in the 1970s (Skylab) and, with other countries, is planning to build a permanent space station beginning in 1997 (see IRSS). Russia has successfully launched 7 space stations since 1971 (see Mir and Salyut); two others failed in orbit before they could be occupied.
SPDM	**Special Purpose Dexterous Manipulator**. Part of Canada's Mobile Servicing System (MSS, see above), it is sometimes described as the "fingers" that attach to the other portion of the MSS, nicknamed the "arm."
SPOT	**Systeme Probatiore d'Obsevation de la Terre.** (French acronym) A series of French remote sensing satellites.
SRB	**Solid Rocket Booster.** Reusable solid-fueled rocket that provides, along with the space shuttle main engines, the initial ascent thrust to lift the shuttle into orbit. The shuttle uses two SRBs that are attached to the external tank during launch. The SRB is made of six subsystems: the solid rocket motor (including case), propellant, thrust vector control, separation, recovery instrumentation, and electrical instrumentation. The SRB are jettisoned after use, recovered at sea, and then refurbished for future launches.
SRM	**Solid Rocket Motor.** Main section of the solid rocket booster that provides the thrust for initial ascent.
SSC	**Stennis Space Center.** NASA field center located in Bay St. Louis, MS. SSC is primarily involved in testing rocket engines. Formerly called NSTL (National Space Technology Laboratories). SSC is NASA's Center of Excellence for propulsion testing systems.
SSME	**Space Shuttle Main Engine.** First reusable liquid-fueled engine that is mounted on the shuttle orbiter. Each orbiter has three SSMEs that are used in conjunction with two solid rocket boosters to provide thrust for launch.
SSTI	**Small Spacecraft Technology Initiative.** NASA program designed to advance the state of technology and reduce the costs and development times associated with the design, integration, launch, and operation of small satellites. The program also actively promotes

	commercial applications of the technology. Lewis and Clark are the first two SSTI satellites (see above).
STARDUST	The fourth of the Discovery program missions, Stardust is a comet sample return mission. Scheduled for launch in 1999, it is designed to collect interstellar dust, and dust from Comet P/Wild 2, returning to Earth in 2006. it may also image the nucleus of the comet.
STEDI	**Student Explorer Development Initiative.** Originally one of the categories of Explorer missions, this has not transitioned into the University Explorer (UNEX) program. Two STEDI missions were selected and are underway: SNOE and Terriers (see those entries). A third had been selected as an alternate, and is now the first UNEX mission (CATSAT, see that entry).
STIS	**Space Telescope Imaging Spectrograph.** Instrument to be placed on the Hubble Space Telescope during the second Hubble servicing mission scheduled for February 1997. STIS will image astronomical objects in the ultraviolet, visible, and near-infrared wavelengths.
STS	**Space Transportation System.** The formal name for the U.S. space shuttle. The abbreviation is used to designate space shuttle missions (STS 1, SS 80, for example).
STSCI	**Space Telescope Science Institute.** Astronomical research center responsible for operating the Hubble Space Telescope as an international observatory that is run by the Association of Universities for Research in Astronomy (AURA). The institute is located in Baltimore, MD.
SWAS	**Submillimeter Wave Astronomy Satellite.** One of the SMEX series of Explorer spacecraft (see SMEX), SWAS is designed to study the chemical composition, energy balance, and structure of interstellar clouds. It is scheduled for launch in 1997.
TDRSS	**Tracking and Data Relay Satellite System.** A NASA satellite system in geostationary orbit for relaying communications between ground controllers and satellites or the space shuttle in low Earth orbit.
TERRIERS	**Tomographic Experiment Using Radiative Recombinative Ionospheric EUV and Radio Sources.** One of the STEDI Explorer missions (see STEDI). Designed by Boston University, it is scheduled for launch in 1997 to study the Earth's ionosphere/upper atmosphere.
TIMED	**Thermosphere, Ionesphere, Mesosphere Energetics and Dynamics.** Scientific mission designed to explore the Earth's mesosphere and lower thermosphere, the least explored and least understood regions of the atmosphere. The mission is cost capped at $100 million and is scheduled for launch in the year 2000. It is managed for NASA by Johns Hopkins University Applied Physics Laboratory.
TITAN	Family of U.S. expendable launch vehicles built by Lockheed Martin.

TOMS	**Total Ozone Mapping Spectrometer.** An instrument designed to measure the horizontal distribution of the ozone layer. NASA has launched several TOMS instruments either as free-flying spacecraft, or on other spacecraft (including U.S., Russian and Japanese spacecraft). A TOMS Earth Probe (see above) was launched in 1996.
TOS	**Transfer Orbit Stage.** An upper stage (see below) developed by Orbital Sciences Corp.
TPS	**Thermal Protection System.** Reusable materials ("Tiles") on the outer surface of the shuttle orbiter that protect it from the high temperatures generated during launch and entry back into the atmosphere from orbit.
TRACE	**Transition Region/Coronal Explorer.** One of the SMEX series of Explorer spacecraft (see SMEX). Scheduled for launch in 1997, TRACE is designed to study the relationship between the Sun's magnetic fields and heating of the solar corona.
TRMM	**Tropical Rainfall Measuring Mission.** U.S./Japanese satellite in the Earth Probe series that will carry a radar for measuring precipitation in the tropics. T RMM is scheduled for launch in 1997.
TSS	**Tethered Satellite System.** U.S./Italian program to develop a method for conducting space experiments using a tether (essentially a long cord) extended as much as 96 kilometers (60 miles) from the shuttle. Instruments for investigating the highest altitudes of Earth's atmosphere could "hang" from the shuttle on the tether to make these regions accessible for scientific study (they are too high for aircraft or balloons, and too low for spacecraft). Two test flights of this system failed (in 1992 and 1996).
UARS	**Upper Atmosphere Research Satellite.** Part of NASA's Mission to Planet Earth, UARS was launched by the shuttle in 1991 to carry out a systematic, comprehensive study of the stratosphere and furnish data on the mesosphere and thermosphere. A major focus of this mission is ozone production and depletion studies.
ULYSSES	An ESA spacecraft launched in 1990 to study the polar regions of the Sun (formerly called the International Solar Polar Mission-ISPM). Originally, both NASA and ESA agreed to build satellites so the Sun could be viewed simultaneously form the North and South poles. A U.S. decision in 1981 to abrogate its agreement and not build the NASA spacecraft (for budgetary reasons) is often cited as a quintessential example of the problems countries have in cooperating with the Untied States in space exploration. (The United States kept its part of the agreement to launch and track ESA's spacecraft and provide the power source for it.)
UPPER STAGE	A separate, small rocket used to take satellites into very high orbits or send them off into the solar system. They are used with both the space shuttle and expendable launch vehicles, which can only take satellites to low Earth orbits. (In some cases, upper stages are replace by "integral propulsion" where the extra rocket power is integrated

	into the satellite itself, but this approach cannot be used for all spacecraft.) Upper stages include Centaur, PAM, TOS, and IUS.
USA	**United Space Alliance.** A company formed by Lockheed Martin and Rockwell International (the latter was subsequently bought by Boeing) to be the "single prime contractor" for the space shuttle program.
UNEX	**University Explorer.** One of the three categories of Explorer programs (see Explorer), this formerly was called the STEDI program (see that entry). UNEX spacecraft are expected to have a design and development cost of less than $6 million (FY 1994 dollars) and take less than 2 years to develop. These costs exclude launch and mission operations and data analysis. One or two launches a year are planned. The first UNEX mission is CATSAT (see that entry).
USGCRP	**U.S. Global Change Research Program.** Interagency program coordinated through the White House Office of Science and Technology Policy to study long tern natural and human-induced changes to the Earth's environment. USGCRP is the U.S. portion of the International Geosphere-Biosphere Program (see IGBP) and Mission to Planet Earth is NASA's contribution to the USGCRP.
VAB	**Vehicle Assembly Building.** Building at Kennedy Space Centers where the external taken, solid rocket boosters, and shuttle orbiter are assembled together and placed on a mobile platform for transport to the launch pad.
VAFB	**Vandenberg Air Force Base.** Launch facility on the coast of California (north of Santa Barbara) from which satellites are placed into polar orbits. The Air Force operates the site for ELV launches. Vandenberg was to be a second launch site for the shuttle, but following the 1986 Challenger accident, shuttle launches from Vandenberg were canceled.
VENTURESTAR	The operational reusable launch vehicle Lockheed Martin plans to build following the X-33 technology development program (see X-33).
VIKING	Viking 1 and 2 are U.S. spacecraft, each consisting of an orbiter and lander, that reached Mars in 1976. The Viking 1 orbiter, which lasted the longest, returned data until 1983.
VOYAGER	Voyager 1 and 2 are U.S. spacecraft launched in 1977 to explore Jupiter and Saturn. Voyager 2 was then sent on to fly past Uranus (in January 1986) and Neptune (in August 1989).
WALLOPS FLIGHT CENTER	(See GSFC)
WCRP	**World Climate Research Program.** Part of the World Climate Program, WCRP is the focus of international climate research efforts. Its activities are guided by the World Meterological Organization and the International Council of Scientific Unions.

WHITE SANDS MISSILE RANGE	(see JSC)
WIND	Part of NASA's GGS initiative (see above), Wind was launched in November 1994 and is studying the interaction between the Earth and Sun as a result of the solar wind.
WIRE	**Wide Field Infrared Explorer.** One of the SMEX series of Explorer spacecraft (see SMEX), WIRE will study galactic evolution. Launch is scheduled for 1998.
X-33	Technology development and flight-test program for a large reusable launch vehicle. The program is funded jointly between NASA and Lockheed Martin. They hope the technologies developed and flight-tested will lead to an operational vehicle, Venturestar, in the first decade of the 21st Century. As currently planned, Venturestar would be funded entirely by the private sector, not NASA.
X-34	NASA technology development and flight-test program for a small reusable launch vehicle. Plans for this program to be a joint government-industry program were terminated when the industrial partners, Orbital Sciences Corp. (OSC) and Rockwell International, withdrew from the program. OSC later was selected by NASA to work as a contractor on the program.
X-38	(See CRV)
ZENIT	Ukrainian expendable launch vehicle (some parts are built in Russia). Ukraine offers commercial launches on Zenit. An international joint venture called Sea Launch, involving companies in the U.S. (Boeing), Ukraine (Yuzhnoye), Russia (Energia), and Norway (Kvaerner), also offers commercial launches of Zenit.

Chapter 8

AN OVERVIEW OF NASA'S MISSION TO PLANET EARTH (MTPE)

David P. Radzanowski and Stephen J. Garber

ABSTRACT

Mission to Planet Earth (MTPE) is the National Aeronautics and Space Administration's (NASA) central contribution to the U.S. Global Change Research Program (USGCRP). The MTPE program provides scientific information so policymakers and scientists can formulate strategies to mitigate human impacts on Earth's environment, such as ozone depletion, deforestation, and possible global warming. MTPE includes a constellation of satellites in various Earth orbits: the program's centerpiece known as the Earth Observing System (EOS) and small satellites known as the Earth Probes. The program also included Landsat 7, ground and aircraft based research, the EOS Data Information System (EOSDIS), and a community of scientists performing research with acquired data from previous or ongoing projects known as phase of missions. The objective of EOS is to acquire a long-term, comprehensive set of environmental measurements about the Earth, particularly those related to global climate change.

In addition, the MTPE program includes data from international Earth observation programs and cooperative projects with Canada, the European Space Agency (ESA), Japan, and Russia. NASA had performed many Earth science programs in the past. This report describes some of the ongoing NASA programs that are considered MTPE phrase one missions. EOS and the Earth Probes were new starts in FY 1991, and in 1993 MTPE was taken out of NASA's Office of Space Science and Applications to become a stand alone program office.

For FY 1995, NASA has $1.340 billion available for MTPE. In February 1995, NASA released its FY 1996 budget request; the proposed figure for MTPE is $1.341 billion.

MTPE Phase One Missions

MTPE's phase one missions include several free-flying satellites for global change observations that are the culmination of planning that began in the late 1970s and 1980s. These missions are: the Upper Atmosphere Research Satellite (UARS) that provides data on global ozone change; the TOPEX/Poseidon mission (with France) that is studying ocean circulation; and the Laser Geodynamics Satellite II (LAGEOSII), a joint U.S. and Italian project to acquire information on crustal deformation, continental drift, and ocean tides.[1] A private firm, Orbital Sciences Corp., also is launching a remote sensing spacecraft called SeaStar in 1995 to collect ocean color data. NASA has contracted with Orbital Sciences to purchase data from the Sea-Viewing Wide Filed Sensor (SeaWiFS) instrument aboard SeaStar.

Earth Observing System (EOS) and Earth Probes

The Earth Observing System (EOS) is the core space-based segment of the MTPE program. The goal of this long-term program is to collect and make available for analysis a wide variety of data to assess trends in global warming, ozone depletion, and deforestation. The three main U.S. (NASA) contributions to EOS are the AM, PM, and Chemistry spacecraft series. The SeaWiFS instrument aboard SeaStar is to serve as the first EOS-Color mission to examine the ecology of Earth's large bodies of water. IF this proves productive, NASA may consider buying more Color data from a private company that would develop, launch, and operate another SeaStar-type spacecraft. Landsat 7, scheduled for launch in 1998, also is considered part of EOS now. In FY 1995 NASA has $591.1 million available for EOS spacecraft, the same amount it is requesting for FY 1996.

Each of the three EOS flight series (AM, PM, and Chemistry) is designed to include up to three spacecraft that would be launched at up to 6-year intervals so that measurements could be taken for each series over an 18-year period. Such an extended time frame is useful to Earth scientists who would like to examine data over a complete solar cycle of 10 to 15 years.

To supplement EOS, NASA plans a series of small satellites called Earth Probes. These satellites will carry instrumentation that requires special orbits or special spacecraft. A table at the end of this report provides details about EOS and Earth Probes Spacecraft.

EOS Data and Information System (EOSDIS)

The data-handling part of MTPE is known as the EOS Data and Information System. EOSDIS is to provide the processing, storage, and distribution of all data collected by EOS as well as the resulting scientific products. EOSDIS also is designed to have spacecraft and instrument command and control capability, and provide data archive, distribution, and information management of all NASA Earth science data. Many feel that the key to success of NASA's entire MTPE program rests with EOSDIS. While NASA certainly has handled

---

[1] UARS was launched in Sept. 1991, TOPEX/Poseidon was launched by the European Space Agency (ESA) in Aug. 1992, and LAGEOS II was launched by the Space Shuttle in Oct. 1992.

very large amounts of information from space, the vast scale of EOS data is likely to be unprecedented.

EOSDIS is to be an evolving system that is continually updated with new advances in computing and networking technology. NASA believes that this "open architecture" approach will provide flexibility to adapt a system of qualitatively new scale in unforeseen ways. Given this open architecture, a reliable, yet flexible plan for EOSDIS funding is an important consideration.

The first major activity in EOSDIS, known as Version 0, is to develop an experience base for handling large Earth science data sets by reworking existing data sets into more user-friendly formats. Version 0 became operational in September 1994. A working prototype for future EOSDIS versions, Version 0 uses existing scientific networks to link research sites across the country. Interfaces will be developed among nine existing Distributive Active Archive Centers (DAACs)[2] to make multiple data bases available to users ina unified format. By providing the DAACs with the capability to process large amounts of data continuously, it is hoped that investigators will be able to identify new scientific variables of interest so that data collection methods could be changed midstream in the program. The next version of EOSDIS, Version 1, is expected in 1996 after the Tropical Rainfall Measuring Mission (TRMM) is launched and begins collecting data.

NASA has $230.6 million available in FY 1995 for EOSDIS. For FY 1996, NASA is requesting $289.8 million for EOSDIS.

SUMMARIES OF SOME KEY ISSUES RELATED TO EOS AND EOSDIS

Cost of EOS

In its original configuration, EOS included two series of polar orbiting spacecraft platforms called EOS-A and EOS-B. Each was to have 10-16 different instruments and was to be replaced every 5 years to achieve a 15-year mission lifetime. Initially, EOS was estimated to cost $17 billion through FY 2000. During 1991 and 1992, NASA responded to growing concerns about the high cost of EOS by breaking up the EOS-A and EOS-B platforms into several smaller spacecraft and by streamlining management and technical configurations. In mid-1992, NASA estimated that the reconfigured EOS program would cost approximately $8 billion through FY 2000.

In fall 1994, NASA took several steps to trim the EOS and EOSDIS program budget through FY 2000 by approximately $750 million to a total of $7.25 billion in response to Federal budget constraints. The EOS-Altimetry mission was split into Laser and Radar components and the Radar Altimetry segment was cost capped. (The chart at the end of this report provides further details.) By delaying the purchase of certain EOSDIS computer equipment until the technology is more mature, NASA believes it can save money.

---

[2] The current DAACs are:" The Alaska Synthetic Aperture Radar (SAR) Facility; the Earth Resources Observation System (EROS) Data Center; the Consortium for International Earth Science Network; NASA's Goddard Space Flight Center DAAC; NASA's Langley Research Center DAAC' the Oak Ridge National Laboratory; NASA's Marshall Space Flight Center/University of Alabama DAAC; the National Snow and Ice Data Center; and NASA's Jet Propulsion Laboratory (JPL) DAAC. There is a related Data Processing Facility in Fairmont, WV.

Reconfiguring payloads also will enable EOS spacecraft after EOS-AM1 to be launched on Delta expendable launch vehicles, instead of the larger and more expensive Atlas rockets.

Although NASA has reduced the cost of EOS and EOSDIS from $17 billion to $7.25 billion through FY 2000, many observers believe that NASA will not have sufficient resources to fund EOS and EOSDIS at their current predicted levels given NASA's overall budget. While NASA has not released an estimate of the total program, cost the currently projected 18 year operational lifespans for each spacecraft series imply that $7.25 billion would be only a fraction of the total runout cost. Once several EOS spacecraft are collecting data after 2000, the operating costs of EOSDIS will likely rise while NASA continues to develop other EOS spacecraft. In an era of constrained budgets, there is concern that in the future NASA may not have sufficient funding for several large programs such as EOS and EOSDIS, the Space Shuttle, the Space Station, and space science.

Complexity of EOSDIS

Outside of NASA, EOSDIS hs been the subject of some controversy. In January 1994, the National Research Council (NRC) released an independent review that found fault with the overly centralized nature of EOSDIS, noting that such an approach was risky, inflexible, and unresponsive to its users. Although this report did not take into consideration changes that EOSDIS managers made in the fall of 1993, NASA made further modifications to address the NRC's criticism. In addition, a 1992 Government Accounting Office (GAO) report, *Earth Observing System: NASA's EOSDIS Development Approach is Risky,* stated that EOSDIS Version 0 did not adequately foster development of the advanced technologies that will eventually be required by EOSDIS. NASA believes that the open architecture of EOSDIS directly allays the GAO's concerns.

Usefulness of EOS Scientific Data

Some scientists have suggested that because Earth system science is such a young and broad field, a wide variety of multi-purpose instruments should be included on EOS so that various measurements can be made to pursue unforeseen paths of inquiry. Some scientists even feel that it is difficult to formulate certain relevant scientific questions. EOS critics, however, object to spending large sums of money for EOS when its objectives do not seem to be defined clearly. EOS managers, however, argue that they have established clear scientific priorities. Advocates also argue that it is science's charter to explore untapped areas of knowledge, especially in relatively new disciplines such as Earth system science.

Viability of Privatization

The issue of privatization or commercialization cuts across many of NASA's space activities including MTPE. Despite the many commercial uses for remotely sensed information. The Government is still the largest consumer of complex scientific information such as MTPE data. With launch vehicle and spacecraft development costs still high, it is difficult for private firms to recoup their investments by selling data that their spacecraft collect. Thus few private businesses have built and launched scientific spacecraft without Government support. While the future cost of access to space potentially could be significantly reduced by using new launch vehicle technology, this is probably 10 to 15 years away. Moreover, it likely will take a considerable amount of time before industry is willing to undertake technical projects with such high financial risks as building and launching spacecraft without significant Government support.

Pricing of Data

Current U.S. Government policy is that EOS and USGCRP data shouldn't cost more than the marginal cost of reproduction for the user. In considering whether to institute user fees, policymakers will likely evaluate whether a billing system would be cost effective for small orders and whether casual users might take advantage of the system by requesting frivolous amounts of data. In addition, the various international participants may need to coordinate their data policies in terms of granting researchers from all countries equal access to equivalently priced data from all instruments. Some European nations have proposed distributing EOS data on a partially commercial basis, while the U.S. is inclined toward providing freedom of access to data.

SUMMARY AND FUTURE OUTLOOK

While the Clinton Administration and many outside observers believe that environmental programs should be a high priority for both the Nation and for NASA in particular, the Republican-controlled 104th Congress may undertake a rigorous examination of MTPE. Chairman Robert Walker of the House Science Committee reportedly has stated that although NASA's budget should keep pace with inflation, the MTPE program will be scrutinized carefully for potential cost savings.[3] In January 1995, the White House directed NASA to cut its budget by $5 billion over the next five years, and it may be difficult for MTPE to continue to fund an EOS program that is to cost $7.3 billion through the year 2000. Despite various restructurings of EOS, many analysts are still concerned about the size and structure of EOS and EOSDIS. Yet many scientists are excited about the potential new discoveries that could be possible through international cooperation on a truly global issue such as the environment.

---

[3] Ferster, Warren. House Panel May Target Mission to Planet Earth, *Space News*, Dec. 19-25, 1994. p.3.

Earth Probes* and EOS Spacecraft

Mission	Launch Date	Scientific Study Goals	Comments
*TOMS (Total Ozone Mapping Spectrometer)	1978, 1991, 5/95, 1996	Atmospheric ozone concentrations.	TOMS are sets of instruments launched on various spacecraft.
*NSCAT (NASA Scatterometer)	1996	Ocean surface winds.	To fly on Japanese ADEOS spacecraft.
*TRMM (Tropical Rainfall Measuring Mission)	1997	Precipitation in tropical regions.	Joint program with Japan.
EOS-AM	6/98, 6/04	Physical properties of clouds; air-land transfers of water, carbon and energy.	Spacecraft to observe same positions on Earth every afternoon.
EOS-PM	12/00, 12/06	Cloud formation; air-sea exchanges of moisture, energy.	Spacecraft to observe same positions on Earth every afternoon.
EOS-Chemistry	12/02	Chemical properties of atmosphere's inner layers.	Third major EOS series.
EOS-Aerosol	12/98, 2000, 2005	Aerosol and gaseous components of the atmosphere.	First launch in '98 is to be SAGE-III aboard Russian vehicle. Second flight is to be aboard Space Station. Third flight is an undetermined flight of opportunity.
EOS-Altimetry	12/99, 7/03, 2004	Sea levels, ocean currents.	Radar altimeter to fly aboard Topex/Poseidon follow-on or Navy's Geosat in '99, laser altimeter to fly in '03, & radar altimeter in '04.
Landsat-7	12/98	Land remote sensing.	Various commercial uses also.

PART III. BIBLIOGRAPHY

BOOKS

1958 NASA/USAF space probes (Able-1); final report. Published/Created: [Washington, National Aeronautics and Space Administration] 1959-Related Authors: United States. Air Force Ballistic Missile Division.Related Titles: NASA/USAF space probes.Description: 3 v. illus., maps. 27 cm. Subjects: Lunar probes Series: United States National Aeronautics and Space Administration. NASA memorandum. LC Classification: TL799.M6 S65

1967 Summer Study of Lunar Science and Exploration: [reports] Directed by Wilmot N. Hess. Published/Created: Washington. Scientific and Technical Information Division, National Aeronautics and Space Administration; [available from the Clearinghouse for Federal Scientific and Technical Information, Springfield, Va.] 1967. Related Authors: California. University, Santa Cruz. Manned Spacecraft Center (U.S.) Description: vi, 398 p. illus. 27 cm. Notes: "Held at the University of California-Santa Cruz, Santa Cruz, California ... under the auspices of Manned Spacecraft Center, NASA." Subjects: Moon--Congresses. Outer space--Exploration--Congresses. Series: United States. National Aeronautics and Space Administration. NASA; SP-157. LC Classification: QB581 .S82 1967c Dewey Class No.: 629.4

A bibliography of adult aerospace books and materials. Compiled for National Aeronautics and Space Administration. Published/Created: Washington [U.S. Govt. Print. Off., 1961] Description: 36p. 26cm. Notes: NASA EP-3. Subjects: Aeronautics--Bibliography. Astronautics--Bibliography. Astronomy--Bibliography. LC Classification: Z5063 .N23 1961

A bibliography of aerospace books and teaching aids for elementary school pupils and teachers. Compiled for Educational Services Branch, National Aeronautics and Space Administration. Published/Created: [Washington, U.S. Govt. Print. Off., 1961] Related Titles: Aeronautics and space bibliography for elementary grades. Description: 26 p. 26 cm. Notes: Cover Aeronautics and space bibliography for elementary grades. Subjects: Astronautics--Juvenile literature--Bibliography. Aeronautics--Juvenile literature--Bibliography. Astronautics--Film catalogs. Aeronautics--Film catalogs. LC

Classification: Z5064.A8 N3 1961 Dewey Class No.: 016.372/8/629/4

A Bibliography on the search for extraterrestrial intelligence / Eugene F. Mallove ... [et al.]. Published/Created: [Washington]: National Aeronautics and Space Administration, Scientific and Technical Information Office; [Springfield, Va.: for sale by the National Technical Information Service], 1978. Related Authors: Mallove, Eugene F. Description: 132 p.; 28 cm. Subjects: Life on other planets-- Bibliography. Interstellar communication--Bibliography. Series: United States. National Aeronautics and Space Administration. Scientific and Technical Information Office. NASA reference publication; 1021. LC Classification: TL521.3 .N17 no. 1021 Z5154.L5 QB54 Dewey Class No.: 629.1/08 s 016.574999

A catalog of NASA special publications. Published/Created: Washington, D.C.: Scientific and Technical Information Branch, National Aeronautics and Space Administration; Springfield, Va.: For sale by the National Technical Information Service], 1981. Description: vii, 104 p.; 26 cm. Subjects: United States. National Aeronautics and Space Administration-- Bibliography--Catalogs. Astronautics-- Bibliography--Catalogs. Series: NASA Sp; 449 LC Classification: Z5061 .U6 1981 TL790 Dewey Class No.: 016.6294 19 Govt. Doc. No.: NAS 1.21:450

A walk in space; Gemini 4 extra-vehicular activity. Published/Created: [Washington, For sale by the Superintendent of Documents, U.S. Govt. Print. Off., 1965] Description: 1 v. (unpaged) illus., ports. 27 cm. Notes: Cover title. On June 3, 1965, during the third revolution of an extended earth orbital mission in space by NASA's Gemini 4, Astronauts James A. McDivitt and Edward H. White II carried out the first Extravehicular Activity in the United States manned space flight program. Subjects: McDivitt, James Alton, 1929- White, Edward Higgins, 1930- Project Gemini (U.S.) Extravehicular activity (Manned space flight) LC Classification: TL789.8.U6 G8

Aeronautical and astronautical events. Published/Created: Washington, National Aeronautics and Space Administration. Related Authors: United States. National Aeronautics and Space Administration. Description: p. cm. Notes: PREMARC/SERLOC merged record Subjects: Aeronautics-- History--Yearbooks. Astronautics-- History--Yearbooks. LC Classification: TL501 .A299

Aeronautics and space bibliography of adult aerospace books and materials. Compiled for National Aeronautics and Space Administration. Edition Information: 2d. ed. Published/Created: Washington, U.S. Govt. Print. Off. [1963] Related Authors: United States. National Aeronautics and Space Administration. Description: iv, 42 p. map. 26 cm. Notes: Published in 1961 under A bibliography of adult aerospace books and materials. Subjects: Aeronautics--Bibliography. Astronautics--Bibliography. Astronomy--Bibliography. LC Classification: Z5063 .N23 1963

Aeronautics in NACA and NASA / Langley Research Center, Ames Research

Center, Lewis Research Center. Published/Created: [Washington, D.C.: NASA, 1991] Related Authors: United States. National Advisory Committee for Aeronautics. United States. National Aeronautics and Space Administration. Langley Research Center. Lewis Research Center. Ames Research Center. Description: 73 p.: ill. (some color); 26 cm. Notes: Includes bibliographical references (p. 72-73). Subjects: United States. National Aeronautics and Space Administration-- History. United States. National Advisory Committee for Aeronautics History. Aeronautics--Research. Astronautics--Research. LC Classification: TL521.312 .A623 1991 Dewey Class No.: 629.13/0072073 20

Aerospace bibliography. Compiled for Educational Programs Division, Office of Public Affairs, National Aeronautics and Space Administration. Edition Information: 3d ed. Published/Created: Washington, For sale by the Supt. of Docs., U.S. Govt. Print. Off. [1966] Related Authors: United States. National Aeronautics and Space Administration. Educational Programs Division. Description: v, 71 p. illus. 26 cm. Notes: The first 2 editions were each issued in 3 pts. and published separately with special titles. "This third edition, now under one cover, includes books published in the period January 1963 through summer 1965." Subjects: Aeronautics--Bibliography. Astronautics--Bibliography. LC Classification: Z5060 .N3 1966 Dewey Class No.: 016.6291

Akens, David S. Historical origins of the George C. Marshall Space Flight Center. Published/Created: Huntsville, Ala., Historical Office, Office of Management Services, George C. Marshall Space Flight Center, National Aeronautics and Space Administration, 1960. Description: 1 v. (various pagings) plates, group ports., maps, diagrs. 27 cm. Subjects: George C. Marshall Space Flight Center. Redstone Arsenal (Ala.) Series: George C. Marshall Space Flight Center. MSFC historical monograph no. 1 LC Classification: TL862.G4 A3 no. 1 Dewey Class No.: 629.42

Alexander, Kent. Countdown to glory: NASA's trials and triumphs in space / Kent Alexander. Published/Created: Los Angeles: Price Stern Sloan, c1989. Description: 192 p.: ill. (some col.); 37 cm. ISBN: 0895867877 : Notes: "A Friedman Group book." Includes bibliographical references (p. [189]). Subjects: United States. National Aeronautics and Space Administration-- History. LC Classification: TL521.312 .A626 1989 Dewey Class No.: 353.0087/78 19

Allaway, Howard. The space shuttle at work / Howard Allaway. Published/Created: Washington: Scientific and Technical Information Branch, and Division of Public Affairs, National Aeronautics and Space Administration: for sale by the Supt. of Docs., U.S. Govt. Print. Off., 1979. Description: 76 p.: ill.; 26 cm. Notes: Includes index. Subjects: Space shuttles. Series: NASA SP; 432. LC Classification: TL795.5 .A39 Dewey Class No.: 629.44

Amending the National aeronautics and space act of 1958 with respect to property rights in inventions; Published/Created: [Washington, U. S. Govt. Print Off., 1962] Related Titles: National aeronautics and space act of

1958. Description: p. cm. Subjects: United States. National Aeronautics and Space Administration. Patents and government-developed inventions--United States. Astronautics--United States--Patents. Series: United States. 87th Congress, 2d session, 1962. House. Report no. 2185. LC Classification: KF32 .S3 1962

Amending the National aeronautics and space act of 1958. Published/Created: [Washington, U. S. Govt. Print. Off.] 1961. Description: p. cm. Subjects: United States. National Aeronautics and Space Council. Series: United States. 87th Congress, 1st session, 1961. Senate. Report no. 175. LC Classification: TL521.312 .A53

Amending various sections of the NASA act of 1958. Published/Created: Washington, U. S. Govt. Print. Off., 1961. Description: p. cm. Subjects: United States. National Aeronautics and Space Administration. LC Classification: TL521 .A5423 1961c

Amendment to the National aeronautics and space act of 1958. Published/Created: Washington, U. S. Govt. Print. Off., 1959. Related Titles: National aeronautics and space act of 1958. Description: p. cm. Subjects: National Aeronautics and Space Administration. LC Classification: TL521 .A5423 1958a

America in space the first five years; a pictorial review. Published/Created: [Washington] National Aeronautics & Space Administration; [for sale by the Superintendent of Documents, U. S. Govt. Print. Off., 1963] Description: p. cm. Subjects: Astronautics--Pictorial works. Astronautics--United States. LC Classification: TL793.5 .U56 1963

America in space; a pictorial review. Published/Created: [Washington] National Aeronautics & Space Administration; [for sale by the Superintendent of Documents, U. S. Govt. Print. Off., 1964] Description: p. cm. Subjects: Astronautics--Pictorial works. Astronautics--United States. LC Classification: TL793.5 .U56 1964

America's spaceport: John F. Kennedy Space Center. Published/Created: [Washington, D.C.?]: National Aeronautics and Space Administration, John F. Kennedy Space Center, [1994] Related Authors: John F. Kennedy Space Center. Description: [44] p.: ill.; 28 cm. Subjects: John F. Kennedy Space Center--History. Launch complexes (Astronautics)--Florida. Astronautics--United States--History. Outer space--Exploration--United States. LC Classification: TL4027.F52 J634 1994 Govt. Doc. No.: NAS 1.2:AM 3/994

An index to NASA tech briefs (briefs 63-10003 through 64-10211) a keyword-in-context index (KWIC) Compiled by Robert E. Booth [and others] Published/Created: Detroit [1965] Related Authors: Booth, Robert E. (Robert Edmond), 1917- United States. National Aeronautics and Space Administration. Technology Utilization Division. NASA tech brief. Related Titles: NASA tech briefs. Description: 53 p. 28 cm. Notes: "CAST-016502." Subjects: United States. National Aeronautical and Space Administration. Technology Utilization Division. NASA tech brief--Indexes. LC Classification: TL521.3.T4 A282

Analysis of Apollo 10; photography and visual observations. Published/Created: Washington, Scientific and Technical Information Office, National Aeronautics and Space Administration; [for sale by the Supt. of Docs., U.S. Govt. Print. Off.] 1971. Description: vii, 226 p. illus., maps (6 fold. col. in pocket) 29 cm. Notes: Includes bibliographical references. Subjects: Project Apollo (U.S.) Moon--Photographs from space. Moon--Observations. Series: NASA SP; 232. LC Classification: QB595 .U53 Dewey Class No.: 523.39 Govt. Doc. No.: NAS 1.21:232

Analysis of Apollo 8; photography and visual observations. Compiled by NASA Manned Spacecraft Center. Published/Created: Washington, Scientific and Technical Information Division, National Aeronautics and Space Administration; [for sale by the Supt. of Docs., U.S. Govt. Print. Off.] 1969. Description: ix, 337 p. illus., maps (4 fold. col. in pocket) 29 cm. Notes: Includes bibliographical references. Subjects: Project Apollo (U.S.) Moon--Photographs from space. Series: NASA SP; 201. LC Classification: TL789.8.U6A598 Dewey Class No.: 523.39

Analysis of cost estimates for the space shuttle and two alternate programs; report to the Congress [on the] National Aeronautics and Space Administration, by the Comptroller General of the United States.
Published/Created: [Washington] 1973. Description: 67 p. illus. 27 cm. Notes: Cover title. "B-173677." Subjects: United States. National Aeronautics and Space Administration. Space shuttles--Costs. LC Classification: TL795.5 .U55 1973 Dewey Class No.: 353.008/778 Govt. Doc. No.: GA 1.13:Sp1/5

Analysis of Surveyor 3 material and photographs returned by Apollo 12. Published/Created: Washington, Scientific and Technical Information Office, National Aeronautics and Space Administration; [for sale by the Supt. of Docs., U.S. Govt. Print. Off.] 1972. Description: viii, 295 p. illus. 27 cm. Notes: Includes bibliographical references. Subjects: Surveyor Program (U.S.) Project Apollo (U.S.) Space vehicles--Materials. Space photography--United States. Series: NASA SP; 284. LC Classification: TL789.8.U6 S92 Dewey Class No.: 629.47/2 Govt. Doc. No.: NAS 1.21:284

Anderson, Frank Walter.
Orders of magnitude: a history of NACA and NASA, 1915-1976 / by Frank W. Anderson, Jr. Published/Created: Washington: National Aeronautics and Space Administration, Scientific and Technical Information Office: for sale by the Supt. of Docs., U.S. Govt. Print. Off., 1976. Description: ix, 100 p.: ill.; 23 cm. Notes: Bibliography: p. 99-100. Subjects: United States. National Advisory Committee for Aeronautics History. United States. National Aeronautics and Space Administration--History. Series: NASA history series. NASA SP; 4403. LC Classification: TL521.312 .A63 Dewey Class No.: 629.1/07/2073

Anderson, John David.
Determination of the masses of the Moon and Venus and the astronomical unit from radio tracking data of the Mariner II spacecraft [by] John D. Anderson.

Published/Created: Pasadena, Calif., Jet Propulsion Laboratory, California Institute of Technology, 1967. Description: viii, 93 p. illus. 28 cm. Notes: "Prepared under contract no. NAS 7-100, National Aeronautics and Space Administration." Subjects: Project Mariner (U.S.) Astronomical unit. Moon--Mass. Venus (Planet)--Mass. Series: California Institute of Technology, Pasadena. Jet Propulsion Laboratory. JPL technical report 32-816. LC Classification: QB501 .A55 Dewey Class No.: 523.3/1

Anderton, David A.
Aeronautics, by David A. Anderton. Published/Created: Washington, National Aeronautics and Space Administration; [for sale by the Supt. of Docs., U.S. Govt. Print. Off., 1971] Description: 24 p. illus. 27 cm. Notes: "EP-85." Subjects: United States. National Aeronautics and Space Administration. Aeronautics--United States. Series: Space in the seventies LC Classification: TL521.312 .A64 1971 Dewey Class No.: 629.13/00973 Govt. Doc. No.: NAS 1.19:85

Anderton, David A.
Aeronautics, by David A. Anderton. Published/Created: Washington, National Aeronautics and Space Administration; [for sale by the Supt. of Docs., U.S. Govt. Print. Off., 1970] Description: 22 p. illus. 27 cm. Notes: "EP 61." Subjects: United States. National Aeronautics and Space Administration. Aeronautics--United States. Series: America in space. LC Classification: TL521.312 .A64 Dewey Class No.: 629.13/00973

Anderton, David A.
Apollo 17 at Taurus Littrow. [Text by David A. Anderton. Published/Created: Washington, Office of Public Affairs, National Aeronautics and Space Administration; for sale by the Supt. of Docs., U.S. Govt. Print. Off., 1973] Related Authors: United States. National Aeronautics and Space Administration. Office of Public Affairs. Description: 32 p. illus. 31 cm. Notes: Cover title. "EP-102." Subjects: Project Apollo (U.S.) Moon--Exploration. LC Classification: TL789.8.U6 A662 Dewey Class No.: 629.45/4 Govt. Doc. No.: NAS 1.19:102

Anderton, David A.
Man in space, by David A. Anderton. Published/Created: Washington, National Aeronautics and Space Administration; [for sale by the Supt. of Docs., U.S. Govt. Print. Off., 1969] Description: 30 p. illus. 27 cm. Notes: "EP-57." Subjects: Manned space flight. Astronautics--United States. Series: America in space. LC Classification: TL789.8.U5 A68 Dewey Class No.: 629.4/0973

Anderton, David A.
Sixty years of aeronautical research, 1917-1977 / by David A. Anderton. Published/Created: Washington: National Aeronautics and Space Administration: for sale by the Supt. of Docs., U.S. Govt. Print. Off., 1978. Related Authors: United States. National Aeronautics and Space Administration. Description: 89 p.: ill.; 26 cm. Notes: "EP145." Subjects: Langley Research Center. LC Classification: TL521.312 .A65 Dewey Class No.: 629.13/007/20755412

Anna, Henry John, 1941-
Task groups and linkages in complex organizations: a case study of NASA /

Henry J. Anna. Published/Created: Beverly Hills, Calif.: Sage Publications, c1976. Description: 64 p.; 22 cm. ISBN: 0803905335 : Notes: Bibliography: p. 64. Subjects: United States. National Aeronautics and Space Administration. Organizational behavior--Case studies. Complex organizations--Case studies. LC Classification: TL521.312 .A66 Dewey Class No.: 301.18/32

Apollo 11 mission report.
Published/Created: Washington, Scientific and Technical Information Office, National Aeronautics and Space Administration; [for sale by the National Technical Information Service, Springfield, Va.] 1971. Related Authors: Apollo 11 (Spacecraft) Description: x, 217 p. illus. 27 cm. Notes: Includes bibliographical references. Subjects: Project Apollo (U.S.) Space flight to the moon. Series: NASA SP; 238. LC Classification: TL789.8.U6 A59916 Dewey Class No.: 629.45/4

Apollo 11: preliminary science report. Published/Created: Washington; [For sale by the Clearinghouse for Federal Scientific and Technical Information, Springfield, Va.] 1969. Description: vii, 204 p. illus., maps. 27 cm. Notes: Includes bibliographies. Subjects: Project Apollo (U.S.) Series: NASA SP; 214. LC Classification: TL789.8.U6 A58 1969 Dewey Class No.: 629.45/4

Apollo 11--man on the moon / NASA. Published/Created: United States: [s.n.], 1969. Related Authors: National Aeronautics and Space Administration. AFI/Triangle Laboratories Estate Collection (Library of Congress) Description: 1 reel of 1 (ca. 749 ft.): si., col.; 16 mm. ref print. Contents: Command module docks with LEM for trip to moon -- Preparing Eagle for landing -- The Eagle separates from Columbia -- The Eagle lands -- Armstrong cautiously moves away from LEM -- Collecting moon rock samples -- Aldrin joins Armstrong on moon's surface -- The U.S. flag implanted -- Aldrin testing mobility in moon's 1/6th gravity -- Kangaroo hop -- Loading moon rock samples -- Armstrong climbs back into Eagle -- The last look before liftoff --Tense and dramatic liftoff from the moon -- Vital rendezvous of Eagle and Columbia -- United again, the astronauts head back to Earth -- Home safe Armstrong, Aldrin and Collins -- President Nixon extends welcome. Notes: Copyright: reg. unknown. 9:32 a.m., E.D.T., July 16, 1969 appears at the beginning. Sources used: NICEM Index to 16 mm educational films, vol 3., 1980, p. 61; Video Source Book, vol. 1, 1997, p. 1002; LCCN record no. 75-701617. Source of Acquisition: Received: 2/2/82; ref print; gift, ATM 133; AFI/Triangle Laboratories Estate Collection. Subjects: Project Apollo (U.S.) Apollo 11 (Spacecraft) Space flight to the moon. LC Classification: FAB 7790 (ref print)

Apollo 12 preliminary science report. Published/Created: Washington, Scientific and Technical Information Division, National Aeronautics and Space Administration; [for sale by the Clearinghouse for Federal Scientific and Technical Information, Springfield, Va.] 1970. Description: xii, 227 p. illus. 27 cm. Notes: Includes bibliographies. Subjects: Project Apollo (U.S.) Space sciences. Moon--Exploration. Series: NASA SP; 235. LC Classification: TL789.8.U6A599 Dewey Class No.:

550

Apollo 13: "Houston, we've got a problem." Published/Created: [Washington; For sale by the Supt. of Docs., U.S. Govt. Print. Off., 1970] Description: 25 p. illus., ports (both part col.) 29 cm. Subjects: Apollo 13 (Spacecraft) Project Apollo (U.S.) Space vehicle accidents. LC Classification: TL789.8.U6 A63 1970 Dewey Class No.: 629.45/4

Apollo 14: preliminary science report. Published/Created: Washington, Scientific amd Technical Information Office, National Aeronautics and Space Administration; [for sale by the Supt. of Docs., U.S. G.P.O.] 1971. Description: xxiv, 309 p. illus. 26 cm. Notes: Includes bibliographical references. Subjects: Project Apollo (U.S.) Space sciences. Series: NASA SP; 272. LC Classification: TL789.8.U6 A59913 Dewey Class No.: 555.9/1 Govt. Doc. No.: NAS1.21:272

Apollo 15 at Hadley Base. Published/Created: [Washington; For sale by the Supt. of Docs., U.S. Govt. Print. Off., 1971] Description: 32 p. illus. (part col.) 23 x 31 cm. Notes: Cover title. "EP-94." Subjects: Project Apollo (U.S.) Space flight to the moon. LC Classification: TL789.8.U6 A63 1971 Dewey Class No.: 559.9/1 Govt. Doc. No.: NAS 1.19:94

Apollo 15: preliminary science report. Published/Created: Washington, Scientific and Technical Information Office, National Aeronautics and Space Administration; [for sale by the Supt. of Docs., U.S. Govt. Print. Off.] 1972. Description: 1 v. (various pagings) illus. 26 cm. Notes: Includes bibliographical references. Subjects: Project Apollo (U.S.) Lunar geology. Moon--Exploration. Series: NASA SP; 289. LC Classification: TL789.8.U6 A59914 Dewey Class No.: 508.99/1 Govt. Doc. No.: NAS 1.21:289

Apollo 16: expedition to Descartes. Published/Created: [Washington, D.C.]: NASA, 1972. Related Authors: United States. National Aeronautics and Space Administration. Description: 8 p.: ill.; 26 cm. Subjects: Moon--Exploration. Apollo 16 (Spacecraft) Astronautics in astronomy. LC Classification: QB581 .A65 1972 Dewey Class No.: 919.9/104 20

Apollo 16: preliminary science report. Published/Created: Washington, Scientific and Technical Information Office, National Aeronautics and Space Administration; [for sale by the Supt. of Docs., U.S. Govt. Print. Off.] 1972. Description: 1 v. (various pagings) illus. 27 cm. Notes: Includes bibliographical references. Subjects: Project Apollo (U.S.) Lunar geology. Moon--Exploration. Series: NASA SP; 315. LC Classification: TL789.8.U6 A59915 Dewey Class No.: 508.99/1 Govt. Doc. No.: NAS 1.21:315

Apollo 17: preliminary science report / prepared by Lyndon B. Johnson Space Center. Published/Created: Washington: Scientific and Technical Information Office, National Aeronautics and Space Administration: for sale by the Supt. of Docs., U.S. Govt. Print. Off., 1973. Description: 650 p. in various pagings: ill.; 26 cm. Notes: Includes bibliographical references. Subjects: Project Apollo (U.S.) Moon. Series: NASA SP; 330. LC Classification: QB581 .L96 1973 Dewey Class No.:

559.9/1 Govt. Doc. No.: NAS 1.21:330

Apollo 8: man around the moon. Published/Created: [Washington, For sale by the Supt. of Docs., U.S. Govt. Print. Off., 1968] Description: 24 p. illus. (part col.), ports. 26 cm. Notes: Cover title. Subjects: Project Apollo (U.S.) LC Classification: TL789.8.U6 A63 1968 Dewey Class No.: 629.45/4

Apollo earth landmark maps. Edition Information: Ed. 1. Published/Created: [S.l.]: NASA, [1968?] Description: 1 atlas (ca. 250 p.): all col. maps; 27 cm. Scale Information: Scale 1:1,000,000. Notes: "Latitude and longitude to .001°; elevation and horizontal uncertainty to .01 NM; elevations referenced to Fischer ellipsoid." "Landmark photographs are available for those landmarks with underlined code numbers." Subjects: Astronautical charts. LC Classification: G1046.P7 U5 1968 Dewey Class No.: 912 19

Apollo expeditions to the moon / edited by Edgar M. Cortright. Published/Created: Washington: Scientific and Technical Information Office, National Aeronautics and Space Administration; for sale by the Supt. of Docs., U.S. Govt. Print. Off., 1975. Related Authors: Cortright, Edgar M. Description: xi, 313 p.: ill. (some col.); 31 cm. Notes: Includes index. Subjects: Project Apollo (U.S.) Space flight to the moon. Series: NASA SP; 350. LC Classification: TL789.8.U6 A513 Dewey Class No.: 629.45/4

Apollo over the moon: a view from orbit / editors: Harold Masursky, G. W. Colton, and Farouk El-Baz, with contributions by Frederick J. Doyle ... [et al.]. Published/Created: Washington: Scientific and Technical Information Office, National Aeronautics and Space Administration: for sale by the Supt. of Docs., U.S. Govt. Print. Off., 1978. Related Authors: Masursky, Harold, 1922- Colton, George Willis, 1920- El-Baz, Farouk. Doyle, Frederick J. Description: vii, 255 p.: ill.; 30 cm. Notes: Bibliography: p. 251-252. Subjects: Project Apollo (U.S.) Moon--Photographs from space. Series: NASA SP; 362. LC Classification: QB595 .A66 Dewey Class No.: 559.9/1/0222

Apollo program summary report. Published/Created: Houston, Tex.: National Aeronautics and Space Administration, Lyndon B. Johnson Space Center, 1975. Description: 508 p. in various pagings: ill.; 27 cm. Notes: "JSC-09423." Includes bibliographical references. Subjects: Project Apollo (U.S.) LC Classification: TL789.8.U6 A5535 1975 Dewey Class No.: 629.45/4

Apollo Soyuz Test Project: first international manned space flight, July 15-24, 1975. Published/Created: [Washington]: National Aeronautics and Space Administration, [1975] Description: 17 p.; 27 cm. Subjects: Apollo Soyuz Test Project. LC Classification: TL788.4 .U57 1975 Dewey Class No.: 629.45/4

Apollo terminology. Prepared for Office of Manned Space Flight. Published/Created: Washington [For sale by Office of Technical Services, Dept. of Commerce] 1963. Related Authors: United States. Office of Manned Space Flight. Description: 109 p. forms. 27 cm. Notes: Published by the Division under its earlier name: Office of Scientific and Technical Information. Subjects: Project Apollo

(U.S.)--Dictionaries. Series: United States. National Aeronautics and Space Administration. NASA SP- 6001 LC Classification: TL789.8.U5 A58 1963a

Apollo translunar/transearth trajectory plotting chart / prepared and published by the Defense Mapping Agency, Aerospace Center for the National Aeronautics and Space Administration. Edition Information: 1st ed., 21 Aug. 1972. Published/Created: St. Louis, Mo.: The Center, [1972] Related Authors: United States. National Aeronautics and Space Administration. Related Titles: National Aeronautics and Space Administration Apollo translunar/transearth trajectory plotting chart (ATT): Apollo Mission 17, December 6, 1972 launch date. Description: 1 map: col.; 45 x 59 cm. Scale Information: Scale not given. Notes: "This chart displays a polar view of the mission profile for Apollo 17 scheduled to be launched December 6, 1972 targeted to Lunar Landing Site Taurus Littrow." In lower right margin: National Aeronautics and Space Administration Apollo translunar/transearth trajectory plotting chart (ATT): Apollo Mission 17, December 6, 1972 launch date. Includes list of events and text. Subjects: Project Apollo (U.S.)--Charts, diagrams, etc. Astronautical charts. Artificial satellites--Moon--Orbits--Charts, diagrams, etc. LC Classification: G3196.P75 1972 .A3 Geographic Class No.: 3196

Apollo-Soyuz Test Project preliminary science report. Published/Created: Washington: U.S. Govt. Print. Off., [1976] Description: 529 p. in various pagings: ill.; 27 cm. Notes: Includes bibliographical references. Subjects: Apollo Soyuz Test Project. Series: NASA technical memorandum; X-58173. LC Classification: TL521.3.T4 A3 no. 58173 TL588.4 Dewey Class No.: 629.1/08 s 500.5

Apollo-Soyuz Test Project: summary science report / prepared by NASA Lyndon B. Johnson Space Center. Published/Created: Washington: Scientific and Technical Information Office, National Aeronautics and Space Administration, 1977-1979. Related Authors: El-Baz, Farouk. Warner, Delia M. Description: 2 v.: ill.; 27 cm. Contents: v. 1. Astronomy, earth atmosphere and gravity field, life sciences, and materials processing.--v. 2. Earth observations and photography. Notes: Vol. 2 edited by F. El-Baz and D. M. Warner. Vol. 2 published by Scientific and Technical Information Branch of NASA. Includes bibliographical references. Subjects: Apollo Soyuz Test Project. Series: NASA SP; 412. LC Classification: TL788.4 .L9 1977 Dewey Class No.: 500.5

Arctic and Antarctic sea ice, 1978-1987: satellite passive-microwave observations and analysis / Per Gloersen ... [et al.]. Published/Created: Washington, D.C.: Scientific and Technical Information Program, National Aeronautics and Space Administration, 1992. Related Authors: Gloersen, Per. Description: xxix, 290 p.: ill. (some col.), maps (some col.); 32 cm. Notes: Includes bibliographical references (p. 213-223) and index. Subjects: Sea ice--Polar regions--Remote sensing. Microwave remote sensing. Series: NASA SP; 511 LC Classification: GB2595 .A72 1992 Dewey Class No.: 551.3/43/0911 20

Arctic sea ice, 1973-1976: satellite passive-microwave observations / Claire L. Parkinson ... [et al.]. Published/Created: Washington, DC: Scientific and Technical Information Branch, National Aeronautics and Space Administration, 1987. Related Authors: Parkinson, Claire L. United States. National Aeronautics and Space Administration. Scientific and Technical Information Branch. Description: xvi, 296 p.: ill., maps (some col.); 30 cm. Notes: Includes index. Bibliography: p. 231-239. Subjects: Nimbus (Artificial satellite) Sea ice--Arctic regions--Remote sensing. Microwave remote sensing. Series: NASA SP; 489 LC Classification: GB2595 .A73 1987 Dewey Class No.: 551.3/43 19

Ariel I: the first international satellite; experimental results. Published/Created: Washington, Scientific and Technical Information Division, National Aeronautics and Space Administration; [for sale by Supt. of Docs., U.S. Govt. Print. Off.] 1966. Description: viii, 158 p. illus. (part col.) 26 cm. Notes: Bibliography: p. 119-124. Subjects: Ariel I (Artificial satellite) Series: NASA SP; 119. LC Classification: TL796 .U47 Dewey Class No.: 629.43/4

Armstrong, Neil, 1930- The first lunar landing: 20th anniversary / as told by the astronauts, Neil Armstrong, Edwin Aldrin, Michael Collins. Published/Created: [Washington, D.C.?]: National Aeronautics and Space Administration: [Supt. of Docs., U.S. G.P.O., distributor, 1989?] Related Authors: Aldrin, Buzz. Collins, Michael, 1930- United States. National Aeronautics and Space Administration. Office of Public Affairs. Description: 24 p.: ill. (some col.); 31 cm. Notes: Cover title. "Produced by the National Aeronautics and Space Administration, Office of Public Affairs"--P. [3] of cover. Subjects: Project Apollo (U.S.) Astronauts--United States--Interviews. Space flight to the moon. Series: NASA EP; 73. LC Classification: TL789.8.U6 A5157 1989

Aroesty, Jerome. Human support issues and systems for the space exploration initiative: results from Project Outreach / J. Aroesty, R. Zimmerman, J. Logan. Published/Created: [Santa Monica, CA (P.O. Box 2138, Santa Monica 90407-2138)]: Rand, [1991] Related Authors: Zimmerman, R. Logan, J. Rand Corporation. United States. Air Force. United States. National Aeronautics and Space Administration. Description: xxii, 115 p.; 28 cm. Notes: Cover title. "Prepared for the United States Air Force, National Aeronautics and Space Administration." Includes bibliographical references (p. 115). Subjects: Life support systems (Space environment) Space Biology. Space flight--Psysiological effect. Manned space flight--Systems engineering. Human engineering. Outer space--Exploration--United States. Series: A Rand note; N-3287-AF/NASA LC Classification: TL1500 .A76 1991 Dewey Class No.: 629.47/7 20

Astronautics and aeronautics; chronology on science, technology, and policy. 1915-60-- Published/Created: Washington, Scientific and Technical Information Division, National Aeronautics and Space Administration [etc.; for sale by the Superintendent of Documents, U. S. Govt. Print. Off.] Related Authors: United States. National Aeronautics and Space Administration. Scientific and Technical Information Division. United

States. National Aeronautics and Space Administration. United States. National Aeronautics and Space Administration. NASA Historical Staff. United States. Congress. House. Committee on Science and Astronautics. Related Titles: Aeronautics and astronautics. Astronautical and aeronautical events. Description: p. cm. Subjects: Astronautics--United States--History. Aeronautics--United States--History. Series: United States. National Aeronautics and Space Administration. NASA SP. LC Classification: TL521.3.A8 A3

Astronomy in space [by] Homer E. Newell [and others] Published/Created: Washington, Scientific and Technical Information Division, Office of Technology Utilization, National Aeronautics and Space Administration; [for sale by the Supt. of Docs., U.S. Govt. Print. Off.] 1967. Related Authors: United States. National Aeronautics and Space Administration. Scientific and Technical Information Division. Description: v, 67 p. illus. 24 cm. Contents: Space astronomy program of the National Aeronautics and Space Administration, by H.E. Newell.--Solar astronomy, by H.J. Smith.--Stellar and galactic astronomy, by N.G. Roman.--Expanding vistas in astronomy, by G.E. Mueller. Notes: Includes bibliographies. Subjects: Astronautics in astronomy. Series: NASA SP; 127. LC Classification: QB136 .A8 Dewey Class No.: 522/.1/0919

Atkinson, Joseph D. The real stuff: a history of NASA's astronaut recruitment program / Joseph D. Atkinson, Jr., Jay M. Shafritz; foreword by Alan B. Shepard and Guion S. Bluford, Jr. Published/Created: New York: Praeger, 1985. Related Authors: Shafritz, Jay M. Description: ix, 227 p.: ill.; 25 cm. ISBN: 0030051878 (alk. paper): 0030051886 (pbk.: alk. paper) : Notes: "Praeger special studies. Praeger scientific." Includes index. Bibliography: p. 195-206. Subjects: United States. National Aeronautics and Space Administration--Officials and employees--Recruiting History. Astronauts--United States--Recruiting--History. LC Classification: TL855 .A85 1985 Dewey Class No.: 629.45/07 19

Atwill, William D. Fire and power: the American space program as postmodern narrative / by William D. Atwill. Published/Created: Athens: University of Georgia Press, c1994. Description: 172 p.; 24 cm. ISBN: 0820316474 (alk. paper) Notes: Includes bibliographical references (p. [157]-168) and index. Subjects: United States. National Aeronautics and Space Administration--History. Astronautics--Social aspects--United States. Astronautics--Political aspects--United States. Literature and technology--United States. American literature--20th century--History and criticism. LC Classification: TL789.8.U5 A88 1994 Dewey Class No.: 303.48/32 20

Baker, David, 1944- Scientific American inventions from outer space: everyday uses for NASA technology / David Baker. Published/Created: New York: Random House, 2000. Related Titles: [Scientific American. Description: p. cm. ISBN: 0375409793 Notes: Includes bibliographical references and index. Subjects: United States. National Aeronautics and Space Administration. Inventions. LC Classification: T212

.B33 2000 Dewey Class No.: 609 21

Baker, Wendy. NASA: America in space / Wendy Baker. Published/Created: New York: Crescent Books: Distributed by Crown Publishers, 1986. Related Titles: America in space. Description: 159 p.: ill. (some col.); 26 cm. ISBN: 0517603640 Notes: "A Quarto book"--T.p. verso. Includes index. Bibliography: p. 154-155. Subjects: United States. National Aeronautics and Space Administration. Astronautics--United States. LC Classification: TL521.312 .B345 1986 Dewey Class No.: 629.4/0973 19

Bates, James R. ALSEP termination report / James R. Bates, W.W. Lauderdale, Harold Kernaghan. Published/Created: Washington, D.C.: National Aeronautics and Space Administration, Scientific and Technical Information Office, 1979. Related Authors: Lauderdale, W. W. Kernaghan, Harold. Related Titles: A.L.S.E.P. termination report. Description: 1 v. (various pagings): ill.; 28 cm. Notes: Includes a bibliography. Subjects: Apollo Lunar Surface Experiments Package. Series: NASA reference publication; 1036 LC Classification: TL789.8.U6 A518 1979 Dewey Class No.: 681/.75 19

Batson, Raymond M. Atlas of Mars: the 1:5,000,000 map series / R.M. Batson, P.M. Bridges, J.L. Inge. Published/Created: Washington, D.C.: Scientific and Technical Information Branch, National Aeronautics and Space Administration: for sale by the Supt. of Docs., U.S. Govt. Print. Off., 1979. Related Authors: Bridges, P. M., joint author. Inge, Jay L., joint author. Description: xi, 146 p.: chiefly ill.; 37 cm. Notes: Bibliography: p. 141-144.
Subjects: Mars (Planet)--Maps. Series: NASA SP; 438. LC Classification: QB641 .B27 Dewey Class No.: 912/.99/23 19

Batson, Raymond M. Atlas of Surveyor 5 television data / by Raymond M. Batson, Raymond Jordan, and Kathleen B. Larson. Published/Created: Washington: Scientific and Technical Information Office, National Aeronautics and Space Administration: for sale by the Supt. of Docs., U.S. Govt. Print. Off., 1974. Related Authors: Jordan, Raymond, joint author. Larson, Kathleen B., joint author. Description: xv, 597 p.: ill.; 30 cm. Notes: Bibliography: p. 593-597. Subjects: Surveyor Program (U.S.) Moon--Photographs from space. Series: NASA SP; 341. LC Classification: QB595 .B37 Dewey Class No.: 523.3/9 Govt. Doc. No.: NAS 1.21:341

Batson, Raymond M. Voyager 1 and 2 atlas of six Saturnian satellites / Raymond M. Batson; image processing, Ella M. Lee, Kevin F. Mullins, Brian A. Skiff; airbrush cartography, Patricia M. Bridges, Jay L. Inge; nomenclature, Harold Masursky, Mary E. Strobell. Published/Created: Washington, DC: Scientific and Technical Information Branch, National Aeronautics and Space Administration, 1984. Description: vii, 175 p.: ill. (some col.); 36 cm. Notes: Includes index. Bibliography: p. 141-142. Subjects: Voyager Project--Atlases. Saturn (Planet)--Satellites--Atlases. Series: NASA SP; 474 LC Classification: QB405 .B37 1984 Dewey Class No.: 523.9/86 19

Before this decade is out--: personal reflections on the Apollo Program / edited by Glen E. Swanson.

Published/Created: Washington, DC: National Aeronautics and Space Administration, NASA History Office, Office of Policy and Plans, 1999. Related Authors: Swanson, Glen E., 1963- Description: xiv, 408 p.: ill.; 25 cm. Notes: Includes bibliographical references (p. 379-385) and index. Subjects: Project Apollo (U.S.)--History. Aerospace engineers--United States--Interviews. Series: NASA SP; 4223 The NASA history series LC Classification: TL789.8.U6 A5187 1999 Dewey Class No.: 629.45/4/0973 21

Belew, Leland F. Skylab: a guidebook, by Leland F. Belew and Ernst Stuhlinger. Published/Created: [Washington, National Aeronautics and Space Administration; for sale by the Supt. of Docs., U.S. Govt. Print. Off., 1973] Related Authors: Stuhlinger, Ernst, 1913- joint author. Description: x, 245 p. illus. 23 cm. Notes: "EP-107." Bibliography: p. 236. Subjects: Skylab Program. LC Classification: TL789.8.U6 S553 Dewey Class No.: 629.44/5 Govt. Doc. No.: NAS 1.19:107

Benedict, Howard. NASA: a quarter century of space achievement / by Howard Benedict; edited by Myrtle Davidson Malone. Published/Created: The Woodlands, Tex. (P.O. Box 7825, The Woodlands 77387): Pioneer Publications, c1984. Related Authors: Malone, Myrtle Davidson. Description: viii, 350 p.: ill. (some col.); 29 cm. Notes: Includes index. Subjects: United States. National Aeronautics and Space Administration. Aerospace industries--United States. LC Classification: TL521 .B46 1984 Dewey Class No.: 629.4/0973 19

Benson, Charles D. Gateway to the moon: building the Kennedy Space Center launch complex / Charles D. Benson and William B. Faherty. Published/Created: Gainesville: University Press of Florida, 2001. Related Authors: Faherty, William Barnaby, 1914- Description: xvi, 316, xvii-xcvii p.: ill.; 23 cm. ISBN: 0813020913 (alk. paper) Notes: First part of a work originally published: Moonport. Washington: Scientific and Technical Information Office, National Aeronautics and Space Administration, 1978, in series: NASA history series. Includes bibliographical references and index. Subjects: John F. Kennedy Space Center--History. LC Classification: TL4027.F52 B46 2001 Dewey Class No.: 629.47/8/0975924 21

Benson, Charles D. Moon launch!: a history of the Saturn-Apollo launch operations / Charles D. Benson and William B. Faherty. Published/Created: Gainesville: University Press of Florida, 2001. Related Authors: Faherty, William Barnaby, 1914- Description: xvi, 317-551, xvii-xcvii p.: ill.; 23 cm. ISBN: 0813020948 (pbk.: alk. paper) Notes: Second part of a work originally published: Moonport. Washington, D.C.: Scientific and Technical Information Office, National Aeronautics and Space Administration, 1978, in series: NASA history series. Includes bibliographical references and index. Subjects: Project Apollo (U.S.) Saturn Project (U.S.) Saturn launch vehicles. LC Classification: TL781.5.S3 B37 2001 Dewey Class No.: 629.45/4/0973 21

Benson, Charles D. Moonport: a history of Apollo launch facilities and operations / Charles D. Benson and William

Barnaby Faherty. Published/Created: Washington: Scientific and Technical Information Office, National Aeronautics and Space Administration: for sale by the Supt. of Docs., U.S. Govt. Print. Off., 1978. Related Authors: Faherty, William Barnaby, 1914- joint author. Description: xx, 633 p.: ill.; 25 cm. Notes: Includes index. Bibliography: p. 599-617. Subjects: John F. Kennedy Space Center--History. Series: NASA history series. NASA SP; 4204. LC Classification: TL4027.F52 J635 Dewey Class No.: 629.47/8/0975927

Bergreen, Laurence. Voyage to Mars: NASA's search for life beyond Earth / Laurence Bergreen. Published/Created: New York: Riverhead Books, 2000. Description: 355 p.; 25 cm. ISBN: 157322166X Notes: Includes bibliographical references (p. [343]-345) and index. Subjects: United States. National Aeronautics and Space Administration. Mars probes. Exobiology. Mars (Planet)--Exploration. LC Classification: QB641 .B467 2000 Dewey Class No.: 629.45/53/00973 21

Beyond the ionosphere: fifty years of satellite communication / Andrew J. Butrica, editor. Published/Created: Washington, D.C.: National Aeronautics and Space Administration, 1997. Related Authors: Butrica, Andrew J. Description: xxxiv, 321 p.: ill.; 25 cm. Notes: Includes bibliographical references (p. 289-292) and index. Subjects: Artificial satellites in telecommunication. Series: NASA history series NASA SP; 4217 LC Classification: TK5104 .B48 1997 Dewey Class No.: 621.382/38 21

Bilstein, Roger E. Orders of magnitude: a history of the NACA and NASA, 1915-1990 / by Roger E. Bilstein. Published/Created: Washington, DC: National Aeronautics and Space Administration, Office of Management, Scientific and Technical Information Division: [For sale by the Supt. of Docs., U.S. G.P.O.], 1989. Related Authors: Anderson, Frank Walter. Orders of magnitude. Description: x, 167 p.: ill.; 23 cm. Notes: Rev. ed. of: Orders of magnitude / by Frank W. Anderson, Jr. Includes index. Includes bibliographical references (p. 151-155) Subjects: United States. National Advisory Committee for Aeronautics History. United States. National Aeronautics and Space Administration--History. Series: The NASA history series NASA SP; 4406 LC Classification: TL521.312 .B56 1989 Dewey Class No.: 353.0087/8 20

Bilstein, Roger E. Stages to Saturn: a technological history of the Apollo/Saturn launch vehicles / Roger E. Bilstein. Published/Created: Washington, DC: National Aeronautics and Space Administration, NASA History Office: For sale by the U.S. G.P.O., Supt. of Docs., 1996. Description: xx, 511 p.: ill.; 25 cm. ISBN: 0160489091 Notes: Includes bibliographical references (p. 493-500) and index. Subjects: Saturn Project (U.S.)--History. Project Apollo (U.S.)--History. Saturn launch vehicles--History. Series: The NASA history series NASA SP; 4206 LC Classification: TL781.5.S3 B54 1996 Dewey Class No.: 629.47/5 21

Bilstein, Roger E. Stages to Saturn: a technological history of the Apollo/Saturn launch vehicles / Roger

E. Bilstein. Published/Created: Washington: Scientific and Technical Information Branch, National Aeronautics and Space Administration, 1980. Description: xx, 511 p.: ill.; 25 cm. Notes: Include index. Bibliography: p. 493-500. Subjects: Saturn Project (U.S.)--History. Project Apollo (U.S.)--History. Saturn rockets--History. Series: NASA history series. NASA SP; 4206. LC Classification: TL781.5.S3 B54 Dewey Class No.: 629.47/522 Govt. Doc. No.: NAS 1.21:4206

Biomedical results from Skylab / edited by Richard S. Johnston and Lawrence F. Dietlein. Published/Created: Washington: Scientific and Technical Information Office, National Aeronautics and Space Administration: [for sale by the Supt. of Docs., U.S. Govt. Print. Off.], 1977. Related Authors: Johnston, Richard S. Dietlein, Lawrence F. Description: xii, 491 p.: ill.; 30 cm. Notes: Includes bibliographies and index. Subjects: Skylab Program. Space medicine. Space flight--Physiological effect. Series: NASA SP; 377. LC Classification: RC1135 .B56 Dewey Class No.: 616.9/80214

Biomedical results of Apollo / managing editors, Richard S. Johnston, Lawrence F. Dietlein, and Charles A. Berry. Published/Created: Washington: Scientific and Technical Information Office, National Aeronautics and Space Administration: for sale by the Supt. of Docs., U.S. Govt. Print. Off., 1975. Related Authors: Johnston, Richard S. Dietlein, Lawrence F. Berry, Charles A. (Charles Alden), 1923- Description: viii, 592 p.: ill.; 26 cm. Notes: Includes bibliographies. Subjects: Project Apollo (U.S.) Space medicine. Space flight--Physiological effect. Series: NASA SP; 368. LC Classification: RC1135 .B57 Dewey Class No.: 616.9/80214

Black Americans in aerospace. Published/Created: Washington, D.C.: Association for the Study of Afro-American Life and History, c1981. Related Authors: Association for the Study of Afro-American Life and History. Description: 19 p.: ports.; 28 cm. Notes: Cover title. Subjects: United States. National Aeronautics and Space Administration--Officials and employees. African Americans in aeronautics. African Americans in astronautics. LC Classification: TL521 .B55 1981 Dewey Class No.: 629.13/0092/2 B 19

Brand, D. O. Mariner Venus 67 dynamic flight data [by] D. O. Brand. Published/Created: Pasadena, Jet Propulsion Laboratory, California Institute of Technology, 1968. Related Authors: United States. National Aeronautics and Space Administration. Description: xi, 64 p. illus. 28 cm. Notes: "Prepared under contract no. NAS 7-100, National Aeronautics & Space Administration." Bibliography: p. 14. Subjects: Project Mariner (U.S.) Series: California Institute of Technology, Pasadena. Jet Propulsion Laboratory. JPL technical report no. 32-1185. LC Classification: TL789.8.U6 M26 Dewey Class No.: 629.46

Bromberg, Joan Lisa. NASA and the space industry / Joan Lisa Bromberg. Published/Created: Baltimore: Johns Hopkins University Press, 1999. Description: x, 247 p.: ill.; 24 cm. ISBN: 0801860504 (acid-free paper) Notes: Includes bibliographical references (p. 229-239) and index.

Subjects: United States. National Aeronautics and Space Administration--History. Aerospace industries--Government policy--United States History. Series: New series in NASA history LC Classification: TL521.312 .B76 1999 Dewey Class No.: 338.4/76291/0973 21

Brooke, Charles W. Development of an electrochemical energy source for the Mariner II spacecraft [by] Charles W. Brooke, Jr. Published/Created: Pasadena, Jet Propulsion Laboratory, California Institute of Technology, 1966. Related Authors: United States. National Aeronautics and Space Administration. Description: v, 52 p. illus. 28 cm. Notes: At head of National Aeronautics and Space Administration. Bibliography: p. 51. Subjects: Project Mariner (U.S.) Space vehicles--Batteries. LC Classification: TL789.8.U6 M28 Dewey Class No.: 629.43/54/2

Brooks, Courtney G. Chariots for Apollo: a history of manned lunar spacecraft / by Courtney G. Brooks, James M. Grimwood, Loyd S. Swenson, Jr. Published/Created: Washington: Scientific and Technical Information Branch, National Aeronautics and Space Administration, 1979. Related Authors: Grimwood, James M., joint author. Swenson, Loyd S., joint author. Description: xvii, 538 p.: ill.; 25 cm. Notes: Includes index. Bibliography: p. 485-502. Subjects: Project Apollo (U.S.)--History. Series: NASA history series. NASA SP; 4205. LC Classification: TL789.8.U6 A5239 Dewey Class No.: 629.45/4

Burgess, Eric. Into the thermosphere: the atmosphere explorers / Eric Burgess, Douglas Torr. Published/Created: Washington, D.C.: Scientific and Technical Information Office, National Aeronautics and Space Administration, 1987. Related Authors: Torr, Douglas. Description: xii, 171 p.: ill. (some col.); 29 cm. Notes: Bibliography: p. 145-169. Subjects: Atmosphere Explorer (Program: U.S.) Thermosphere--Research--United States. Explorer (Artificial satellite) Orbiting geophysical observatories. Series: NASA SP; 490 LC Classification: QC881.2.T4 B87 1987 Dewey Class No.: 551.5/14/072073 19

Burrough, Bryan, 1961- Dragonfly: NASA and the crisis aboard the MIR / Bryan Burrough. Edition Information: 1st ed. Published/Created: New York, NY: HarperCollinsPublishers, c1998. Description: 528 p.: ill. (some col.); 25 cm. ISBN: 0887307833 Notes: Includes index. Subjects: Mir (Space station)--Accidents. United States. National Aeronautics and Space Administration. Space vehicle accidents. Astronautics--Government policy--United States. LC Classification: TL867 .B87 1998 Dewey Class No.: 363.12/492 21

Burrows, William E., 1937- The infinite journey: eyewitness accounts of NASA and the age of space / written by William E. Burrows; edited by Mary Kalamaras. Edition Information: 1st ed. Published/Created: New York: Discovery Books, c2000. Related Authors: Kalamaras, Mary. Description: 240 p.: col. ill.; 31 cm. ISBN: 1563319241 Notes: Includes index. Subjects: United States. National Aeronautics and Space Administration--History. Astronautics--United States--History. Outer space--Exploration--United States--History. LC

Classification: TL789.8.U5 B87 2000 Dewey Class No.: 629.4/0973 21

Butrica, Andrew J. To see the unseen: a history of planetary radar astronomy / by Andrew J. Butrica. Published/Created: Washington, DC: National Aeronautics and Space Administration, 1996. Description: xiii, 301 p.: ill.; 25 cm. Notes: Includes bibliographical references and index. Subjects: Planetology--United States. Planets--Exploration. Radar in astronomy. Series: The NASA history series NASA SP; 4218 LC Classification: QB602.9 .B87 1996 Dewey Class No.: 523.2/028 20

Byrnes, Mark E. Politics and space: image making by NASA / Mark E. Byrnes. Published/Created: Westport, Conn.: Praeger, 1994. Description: 212 p.: ill.; 25 cm. ISBN: 0275949508 (alk. paper) Notes: Includes bibliographical references (p. [187]-206) and index. Subjects: United States. National Aeronautics and Space Administration-- Public opinion. Mass media--United States--Influence. Astronautics--United States--Public opinion. United States-- Politics and government. LC Classification: TL521.312 .B95 1994 Dewey Class No.: 353.0087/8 20

Carroll, William F. Design, test, and performance of the Mariner V temperature control reference [by] W. F. Carroll. Published/Created: Pasadena, Jet Propulsion Laboratory, California Institute of Technology, 1968. Description: v, 25 p. illus. 28 cm. Notes: "Prepared under contract no. NAS 7-100, National Aeronautics & Space Administration." Bibliography: p. 25. Subjects: Project Mariner (U.S.) Remote sensing--Equipment and supplies. Space vehicles--Materials. Artificial satellites in remote sensing. Series: California Institute of Technology, Pasadena. Jet Propulsion Laboratory. JPL technical report no. 32-1250. LC Classification: TL900 .C3 Dewey Class No.: 629.46

Carroll, William F. Mariner Mars 1964 temperature control hardware; design and development [by] W. Carroll, G. G. Coyle [and] H. von Delden. Published/Created: Pasadena, Calif., Jet Propulsion Laboratory, California Institute of Technology, 1967. Related Authors: Coyle, G. G., joint author. Von Delden, H., joint author. Description: vi, 22 p. illus. 28 cm. Notes: "Prepared under contract no. NAS 7-100, National Aeronautics & Space Administration." Bibliography: p. 22. Subjects: Project Mariner (U.S.) Space vehicles-- Thermodynamics. Series: California Institute of Technology, Pasadena. Jet Propulsion Laboratory. JPL technical report no. 32-955. LC Classification: TL789.8.U6 M452 Dewey Class No.: 629.45/5/3

Chambers, Joseph R. Partners in freedom: contributions of the Langley Research Center to U.S. military aircraft of the 1990's / by Joseph R. Chambers. Published/Created: Washington, D.C.: National Aeronautics and Space Administration,Office of Policy and Plans, NASA History Division, 2000. Description: xii, 256 p.: ill. (some col.); 29 cm. Notes: "NASA SP-2000-4519." Includes bibliographical references (p. 233-242) and index. Subjects: Langley Research Center. Aeronautics, Military- -Research--United States--History. Airplanes, Military--United States-- Design and construction. Series: The NASA history series Monographs in aerospace history; no. 19 LC

Classification: UG644.H36 .C43 2000
Dewey Class No.: 623.7/46/072073 21

Chapman, Richard L. Project management in NASA; the system and the men, by Richard L. Chapman, with the assistance of Robert H. Pontious and Lewis B. Barnes. Published/Created: Washington, Scientific and Technical Information Office, National Aeronautics and Space Administration; [for sale by the Supt. of Docs., U.S. Govt. Print. Off.] 1973. Description: x, 128 p. illus. 24 cm. Notes: Bibliography: p. 123-127. Subjects: United States. National Aeronautics and Space Administration--Management. Project management. Series: NASA SP; 324. LC Classification: TL521.312 .C45 Dewey Class No.: 658.4/04 Govt. Doc. No.: NAS 1.21:324

Chappell, Russell E. Apollo / text by Russell E. Chappell. Published/Created: Washington: [Office of Public Affairs, National Aeronautics and Space Administration]: for sale by the Supt. of Docs., U.S. Govt. Print. Off., [1974] Related Authors: United States. National Aeronautics and Space Administration. Office of Public Affairs. Description: 63 p.: ill.; 35 cm. Notes: "EP-100." Subjects: Project Apollo (U.S.) LC Classification: TL789.8.U6 A5246 Dewey Class No.: 629.45/4

Chronology of space shuttle flights, 1981-2000 / compiled by Judy A. Rumerman with Stephen J. Garber. Published/Created: Washington, D.C.: NASA History Division, Office of Policy and Plans, NASA Headquarters, [2000] Related Authors: Rumerman, Judy A. Garber, Stephen J. United States. National Aeronautics and Space Administration. History Office. Description: 34 p.: ill.; 28 cm. Notes: "October 2000." "HHR-70." Includes bibliographical references (p. 4). Subjects: Space shuttles--United States--History--Chronology. Space flights--United States--History--Chronology. LC Classification: TL795.5 .C49 2000 Dewey Class No.: 629.44/1/0973 21

Clemmons, Dewey L. The Echo I inflation system, Published/Created: Washington, National Aeronautics and Space Administration; [1964] Description: 1 l., 53 p. illus. 27 cm. Subjects: Project Echo. Expandable space structures. Series: United States. National Aeronautics and Space Administration. NASA technical note D-2194. LC Classification: TL521 .A3525 no. 2194

Collins, Stewart A. The Mariner 6 and 7 pictures of Mars, by Stewart A. Collins. Prepared at Jet Propulsion Laboratory. Published/Created: Washington, Scientific and Technical Information Office, National Aeronautics and Space Administration; [for sale by the Supt. of Docs., U.S. Govt. Print. Off.] 1971. Related Authors: Jet Propulsion Laboratory (U.S.) Description: ix, 159 p. illus. 30 cm. Notes: Bibliography: p. 148. Subjects: Project Mariner (U.S.) Mars (Planet)--Photographs from space. Series: United States. National Aeronautics and Space Administration. NASA SP -263. LC Classification: QB641 .C65 Dewey Class No.: 523.4/3/9 Govt. Doc. No.: NAS 1.21:263

Compositions of major and minor minerals in five Apollo 12 crystalline rocks [by] Bevan M. French [and others] Published/Created: Washington,

Scientific and Technical Information Office, National Aeronautics and Space Administration; [for sale by the National Technical Information Service, Springfield, Va.] 1972. Related Authors: French, Bevan M. Description: 142 p. illus. 27 cm. Notes: Bibliography: p. 141. Subjects: Project Apollo (U.S.) Lunar petrology. Lunar mineralogy. Series: NASA SP; 306. LC Classification: QB592 .U53 Dewey Class No.: 552/.09/991 Govt. Doc. No.: NAS 1.21:306

Compton, William David, 1927- Living and working in space: a history of Skylab / W. David Compton and Charles D. Benson. Published/Created: Washington, D.C.: Scientific and Technical Information Branch, National Aeronautics and Space Administration: For sale by the Supt. of Docs., U.S. G.P.O., 1983. Related Authors: Benson, Charles D. Description: xiii, 449 p.: ill. (some col.); 25 cm. Notes: Includes index. Bibliography: p. 397-442. Subjects: Skylab Program. Series: The NASA history series NASA SP; 4208 LC Classification: TL789.8.U6 S5546 1983 Dewey Class No.: 629.44/5 19

Compton, William David, 1927- Where no man has gone before: a history of Apollo lunar exploration missions / William David Compton. Published/Created: Washington, DC: National Aeronautics and Space Administration, Office of Management, Scientific and Technical Information Division, 1989. Related Authors: United States. National Aeronautics and Space Administration. Scientific and Technical Information Division. Description: xiii, 415 p.: ill. (1 col.); 26 cm. Notes: Includes bibliographical references. Subjects: Project Apollo (U.S.)-- History. Series: The NASA history series NASA SP; 4214 LC Classification: TL789.8.U6 A528 1989 Dewey Class No.: 919.9/104 19 Govt. Doc. No.: NAS 1.21:4214

Corliss, William R. Exploring the moon and planets, by William R. Corliss. Published/Created: Washington, National Aeronautics and Space Administration; [for sale by the Supt. of Docs., U.S. Govt. Print. Off., 1969] Related Authors: United States. National Aeronautics and Space Administration. Description: 26 p. illus. 27 cm. Subjects: Planets--Exploration. Moon--Exploration. Series: America in space. LC Classification: QB581 .C67 Dewey Class No.: 523.4

Corliss, William R. Linking man and spacecraft, by William R. Corliss. Published/Created: Washington, National Aeronautics and Space Administration; [for sale by the Supt. of Docs., U.S. Govt. Print. Off., 1969] Description: 18 p. illus. 27 cm. Notes: "EP-56." Subjects: Astronautics-- Communication systems. Series: America in space. LC Classification: TL3025 .C68 Dewey Class No.: 629.47

Corliss, William R. NASA sounding rockets, 1958-1968; a historical summary [by] William R. Corliss. Published/Created: Washington, Scientific and Technical Information Office, National Aeronautics and Space Administration; for sale by the Supt. of Docs., U.S. Govt. Print. Off., 1971. Description: vii, 158 p. illus. 25 cm. Notes: Bibliography: p. 77-78. Subjects: Rockets, Sounding--History. Series: United States. National Aeronautics and Space Administration. NASA historical report series. NASA SP; 4401. LC

Classification: TL785.8.S6 C67 Dewey Class No.: 551.5/14/028 Govt. Doc. No.: NAS 1.21:4401

Corliss, William R. NASA spacecraft, by William R. Corliss. Published/Created: Washington, National Aeronautics and Space Administration; [for sale by the Supt. of Docs., U.S. Govt. Print. Off., 1969] Description: 26 p. illus. 27 cm. Notes: "EP-54." Subjects: Artificial satellites. Space vehicles. Series: America in space. LC Classification: TL796.5.U5 C6 Dewey Class No.: 629.43/4/0973

Corliss, William R. Planetary exploration, by William R. Corliss. Published/Created: Washington, National Aeronautics and Space Administration; [for sale by the Supt. of Docs., U.S. Govt. Print. Off., 1971] Description: 28 p. illus. 27 cm. Subjects: Space flight to Mars. Space probes. Planets--Exploration. Astronautics--United States. Series: Space in the seventies LC Classification: TL799.M3 C67 Dewey Class No.: 629.43/54 Govt. Doc. No.: NAS 1.19:82

Corliss, William R. Putting satellites to work, by William R. Corliss. Published/Created: Washington, National Aeronautics and Space Administration; [for sale by the Supt. of Docs., U.S. Govt. Print. Off., 1969] Description: 26 p. illus. 27 cm. Notes: "EP-53." Subjects: Scientific satellites. Series: America in space. LC Classification: TL796 .C6 Dewey Class No.: 629.43/5

Corliss, William R. Satellites at work in communications, meteorology, geodesy, navigation, air traffic control, and earth resources technology, by William R. Corliss. Published/Created: Washington, National Aeronautics and Space Administration [1971] Description: 28 p. illus. (part col.) 27 cm. Notes: "EP-84." Subjects: Scientific satellites. Series: Space in the seventies LC Classification: TL798.S3 C59 Dewey Class No.: 629.43/52 Govt. Doc. No.: NAS 1.19:84

Corliss, William R. Scientific satellites [by] William R. Corliss. Published/Created: Washington, Scientific and Technical Information Division, National Aeronautics and Space Administration; [for sale by the Supt. of Docs., U.S. Govt. Print. Off.] 1967. Description: vii, 822 p. illus. 24 cm. Notes: Includes bibliographies. Subjects: Scientific satellites. Series: NASA SP; 133. LC Classification: TL798.S3 C6 Dewey Class No.: 629.43/4

Corliss, William R. Space physics and astronomy, by William R. Corliss. Published/Created: Washington, National Aeronautics and Space Administration; [for sale by the Supt. of Docs., U.S. Govt. Print. Off., 1969] Description: 22 p. illus. 27 cm. Notes: "EP-51." Subjects: Cosmic physics. Astronomy. Series: America in space. LC Classification: QC806 .C67 Dewey Class No.: 530

Corliss, William R. Spacecraft power, by William R. Corliss. Published/Created: Washington, National Aeronautics and Space Administration; [for sale by the Supt. of Docs., U.S. Govt. Print. Off., 1970] Description: 18 p. illus. 26 cm. Notes: "EP-59." Subjects: Space vehicles--Auxiliary power supply. Series: America in space. LC Classification: TL1100 .C67 Dewey

Class No.: 629.47/44

Corliss, William R. Spacecraft tracking, by William R. Corliss. Published/Created: Washington, National Aeronautics and Space Administration; [for sale by the Supt. of Docs., U.S. Govt. Print. Off., 1969] Related Authors: United States. National Aeronautics and Space Administration. Description: 18 p. illus. 26 cm. Subjects: Space vehicles--Tracking. Artificial satellites--Tracking. Series: America in space. LC Classification: TL4030 .C67 Dewey Class No.: 629.45/7

Corliss, William R. The interplanetary pioneers, by William R. Corliss. Published/Created: Washington, Scientific and Technical Information Office, National Aeronautics and Space Administration; [for sale by the Supt. of Docs., U.S. Govt. Print. Off.] 1972-[73] Description: 3 v. illus. 24 cm. Contents: v. 1. Summary.--v. 2. System design and development.--v. 3. Operations. Notes: Includes bibliographies. Subjects: Pioneer (Space probes) Outer space--Exploration. Series: NASA SP; 278-280. LC Classification: TL789.8.U6 P76 Dewey Class No.: 629.43/54 Govt. Doc. No.: NAS 1.21:278

Corliss, William R. The Viking mission to Mars / William R. Corliss. Edition Information: Rev. ed. Published/Created: Washington: Scientific and Technical Information Office, National Aeronautics and Space Administration: for sale by the Supt. of Docs., U.S. Govt. Print. Off., 1975 i.e. 1976. Description: vii, 77 p.: ill.; 23 cm. Notes: Bibliography: p. 77. Subjects: Viking Mars Program (U.S.) Mars (Planet)--Exploration. Series: NASA SP; 334. LC Classification: QB641 .C67 1976 Dewey Class No.: 559.9/23 Govt. Doc. No.: NAS 1.21:334/2

Coyle, G. G. Mariner IV science platform structure and actuator design, development and flight performance [by] G. Coyle. Published/Created: Pasadena, Jet Propulsion Laboratory, California Institute of Technology, 1965. Related Authors: United States. National Aeronautics and Space Administration. Description: iv, 18 p. illus. 28 cm. Notes: At head of National Aeronautics and Space Administration. "Prepared under contract no. NAS 7-100, National Aeronautics & Space Administration." Subjects: Project Mariner (U.S.) Space vehicles--Optical equipment. Television in astronautics. Series: California Institute of Technology, Pasadena. Jet Propulsion Laboratory JPL technical report no. 32-832. LC Classification: TL789.8.U6 M3455

Crouch, Tom D. The National Aeronautics and Space Administration / Tom D. Crouch. Published/Created: New York: Chelsea House Publishers, c1990. Description: 140 p.: ill.; 24 cm. ISBN: 1555461204 0791009017 (pbk.) Summary: Surveys the history of the National Aeronautics and Space Administration and describes its structure, current functions, and influence on our society. Notes: Includes bibliographical references (p. 134-135) and index. Subjects: United States. National Aeronautics and Space Administration--Juvenile literature. United States. National Aeronautics and Space Administration. Series: Know your government (New York, N.Y.) LC Classification: TL521 .C655 1990

Dewey Class No.: 353.0087/78 20

Cruise to Jupiter: contributions of the Galileo Imaging Science Team / edited by Michael J.S. Belton. Published/Created: [S.l.: s.n., 1997?] Related Authors: Belton, M. J. S. Galileo Imaging Science Team. Description: 1 v. (various pagings): ill. (some col.); 29 cm. Notes: "This work constitutes the final report of the Galileo Imaging Science Team to the Galileo Project and the National Aeronautics and Space Administration on the cruise phase of the mission." "Galileo project document 625-800." Consists mainly of reprints of articles from various scientific journals published between 1984 and 1996. Includes bibliographical references. Subjects: Galileo Project. Jupiter probes. Space flight to Jupiter. Jupiter (Planet)--Exploration. Outer space--Exploration. LC Classification: QB661 .C78 1997 Dewey Class No.: 523.4 21

Cunningham, Walter, 1932- The all-American boys / Walter Cunningham, with assistance by Mickey Herskowitz. Published/Created: New York: Macmillan, c1977. Related Authors: Herskowitz, Mickey, joint author. Description: x, 321 p., [4] leaves of plates: ill.; 24 cm. ISBN: 0025292404 Notes: Includes index. Subjects: United States. National Aeronautics and Space Administration. Astronauts--United States. LC Classification: TL521.312 .C86 Dewey Class No.: 387.8

Daniels, Lynne L. Statements of prominent Americans on the opening of the space age; a chronology of select statements of October 4, 1957 to November 13, 1958. Published/Created: Washington, National Aeronautics and Space Administration, NASA Historical Staff, 1963. Description: ii, 29 l. 27 cm. Subjects: Astronautics--History. Series: United States. National Aeronautics and Space Administration. NASA Historical Staff. NASA historical note no. 21 cm. LC Classification: TL521.3.H5 A3 no. 21

Data bases and data base systems, related to NASA's Aerospace Program: a bibliography with indexes. Published/Created: Washington, D.C.: Scientific and Technical Information Branch, National Aeronautics and Space Administration; Springfield, Va.: Available from the National Technical Information Service, 1981. Description: 511 p. in various pagings; 27 cm. Notes: "A selection of annotated references to unclassified reports and journal articles entered into the NASA scientific and technical information system from 1975 through 1980 in Scientific and Technical Aerospace Reports (STAR), International Aerospace Abstracts (IAA)." Subjects: Database management--Bibliography. Astronautics--Databases--Bibliography. Series: NASA SP; 449 LC Classification: Z5643.D36 U54 1981 QA76.9.D3 Dewey Class No.: 016.6294 19 Govt. Doc. No.: NAS 1.21:7045

Davies, Merton E. A preliminary control net of Mars [by] M. E. Davies and R. A. Berg. Published/Created: Santa Monica, Calif., Rand, 1970. Related Authors: Berg, Robert Alvin, 1936- joint author. United States. National Aeronautics and Space Administration. Description: ix, 51 p. illus. 28 cm. Notes: Prepared for National Aeronautics and Space Administration, under contract no. NAW-12402. Includes bibliographical references. Subjects: Project Mariner

(U.S.) Nets (Geodesy)--Mars (Planet) Mars (Planet)--Surface. Series: [Rand Corporation. Research memorandum] RM-6381-JPL LC Classification: Q180.A1 R36 no. 6381 QB641 Dewey Class No.: 081 s 526.3/0999/23

Davies, Merton E. Control net of Mars: February 1978 / Merton E. Davies, Frank Y. Katayama, James A. Roth. Published/Created: Santa Monica, CA: Rand, [1978] Related Authors: Katayama, Frank Y., 1924- Roth, James A., 1937- United States. National Aeronautics and Space Administration. Rand Corporation. Description: vii, 91 p.: ill.; 28 cm. ISBN: 0833000632 Notes: "R-2309-NASA." "February 1978." "Sponsored by the National Aeronautics and Space Administration." Includes bibliographical references (p. 91). Subjects: Project Mariner (U.S.) Viking Mars Program (U.S.) Nets (Geodesy)--Mars (Planet) Mars (Planet)--Surface. LC Classification: QB641 .D39 1978 Dewey Class No.: 081 s 526.6/0999/23

Davies, Merton E. Coordinates of features on the Mariner 6 and 7 pictures of Mars [by] Merton E. Davies. Published/Created: Santa Monica, Calif., Rand, 1972. Description: xiii, 57 p. illus. 28 cm. Notes: "Research ... supported by the National Aeronautics and Space Administration under Contract no. NASA-2143." Bibliography: p. 57. Subjects: Project Mariner (U.S.) Coordinates. Mars (Planet)--Surface. Mars (Planet)--Photographs from space. Series: [Rand Corporation. Rand report] R-896-NASA LC Classification: AS36.R3 R-896 QB641 Dewey Class No.: 081 s 526.3/0999/23

Davies, Merton E. Improved coordinates of features in the vicinity of the Viking 1 lander site on Mars / Merton E. Davies, Stephen H. Dole. Published/Created: Santa Monica, CA: Rand, 1980. Related Authors: Dole, Stephen H., joint author. United States. National Aeronautics and Space Administration. Description: ix, 16 p.: ill.; 28 cm. ISBN: 0833002368 Notes: "Prepared for the National Aeronautics and Space Administration." Bibliography: p. 15-16. Subjects: Viking Mars Program (U.S.) Mars (Planet)--Surface. Series: Rand Corporation. Rand report; R-2600-NASA. LC Classification: AS36 .R3 R-2600 QB641 Dewey Class No.: 081 s 523.1/3

Dawson, Virginia P. (Virginia Parker) Engines and innovation: Lewis laboratory and American propulsion technology / Virginia P. Dawson. Published/Created: Washington, DC: National Aeronautics and Space Administration, Office of Management, Scientific and Technical Information Division: For sale by the Supt. of Docs., U.S. G.P.O., 1991. Description: x, 276 p.: ill.; 26 cm. Notes: Includes bibliographical references (p. 219-223) and index. Subjects: Propulsion Laboratory (U.S.)--History. Series: NASA SP; 4306 The NASA history series LC Classification: TL568.P76 D38 1991 Dewey Class No.: 629.4/06/079493 20 Govt. Doc. No.: NAS 1.21:4306

Deerwester, Jerry M. Reference system characteristics for manned stopover missions to Mars and Venus, by Jerry M. Deerwester and Susan M. Norman. Published/Created: Washington, National Aeronautics and Space Administration; [for sale by the

National Technical Information Service, Springfield, Va.] 1971. Related Authors: Norman, Susan M., joint author. Description: iii, 72 p. illus. 27 cm. Notes: Cover title. Includes bibliographical references. Subjects: Space flight to Mars. Space trajectories. Interplanetary voyages. Series: United States. National Aeronautics and Space Administration. NASA technical note, NASA TN D-6226 LC Classification: TL521 .A3525 no. 6226 Dewey Class No.: 629.1/08 s

Dickson, Katherine Murphy. History of aeronautics and astronautics; a preliminary bibliography. Published/Created: Washington, National Aeronautics and Space Administration, 1967. Related Authors: United States. National Aeronautics and Space Administration. NASA Historical Staff. Description: vi, 117 l. 27 cm. Notes: "Sponsored by NASA Historical Staff, Office of Policy Analysis." "HHR-24." Subjects: Aeronautics--Bibliography of bibliographies. Astronautics--Bibliography of bibliographies. LC Classification: Z5063.A1 D5 Dewey Class No.: 016.01662913

Dictionary of technical terms for aerospace use. William H. Allen, editor. Edition Information: 1st ed. Published/Created: Washington; [For sale by the Superintendent of Documents, U.S. Govt. Print. Off.] 1965. Related Authors: Allen, William Hubert, 1916- Description: xi, 314 p. 25 cm. Notes: Bibliography: p. 313-314. Subjects: Aeronautics--Dictionaries. Astronautics--Dictionaries. Series: NASA SP; 7. LC Classification: TL521 .A333 no. 7

Doctors, Samuel I. The NASA technology transfer program; an evaluation of the dissemination system [by] Samuel I. Doctors. Published/Created: New York, Praeger Publishers [1971] Description: xvi, 226 p. 25 cm. Notes: Bibliography: p. 201-220. Subjects: United States. National Aeronautics and Space Administration. Technology transfer. LC Classification: T174.3 .D58 Dewey Class No.: 029.9/6

Doctors, Samuel I. The role of federal agencies in technology transfer [by] Samuel I. Doctors. Published/Created: Cambridge, Mass., M.I.T. Press [1969] Description: xxvi, 230 p. 24 cm. Notes: Bibliography: p. 206-223. Subjects: United States. National Aeronautics and Space Administration. Technology Utilization Division. Technology transfer. LC Classification: T174.3 .D6 Dewey Class No.: 607/.2/73

Donn, Bertram. Atlas of Comet Halley 1910 II / Bertram Donn, Jürgen Rahe, John C. Brandt. Published/Created: Washington, D.C.: Scientific and Technical Information Branch, National Aeronautics and Space Administration: For sale by the Supt. of Docs., U.S. G.P.O., 1986. Related Authors: Rahe, Jürgen. Description: xi, 600 p.: ill.; 31 cm. Notes: Bibliography: p. 599-600. Subjects: Halley's comet--Atlases. Series: NASA SP; 488 LC Classification: QB723.H2 D66 1986 Dewey Class No.: 523.6/4 19

Dunar, Andrew J. Power to explore: a history of Marshall Space Flight Center, 1960-1990 / Andrew J. Dunar and Stephen P. Waring. Published/Created: Washington, DC: National Aeronautics and Space Administration, NASA History Office, Office of Policy and

Plans, 1999. Related Authors: Waring, Stephen P. Description: x, 713 p.: ill.; 25 cm. Notes: Includes bibliographical references and index. Subjects: George C. Marshall Space Flight Center--History. Series: NASA historical series LC Classification: TL862.G4 D86 1999 Dewey Class No.: 629.4/0973 21

Dunne, James A. The voyage of Mariner 10: mission to Venus and Mercury / by James A. Dunne and Eric Burgess; prepared by Jet Propulsion Laboratory, California Institute of Technology. Published/Created: Washington: Scientific and Technical Information Division, National Aeronautics and Space Administration: for sale by the Supt. of Docs., U.S. Govt. Print Off., 1978. Related Authors: Burgess, Eric, joint author. Jet Propulsion Laboratory (U.S.) Description: vii, 224 p.: ill.; 30 cm. Notes: Includes index. Bibliography: p. 219-222. Subjects: Project Mariner (U.S.) Venus (Planet)--Observations. Mercury (Planet)--Observations. Series: NASA SP; 424. LC Classification: QB621 .D86 Dewey Class No.: 523.4/2

Earth observations and global change decision making: a special bibliography / National Aeronautics and Space Administration. Published/Created: Washington, DC: National Aeronautics and Space Administration, Office of Management, Scientific and Technical Information Program, 1991. Related Authors: United States. National Aeronautics and Space Administration. Scientific and Technical Information Program. Description: [93] p.; 28 cm. Notes: This bibliography lists 294 reports, articles, and other documents introduced in the NASA Scientific and Technical Information Database in 1990. "NASA STI program, scientific & technical information"--Cover. Includes indexes. Subjects: Climatic changes--Bibliography--Catalogs. Environmental monitoring--Bibliography--Catalogs. Global warming--Bibliography--Catalogs. Earth sciences--Remote sensing--Bibliography--Catalogs. Series: NASA SP; 7092 LC Classification: Z6683.C5 E27 1991 QC981.8.C5 Dewey Class No.: 016.5516 20 Govt. Doc. No.: NAS 1.21:7092

Earth photographs from Gemini III, IV, and V. Published/Created: Washington [For sale by the Supt. of Docs, U.S. Govt., Print. Off.] 1967. Description: ix, 266 p. (chiefly col. illus.) 30 cm. Subjects: Earth--Photographs from space. Series: NASA SP; 129. LC Classification: QB637 .U55 Dewey Class No.: 525/.022/2

Earth photographs from Gemini VI through XII. Published/Created: Washington [For sale by the Supt. of Docs., U.S. Govt. Print. Off.] 1968. Description: x, 327 p. (chiefly col. illus.) 30 cm. Notes: "A companion volume to Earth photographs from Gemini III, IV, and V, which was issued in 1967 as Special publication 129." Bibliography: p. 327. Subjects: Earth--Photographs from space. Series: NASA SP; 171. LC Classification: QB637 .U56 Dewey Class No.: 525/.022/2

Eddy, John A. A new sun: the solar results from Skylab / by John A. Eddy; edited by Rein Ise; prepared by George C. Marshall Space Flight Center. Published/Created: Washington: Scientific and Technical Information Office, National Aeronautics and Space Administration: for sale by the Supt. of

Docs., U.S. Govt. Print. Off., 1979. Related Authors: Ise, Rein. George C. Marshall Space Flight Center. Description: xix, 198 p.: ill.; 30 cm. Notes: Includes index. Subjects: Skylab Program. Sun. Series: NASA SP; 402. LC Classification: QB521 .E33 Dewey Class No.: 523.7 Govt. Doc. No.: NAS 1.21:402

Educational horizons / National Aeronautics and Space Administration, Education Division, Education Publications Branch. Published/Created: Washington, D.C.: The Branch, [1992- Related Authors: United States. National Aeronautics and Space Administration. Education Publications Branch. United States. National Aeronautics and Space Administration. Education Division. Description: v.: ill.; 44 x 28 cm. Vols. for spring 1992- published as if quarterly with summer issue purposefully omitted. Vol. 1, no. 1 (Feb. 1992)- Continues: Report to educators 0883-0983 (DLC) 85645911 (OCoLC)10864699 Continued by: Educational horizons (Washington, D.C.: 1995) (DLC)sn 97032448 (OCoLC)34331163 Cancel/Invalid LCCN: sn 92025086 Notes: Title from caption. Issued by: NASA, Education Division, Office of Human Resources and Education, fall 1992- SERBIB/SERLOC merged record Subjects: United States. National Aeronautics and Space Administration-- Periodicals. Astronautics--Study and teaching--United States Periodicals. Aeronautics--Study and teaching-- United States Periodicals. LC Classification: TL521.3.R4 U5a Dewey Class No.: 629.4/071073 20 Govt. Doc. No.: NAS 1.49/2:

Electrical power management standard. Published/Created: Washington [For sale by the Clearinghouse for Federal Scientific and Technical Information, Springfield, Va.] 1965. Related Authors: United States. Office of Manned Space Flight. Related Titles: Apollo program. Description: v, 31 p. 27 cm. Notes: "CMO 19-000-1." At head of Office of Manned Space Flight, Apollo Program. Subjects: Project Apollo (U.S.) Space vehicles--Electric equipment. Series: NASA SP; 6005. LC Classification: TL1100 .U53

Energy-related research and development. Prepared at the request of Frank E. Moss, chairman, Committee on Aeronautical and Space Sciences, United States Senate. Published/Created: Washington, U.S. Govt. Print. Off., 1974. Related Authors: United States. Congress. Senate. Committee on Aeronautical and Space Sciences. Description: vii, 90 p. 24 cm. Notes: At head of 93d Congress, 2d session. Committee print. Subjects: United States. National Aeronautics and Space Administration. Power resources- -Research--United States. LC Classification: TJ153 .U538 1974 Dewey Class No.: 621

Engineering challenges to the long-term operation of the International Space Station / Committee on the Engineering Challenges to the Long-Term Operation of the International Space Station, Aeronautics and Space Engineering Board, Commission on Engineering and Technical Systems, National Research Council. Published/Created: Washington, D.C.: National Academy Press, c2000. Related Authors: United States. National Aeronautics and Space Administration. National Research

Council (U.S.). Aeronautics and Space Engineering Board. National Research Council (U.S.). Commission on Engineering and Technical Systems. Description: x, 43 p., [2] p. of plates: ill.; 28 cm. ISBN: 0309069386 Notes: "This study was supported by the National Aeronautics and Space Administration under contract No. NASW 4938"--P. [ii] Includes bibliographical references. Additional Form Avail.: Also issued on the World Wide Web. Subjects: International Space Station. Space stations--Maintenance and repair. Maintainability (Engineering) Service life (Engineering) MANNED ORBITAL LABORATORIES. SPACE LABORATORIES. SPACECRAFT MAINTENANCE. Series: Compass series (Washington, D.C.) LC Classification: TL797 .N37 2000 Dewey Class No.: 629.44/2 21

EOS science plan: the state of science in the EOS Program / editor Michael D. King; technical editors Reynold Greenstone, William Bandeen. Published/Created: Greenbelt, Md.: National Aeronautics and Space Administration, [1999] Related Authors: King, Michael D. Description: 397 p.: ill., maps (some col.); 29 cm. Notes: "January 1999." Includes bibliographical references and index. Subjects: Earth Observing System (Program) Atmospheric physics. Atmospheric chemistry. LC Classification: QC861.2 .E57 1999 Dewey Class No.: 551.51/07/02 21

Establishment of the National space program; report on H. R. 12575. Published/Created: Washington, U.S. Govt. Print. Off., 1958. Related Authors: National space program. Description: iii, 39 p. 24 cm. Subjects: United States. National Aeronautics and Space Administration. Astronautics. Series: 85th Cong., 2nd sess. House report no. 1770 United States. 85th Cong., 2nd sess., 1958. House. Report; no. 1770. LC Classification: TL521 .A54147 1958c Dewey Class No.: 629.1388

Ethell, Jeffrey L. Fuel economy in aviation / Jeffrey L. Ethell. Published/Created: Washington, DC: Scientific and Technical Information Branch, National Aeronautics and Space Administration, 1983. Description: ix, 111 p.: ill. (some col.); 24 cm. Subjects: Aircraft Energy Efficiency Program (U.S.) Airplanes--Fuel consumption. Aeronautics--Research--United States. Series: NASA SP; 462 LC Classification: TL704.7 .E78 1983 Dewey Class No.: 629.134/351 19

Etzioni, Amitai. The moon-doggle; domestic and international implications of the space race. Edition Information: [1st ed.] Published/Created: Garden City, N.Y., Doubleday, 1964. Related Titles: Domestic and international implications of the space race. Description: xv, 198 p. 22 cm. Subjects: United States. National Aeronautics and Space Administration. Astronautics and state--United States. Science and state--United States. LC Classification: TL789.8.U5 E8 Dewey Class No.: 353

Evans, David Donald, 1930- Development of the Ranger Block III spacecraft propulsion system [by] D. D. Evans, T. A. Groudle [and] R. F. Mattson. Published/Created: Pasadena, Jet Propulsion Laboratory, California Institute of Technology, 1966. Related Authors: Groudle, T. A., joint author. Mattson, R. F., joint author. United

States. National Aeronautics and Space Administration. Description: v, 39 p. illus. 28 cm. Notes: At head of National Aeronautics and Space Administration. "Prepared under contract no. NAS 7-100, National Aeronautics & Space Administration." Bibliography: p. 38. Subjects: Project Ranger. Space vehicles--Propulsion systems. Series: California Institute of Technology, Pasadena. Jet Propulsion Laboratory. JPL technical report no. 32-829 LC Classification: TL789.8.U6R353 Dewey Class No.: 629.43/53

Exobiology in solar system exploration: the proceedings of a symposium held in August 1988 / sponsored by the Exobiology Program of NASA's Division of Life Sciences; edited by Glenn C. Carle, Deborah E. Schwartz, and Judith L. Huntington. Published/Created: Moffett Field, CA: National Aeronautics and Space Administration, Ames Research Center, 1992. Related Authors: Carle, Glenn C., 1936- Schwartz, Deborah E. Huntington, Judith L., 1960- Exobiology Program (U.S.) Description: v, 297 p.: ill.; 27 cm. Notes: Includes bibliographical references. Subjects: Life on other planets. Exobiology. Solar system Astronautics in astronomy. Outer space--Exploration. Series: NASA SP; 512 LC Classification: QB54 .E87 1992 Dewey Class No.: 574.999 20

Explorer vi and Pioneer v data. Published/Created: [Washington, 1963] Related Authors: Rosen, Alan. United States. National Aeronautics and Space Administration. Description: 2 v. illus., map, diagrs. 27 cm. Subjects: Geophysics--Observations. Cosmic physics. Explorer (Artificial satellite) Pioneer (Space probes) Series: United States. National Aeronautics and Space Administration. NASA contractor report CR-3-4. LC Classification: TL521.3.C6 A3 no. 3-4

Eyewitness to space; paintings and drawings related to the Apollo mission to the moon, selected, with a few exceptions, from the art program of the National Aeronautics and Space Administration (1963 to 1969) Written by Hereward Lester Cooke, with the collaboration of James D. Dean. Foreword by J. Carter Brown. Pref. by Thomas O. Paine. Published/Created: New York, H. N. Abrams [1971] Related Authors: Cooke, Hereward Lester. Dean, James D. United States. National Aeronautics and Space Administration. Description: 227 p. illus. (part col.) 32 x 43 cm. ISBN: 0810901129 Subjects: Project Apollo (U.S.) Art, American--20th century--Catalogs. LC Classification: N6512 .E9 Dewey Class No.: 760/.074/0153

Ezell, Edward Clinton. On Mars: exploration of the Red Planet, 1958-1978 / Edward Clinton Ezell, Linda Neuman Ezell. Published/Created: Washington, D.C.: Scientific and Technical Information Branch, National Aeronautics and Space Administration: For sale by the Supt. of Docs., U.S. G.P.O., 1984. Related Authors: Ezell, Linda Neuman. Description: xvi, 535 p.: ill. (some col.; 25 cm. Notes: Includes index. Bibliography: p. 481-524. Subjects: Viking Mars Program (U.S.)--History. Series: The NASA history series NASA SP; 4212 LC Classification: TL789.8.U6 V524 1984 Dewey Class No.: 629.43/543 19

Ezell, Edward Clinton. The partnership: a history of the Apollo-Soyuz test project

/ Edward Clinton Ezell and Linda Neuman Ezell. Published/Created: Washington: National Aeronautics and Space Administration: [for sale by the Supt. of Docs., U.S. Govt. Print. Off.] 1978. Related Authors: Ezell, Linda Neuman, joint author. Description: xx, 560 p.: ill.; 25 cm. Notes: Includes bibliographical references and index. Subjects: Apollo Soyuz Test Project. Series: NASA history series. NASA SP; 4209. LC Classification: TL788.4 .E95 Dewey Class No.: 629.45/4

Facilities data / National Aeronautics and Space Administration, Office of Facilities. Published/Created: Washington: The Office: for sale by the Supt. of Docs., U.S. Govt. Print. Off., 1974. Description: 120 p.: ill. (some col.); 20 x 26 cm. Subjects: United States. National Aeronautics and Space Administration--Equipment and supplies. LC Classification: TL521.312 .U54 1974 Dewey Class No.: 629.1/07/2073

Fallaci, Oriana. If the sun dies / Oriana Fallaci; translated from the Italian by Pamela Swinglehurst. Edition Information: 1st Atheneum pbk. ed. Published/Created: New York: Atheneum, 1981, c1966. Description: p. cm. ISBN: 0689706103 (pbk.) : Notes: Cataloging based on CIP information. Translation of: Se il sole muore. Includes index. Subjects: United States. National Aeronautics and Space Administration. Outer space--Exploration. LC Classification: TL793 .F2913 1981 Dewey Class No.: 500.5/072 19

Fallaci, Oriana. If the sun dies. Translated from the Italian by Pamela Swinglehurst. Edition Information: [1st ed.] Published/Created: New York, Atheneum, 1966. Related Authors: United States. National Aeronautics and Space Administration Officials and employees. Description: ix, 403 p. 25 cm. Notes: Translation of Se il sole muore. LC Classification: PQ4866.A4 S413

Fallaci, Oriana. If the sun dies; translated from the Italian by Pamela Swinglehurst. Published/Created: London, Collins, 1967. Description: 415 p. 22 cm. Notes: Translation of Se il sole muore. Subjects: United States. National Aeronautics and Space Administration. Outer space--Exploration. LC Classification: TL793 .F2913 1967 Dewey Class No.: 858/.9/1403

Faller, Kenneth H. Remote sensing of oceanic parameters during the Skylab/Gamefish Experiment / Kenneth H. Faller. Published/Created: Washington, D.C.: National Aeronautics and Space Administration, Scientific and Technical Information Office; Springfield, Va.: For sale by the National Technical Information Service, 1977. Description: 38 p.: ill.; 27 cm. Notes: "November 1977"--Cover. Includes bibliographical references. Subjects: Skylab Program. Oceanography--Remote sensing. Series: NASA reference publication; 1012 LC Classification: GC10.4.R4 F34 1977 Dewey Class No.: 551.46/0028 19

Fimmel, Richard O. Pioneer odyssey / Richard O. Fimmel, William Swindell, Eric Burgess; prepared at Ames Research Center. Edition Information: Rev. ed. Published/Created: Washington: Scientific and Technical Information Office, National

Aeronautics and Space Administration: for sale by the Supt. of Docs., U.S. Govt. Print. Off., 1977. Related Authors: Swindell, William, joint author. Burgess, Eric, joint author. Ames Research Center. Description: xi, 217 p.: ill.; 30 cm. Notes: On cover and spine: NASA SP-349. Includes index. Bibliography: p. 209-212. Subjects: Pioneer (Space probes) Jupiter (Planet)--Exploration. Series: NASA SP; 396. LC Classification: QB661 .F53 1977 Dewey Class No.: 523.4/5 Govt. Doc. No.: NAS 1.21:396

Fimmel, Richard O. Pioneer odyssey: encounter with a giant / by Richard O. Fimmel, William Swindell, and Eric Burgess, prepared at Ames Research Center. Published/Created: Washington: Scientific and Technical Information Division, National Aeronautics and Space Administration: for sale by the Supt. of Docs., U.S. Govt. Print. Off., 1974. Related Authors: Swindell, William, joint author. Burgess, Eric, joint author. Ames Research Center. United States. National Aeronautics and Space Administration. Scientific and Technical Information Division. Description: x, 171 p.: ill. (some col.); 29 cm. Notes: Includes index. Bibliography: p. 163-164. Subjects: Pioneer (Space probes) Jupiter (Planet)--Exploration. Series: NASA SP; 349. LC Classification: QB661 .F54 Dewey Class No.: 523.4/5 Govt. Doc. No.: NAS 1.21:349.

Fimmel, Richard O. Pioneer Venus / prepared at Ames Research Center [by] Richard O. Fimmel, Lawrence Colin, Eric Burgess. Published/Created: Washington, DC: Scientific and Technical Information Branch, National Aeronautics and Space Administration: For sale by the Supt. of Docs., U.S. G.P.O., 1983. Related Authors: Colin, Lawrence. Burgess, Eric. Description: xi, 253 p.: ill. (some col.); 28 cm. Notes: Includes index. Bibliography: p. 221-237. Subjects: Venus probes. Pioneer (Space probes) Series: NASA SP; 461 LC Classification: QB621 .F55 1983 Dewey Class No.: 559.9/22 19

Fimmel, Richard O. Pioneer, first to Jupiter, Saturn, and beyond / Richard O. Fimmel, James Van Allen, Eric Burgess; prepared at Ames Research Center. Published/Created: Washington, D.C.: Scientific and Technical Information Office, National Aeronautics and Space Administration: For sale by the Supt. of Docs., U.S. G.P.O., 1980. Related Authors: Van Allen, James Alfred, 1914- Burgess, Eric. Ames Research Center. United States. National Aeronautics and Space Administration. Scientific and Technical Information Office. Description: x, 285 p.: ill. (some col.); 29 cm. Notes: Includes index. S/N 033-000-00805-8 Item 830-I Subjects: Pioneer (Space probes) Outer space--Exploration. Series: NASA SP; 446 LC Classification: QB661 .F52 Dewey Class No.: 919.9/25/04 19 Govt. Doc. No.: NAS 1.21:446

Fimmel, Richard O. Pioneering Venus: a planet unveiled / Richard O. Fimmel, Lawrence Colin, Eric Burgess; prepared at NASA Ames Research Center. Published/Created: Washington, D.C.: National Aeronautics an Space Administration, [1995] Related Authors: Colin, Lawrence. Burgess, Eric. Description: xi, 358 p.: ill. (some col.); 30 cm. ISBN: 0964553708 (hardcover) 0964553716 (pbk.) Notes: Includes bibliographical references (p. 305-310) and index. Subjects: Venus probes. Pioneer (Space probes) LC Classification: QB621 .F55 1995 Dewey Class No.: 523.4/2 20

Fisher, Allan C. The great venture; man's first steps into space. Published/Created: [Washington, U. S. Information Service, 1961?] Description: 44 p. illus. 27 cm. Subjects: United States. National Aeronautics and Space Administration. LC Classification: TL798.8.U5 F5

Forecasts and appraisals for management evaluation. Published/Created: Washington, National Aeronautics and Space Administration; [for sale by the Clearinghouse for Federal Scientific and Technical Information, Springfield, Va., 1966- Description: v. illus. 27 cm. Notes: Includes bibliographies. Subjects: Project Apollo (U.S.) Astronautics--Data processing. Series: NASA SP; 6009. LC Classification: TL789.8.U6 A585 Dewey Class No.: 658.5/02

French, Bevan M. Mars: the Viking discoveries / by Bevan M. French. Published/Created: [Washington]: National Aeronautics and Space Administration: for sale by the Supt. of Docs., U.S. Govt. Print. Off., 1977. Description: 36 p.: ill.; 28 cm. Notes: "EP-146." Bibliography: p. 32-33. Filmography: p. 36. Subjects: Viking Mars Program (U.S.) Mars (Planet)-- Exploration. LC Classification: QB641 .F73 Dewey Class No.: 919.9/23/04

French, Bevan M. Return to the red planet: the Mars Observer mission. Published/Created: Pasadena, Calif.: National Aeronautics and Space Administration, Jet Propulsion Laboratory, California Institute of Technology; [Washington, D.C.: For sale by the Supt. of Docs., U.S. G.P.O., 1993] Related Authors: Jet Propulsion Laboratory (U.S.) Related Titles: Mars Observer mission. Description: 59 p.: ill.; 26 cm. ISBN: 0160424267 : Notes: Cover title. "Written by Bevan M. French"--Acknowledgments. Shipping list no.: 93-0530-P. "JPL 400-504 8/93"--P. [4] of cover. Subjects: Mars Observer (Spacecraft) Space flight to Mars. Mars probes. Mars (Planet)-- Exploration. Mars (Planet)-- Observations. LC Classification: TL799.M3 F74 1993 Dewey Class No.: 629.43/543 20 Govt. Doc. No.: NAS 1.12/7:400-504

Frierson, J. Q., 1919- Rocketed into history: NASA claims a paradise / J.Q. Frierson. Published/Created: New York: Rivercross Pub., c1996. Description: x, 180 p.: map; 23 cm. ISBN: 094495782X (hc) Subjects: United States. National Aeronautics and Space Administration-- History--Fiction. Launch complexes (Astronautics)--Location--Mississippi Hancock County--Fiction. Hancock County (Miss.)--History--Fiction. LC Classification: PS3556.R5682 R63 1996 Dewey Class No.: 813/.54 20

Fries, Sylvia Doughty. NASA engineers and the age of Apollo / Sylvia Doughty Fries. Published/Created: Washington, DC: National Aeronautics and Space Administration, Scientific and Technical Information Program, 1992. Related Authors: United States. National Aeronautics and Space Administration. Scientific and Technical Information Program. Description: xx, 216 p.; 25 cm. Notes: Includes bibliographical references and index. Subjects: United States. National Aeronautics and Space

Administration--Officials and employees--Biography. Project Apollo (U.S.) Aeronautical engineers--United States--Biography. Series: NASA SP; 4104 The NASA history series LC Classification: TL521.312 .N365 1992 Dewey Class No.: 629.1/092/273 B 20

Froehlich, Walter, 1921- Space station: the next logical step / [by Walter Froehlich]. Published/Created: Washington, DC: National Aeronautics and Space Administration: For sale by the Supt. of Docs., G.P.O., [1984?] Related Authors: United States. National Aeronautics and Space Administration. Description: 51 p.: col. ill.; 28 cm. Notes: S/N 033-000-00932-1 Item 830-G Subjects: Space stations. Series: NASA EP; 213. LC Classification: TL797 .F758 1984 Dewey Class No.: 629.44/2 19 Govt. Doc. No.: NAS 1.19:213

Froehlich, Walter, 1921- Apollo 14: science at Fra Mauro. Published/Created: [Washington, Office of Public Affairs, National Aeronautics and Space Administration; for sale by the Supt. of Docs., U.S. Govt. Print. Off., 1971] Related Authors: United States. National Aeronautics and Space Administration. Office of Public Affairs. Description: 48 p. illus. (part col.) 31 cm. Notes: Cover title. "EP-91." Subjects: Project Apollo (U.S.) Space flight to the moon. LC Classification: TL789.8.U6 A5393 Dewey Class No.: 629.45/4

Froehlich, Walter, 1921- Apollo 16 at Descartes. [Text by Walter Froehlich. Published/Created: Washington, Office of Public Affairs, National Aeronautics and Space Administration; for sale by the Supt. of Docs., U.S. Govt. Print. Off., 1972] Related Authors: United States. National Aeronautics and Space Administration. Office of Public Affairs. Description: 32 p. col. illus. 23 x 31 cm. Notes: Cover title. Subjects: Project Apollo (U.S.) Moon--Exploration. Moon--Exploration. LC Classification: TL789.8.U6 A5339 Dewey Class No.: 559.9/1 Govt. Doc. No.: NAS 1.10:97

Froehlich, Walter, 1921- Apollo Soyuz / by Walter Froehlich. Published/Created: Washington: National Aeronautics and Space Administration: for sale by the Supt. of Docs., U.S. Govt. Print. Off., 1976. Description: 131 p.: ill.; 23 cm. Notes: "EP-109." "Stock number 033-000-00652-7." Subjects: Apollo Soyuz Test Project. LC Classification: TL788.4 .F76 Dewey Class No.: 629.44/2

Froehlich, Walter, 1921- Man in space. Published/Created: Washington, National Aeronautics and Space Administration; [for sale by the Supt. of Docs., U.S. Govt. Print. Off., 1971] Description: v, 21 p. illus. (part col.) 27 cm. Notes: "EP-81." Subjects: Astronautics--United States. Outer space--Exploration. Series: Space in the seventies LC Classification: TL789.8.U5 F76 Dewey Class No.: 629.45 Govt. Doc. No.: NAS 1.19:81

Froehlich, Walter, 1921- Spacelab, an international short-stay orbiting laboratory. Published/Created: Washington, D.C.: National Aeronautics and Space Administration: For sale by the Supt. of Docs., U.S. G.P.O., 1983. Related Authors: United States. National Aeronautics and Space Administration. Description: v, 77 p.: col. ill.; 28 cm. Notes: S/N 033-000-

00895-3 Item 830-G Subjects: Spacelab Program. Space stations. Series: NASA EP; 165. LC Classification: TL797 .F76 1983 Dewey Class No.: 629.44/5 19 Govt. Doc. No.: NAS 1.19:165

From engineering science to big science: the NACA and NASA Collier Trophy research project winners / edited by Pamela E. Mack. Published/Created: Washington, D.C.: National Aeronautics and Space Administration, 1998. Related Authors: Mack, Pamela Etter. Description: xxiii, 427 p.: ill.; 25 cm. Notes: Includes bibliographical references and index. Subjects: Collier Trophy. Aerospace engineering--United States--History. Airplanes--United States--Design and construction History. Aeronautics--Research--United States--History. Series: NASA SP; 4219 The NASA history series LC Classification: TL537 .F76 1998 Dewey Class No.: 629.1/0973 21

Future aeronautics and space opportunities. Published/Created: [Washington: National Aeronautics and Space Administration], 1974- Related Authors: Fleming, William A. Description: v.: ill.; 22 x 28 cm. Incomplete Contents: v. 1. Space. Notes: Cover title. Vol. 1 composed and edited by William A. Fleming. Subjects: United States. National Aeronautics and Space Administration. Astronautics. Aeronautics. Space sciences. LC Classification: TL790 .U58 1974 Dewey Class No.: 629.4 Govt. Doc. No.: NAS 1.2:Op5/v.1

Future National space objectives; staff study, Eighty-ninth Congress, second session. Published/Created: Washington, U.S. Govt. Print. Off., 1966. Description: xvi, 439 p. illus., maps. 24 cm. Notes: At head of Committee print. "Serial O." Subjects: U. S. National Aeronautics and Space Administration. Astronautics--United States. LC Classification: TL789.8.U5 A52 1966 Dewey Class No.: 629.40973

Galileo to Jupiter: probing the planet and mapping its moons. Published/Created: [Washington: National Aeronautics and Space Administration; Pasadena, Calif.: Jet Propulsion Laboratory, California Institute of Technology, 1979] Related Authors: Jet Propulsion Laboratory (U.S.) Description: 20 p.: ill.; 28 cm. Notes: Cover title. Subjects: Satellites--Jupiter--Exploration. Jupiter (Planet)--Exploration. LC Classification: QB661 .U54 1979 Dewey Class No.: 523.4/5/072

Galileo: exploration of Jupiter's system / C.M. Yeates ... [et al.]. Published/Created: Washington, D.C.: Scientific and Technical Information Branch, National Aeronautics and Space Administration: For sale by the Supt. of Docs., U.S. G.P.O., 1985. Related Authors: Yeates, C. M. Galileo Project. Description: viii, 175 p.: ill. (some col.); 28 cm. Notes: Bibliography: p. 175. Subjects: Galileo Project. Jupiter (Planet)--Exploration. Series: NASA SP; 479 LC Classification: QB661 .G35 1985 Dewey Class No.: 523.4/5 19

Gawdiak, Ihor, 1935- Astronautics and aeronautics, 1986-1990: a chronology / by Ihor Y. Gawdiak, Ramón J. Miro, and Sam Stueland. Published/Created: Washington, DC: National Aeronautics and Space Administration: For sale by the U.S. G.P.O., 1997. Related Authors: Miró, Ramón, 1968- Stueland, Sam, 1955- Description: v, 370 p.; 23 cm. Notes: Includes index. Subjects:

Astronautics--Chronology. Aeronautics--Chronology. Series: The NASA history series NASA SP; 4027. LC Classification: TL788.5 .G38 1997 Dewey Class No.: 629.1/09 21

Gawdiak, Ihor, 1935- Astronautics and aeronautics, 1991-1995: a chronology / compiled by Ihor Y. Gawdiak and Charles Shetler. Published/Created: Washington, DC: National Aeronautics & Space Administration, Office of Policy and Plans, NASA History Division, 2000. Related Authors: Shetler, Charles, 1941- Description: v, 773 p.; 23 cm. Notes: Includes bibliographical references and index. Subjects: United States. National Aeronautics and Space Administration--History. Aeronautics--United States--Chronology. Astronautics--United States--Chronology. Series: The NASA history series NASA SP; 4028 LC Classification: TL521 .G38 1999 Dewey Class No.: 629.1/09 21

Glennan, Thomas Keith, 1905- The birth of NASA: the diary of T. Keith Glennan / with an introduction by Roger D. Launius; edited by J.D. Hunley. Published/Created: Washington, DC: NASA History Office, National Aeronautics and Space Administration: For sale by U.S. G.P.O., 1993. Related Authors: Hunley, J.D., 1941- Description: xxxii, 389 p.: ill., map; 25 cm. ISBN: 0160419360 Notes: Includes bibliographical references and index. Subjects: Glennan, Thomas Keith, 1905- --Diaries. United States. National Aeronautics and Space Administration--History--Sources. Space sciences--Research--United States--History. Series: NASA SP; 4105 The NASA history series LC Classification: TL521.312 .G58 1993 Dewey Class No.: 353.0087/77 20

Goddard Space Flight Center/Johnson Space Center LACIE interface control document LANDSAT data/LACIE 2. Edition Information: Rev. A. Published/Created: Houston, Tex.: National Aeronautics and Space Administration, Lyndon B. Johnson Space Center, [1977] Related Authors: Goddard Space Flight Center. Lyndon B. Johnson Space Center. Related Titles: LACIE interface control document LANDSAT data/LACIE 2. Large Area Crop Inventory Experiment (LACIE) level 3 baseline. Description: 35 p. in various pagings; 28 cm. Notes: Cover title. At head of Large Area Crop Inventory Experiment (LACIE) level 3 baseline. "January 1977." "LACIE-C00701." Subjects: LACIE (Project: U.S.) Crops--Remote sensing--Data processing. Artificial satellites in agriculture--United States. Series: JSC (Series); 11670. LC Classification: IN PROCESS (UTILITY LOAD)

Goldstein, Richard M. The superior conjunction of Mariner IV [by] R. M. Goldstein, et al. Published/Created: Pasadena, Jet Propulsion Laboratory, California Institute of Technology, 1967. Related Authors: United States. National Aeronautics and Space Administration. Description: vi, 56 p. illus. 28 cm. Notes: "Prepared under contract no. NAS 7-100, National Aeronautics & Space Administration." Bibliography: p. 55-56. Subjects: Astronautics--Communication systems. Interstellar communication. Project Mariner (U.S.) Sun--Corona. Series: California Institute of Technology, Pasadena. Jet Propulsion Laboratory. JPL technical report no. 32-1092. LC

Classification: TL3025 .G6

Golovin, Nicholas E. (Nicholas Erasmus), 1912-1969. Papers of Nicholas E. Golovin, 1943-1970 (bulk 1958-1969) Related Authors: Coler, Myron A. (Myron Abraham), 1913- Essays on creativity in the sciences (1963) Description: 10,125 items. 27 containers. 10.8 linear feet. Biog./History Note: Physicist, government official, and presidential advisor for aviation and space science. Summary: Correspondence, memoranda, speeches and writings, reports, notes, working diaries, agenda and meeting summaries, academic, employment, and military records, clippings, photographs, and other papers pertaining chiefly to Golovin's administrative positions at the National Bureau of Standards, National Aeronautics and Space Administration (NASA), and the U.S. Office of Science and Technology. Documents his roles as director of the Large Launch Vehicle Planning Group, a program sponsored by NASA and the Dept. of Defense, in the planning, programming, and budgeting system at the Office of Science and Technology, and as a member of the President's Science Advisory Committee during the John F. Kennedy and Lyndon B. Johnson administrations. Also documents his participation in several International Astronautical Congresses, his work with the White Sands Missile Range, Advanced Research Projects Agency, and Myron Coler's Creative Science Program at New York University, and his various technical writing projects. Includes Golovin's reports and recommendations to Kennedy and Johnson on topics such as the Project Apollo space program, manned space flight, space vehicles and vehicle propulsion systems, and supersonic transport noise problems; his paper titled, A Fourth Branch of Government, in which he proposed the creation of a branch of government employing physicists and social scientists to make technological choices in public policy; and chapter drafts and a published copy of Essays on Creativity in the Sciences (1963) written by Creative Science Program associates. Correspondents include Edward U. Condon, Anne Golovin, Donald F. Hornig, Charles S. Murphy, and Jerome B. Wiesner. Notes: MSS57265 Source of Acquisition: Gift, Anne Golovin, 1977. Subjects: Coler, Myron A. (Myron Abraham), 1913- Condon, Edward Uhler, 1902-1974. Golovin, Anne Castrodale. Hornig, Donald F., 1920- Johnson, Lyndon B. (Lyndon Baines), 1908-1973. Kennedy, John F. (John Fitzgerald), 1917-1963. Murphy, Charles S. (Charles Springs), 1909- Wiesner, Jerome B. (Jerome Bert), 1915- United States. Advanced Research Projects Agency. United States. National Aeronautics and Space Administration. United States. National Bureau of Standards. United States. Office of Science and Technology. United States. President's Science Advisory Committee. Large Launch Vehicle Planning Group. New York University. Creative Science Program. Project Apollo (U.S.) International Astronautical Congress. Aeronautics. Launch vehicles (Astronautics) Manned space flight. Program budgeting. Science--Study and technology. Science and state--United States. Separation of powers--United States. Social sciences and state--United States. Space sciences. Space vehicles. Space vehicles--Propulsion systems. Supersonic transport planes. Technology and state--

United States. United States--Politics and government--1961-1963. United States--Politics and government--1963-1969. White Sands Missile Range (N.M.) Physicists. Public officials. Local Call/Shelving: 0806M

Governmental organization for space activities: report of the Committee on Aeronautical and Space Sciences, United States Senate, made by its Subcommittee on Governmental Organization for Space Activities. Published/Created: [Washington, D.C.?: U.S. G.P.O., 1959] Description: v, 58 p.; 23 cm. Notes: "August 25, 1959." Subjects: United States. National Aeronautics and Space Act of 1958. United States. Advanced Research Projects Agency. United States. National Aeronautics and Space Administration. Federal Council for Science and Technology (U.S.) Series: United States. Congress. Senate. Report; 86th Congress, no. 806. LC Classification: TL789.8.U5 A53 1959a

Green, Constance McLaughlin, 1897- Vanguard; a history, by Constance McLaughlin Green and Milton Lomask. Published/Created: Washington, Scientific and Technical Information Division, National Aeronautics and Space Administration; [for sale by the National Technical Information Service, Springfield, Va.] 1970. Related Authors: Lomask, Milton, joint author. Description: xvi, 308 p. illus., ports. 25 cm. Notes: Includes bibliographical references. Subjects: Project Vanguard--History. Astronautics--United States--History. Series: United States. National Aeronautics and Space Administration. NASA historical series. NASA SP; 4202. LC Classification: TL796.5.U6 V35 Dewey Class No.: 629.43/2

Grimwood, James M. Project Gemini: technology and operations; a chronology. Prepared by James M. Grimwood and Barton C. Hacker with Peter J. Vorzimmer. Published/Created: Washington, Scientific and Technical Information Division, National Aeronautics and Space Administration; [for sale by the Supt. of Docs., U.S. Govt. Print. Off.] 1969. Related Authors: Hacker, Barton C., 1935- joint author. Vorzimmer, Peter J., 1937- joint author. Description: xvi, 308 p. illus., maps. 27 cm. Subjects: Project Gemini (U.S.) Series: United States. National Aeronautics and Space Administration. NASA historical series. NASA SP; 4002. LC Classification: TL789.8.U6 G36 Dewey Class No.: 629.45/4

Grimwood, James M. Project Mercury; a chronology. Published/Created: Washington, Office of Scientific and Technical Information, National Aeronautics and Space Administration; [for sale by the Superintendent of Documents, U.S. Govt. Print. Off.] 1963. Description: xiv, 238 p. illus., ports., maps. 26 cm. Notes: Bibliographical footnotes. Subjects: Project Mercury (U.S.) Series: Manned Spacecraft Center (U.S.) MSC publication HR-1 United States. National Aeronautics and Space Administration. NASA SP-; 4001. LC Classification: TL789.8.U6 M465

Guide to Magellan image interpretation / John P. Ford ... [et. al]. Published/Created: Pasadena, Calif.: National Aeronautics and Space Administration: Jet Propulsion Laboratory, California Institute of Technology, [1993] Related Authors:

Ford, John P. United States. National Aeronautics and Space Administration. Jet Propulsion Laboratory (U.S.) Description: vii, 148 p.: ill. (some col.), maps (some col.); 28 cm. Notes: Shipping list no.: 93-0662-P. "November 1, 1993." "JPL publication 93-24." Includes bibliographical references and index. Subjects: Magellan (Spacecraft) Space flight to Venus. Venus probes. LC Classification: QB621 .G85 1993 Dewey Class No.: 523.4/2 20 Govt. Doc. No.: NAS 1.12/7:93-24

Gump, David. Space enterprise: beyond NASA / David P. Gump. Published/Created: New York: Praeger, 1990. Description: 220 p.: ill.; 24 cm. ISBN: 0275933148 (alk. paper) : Notes: Includes bibliographical references (p. [205]-215). Subjects: United States. National Aeronautics and Space Administration. Space stations. Space industrialization. LC Classification: TL797 .G86 1990 Dewey Class No.: 333.9/4 20

Hacker, Barton C., 1935- On the shoulders of Titans: a history of Project Gemini / by Barton C. Hacker and James M. Grimwood. Published/Created: Washington: Scientific and Technical Information Office, National Aeronautics and Space Administration: for sale by the Supt. of Docs., U.S. Govt. Print. Off., 1977 [i.e. 1978] Related Authors: Grimwood, James M., joint author. Description: xx, 625 p.: ill.; 25 cm. Notes: Includes index. Bibliography: p. 502-513. Subjects: Project Gemini (U.S.)--History. Series: NASA SP; 4203. LC Classification: TL789.8.U6 G394 Dewey Class No.: 629.45/4/09046

Hakkila, Jon Eric, 1957- NASA / by Jon Hakkila and Adele D. Richardson. Published/Created: Mankato, Minn.: Smart Apple Media, 1999. Related Authors: Richardson, Adele, 1966- Description: p. cm. ISBN: 1583400508 (alk. paper) Summary: Examines the function of NASA and the impact of its various space exploration programs. Notes: Includes index. Subjects: United States. National Aeronautics and Space Administration--Juvenile literature. United States. National Aeronautics and Space Administration. Astronautics--United States--Juvenile literature. Outer space--Exploration--United States--Juvenile literature. Series: Above & beyond. LC Classification: TL521.312 .H32 1999 Dewey Class No.: 629.4/0973 21

Hall, Harvey, 1904- History of a physicist: joys and woes / Harvey Hall. Published/Created: [Fairfax, Va.?]: H. Hall, c2000. Description: vi, 272 p.: ill., ports.; 23 cm. Notes: Includes bibliographical references. Subjects: Hall, Harvey, 1904- United States. National Aeronautics and Space Administration--Officials and employees--Biography. Physicists--United States--Biography. Physics--United States--History--20th century. LC Classification: QC16.H25 H35 2000 Dewey Class No.: 530/.092 B 21

Hall, R. Cargill. Lunar impact: a history of Project Ranger / R. Cargill Hall. Published/Created: Washington: Scientific and Technical Information Office, National Aeronautics and Space Administration: [for sale by the Supt. of Docs., U.S. Govt. Print. Off.], 1977. Description: xvii, 450 p.: ill.; 25 cm. Notes: Includes index. Bibliography: p. 359-365. Subjects: Project Ranger.

Series: NASA history series. NASA SP; 4210. LC Classification: TL789.8.U6 R35314 Dewey Class No.: 629.43/53

Hallion, Richard. On the frontier: flight research at Dryden, 1946-1981 / Richard P. Hallion. Published/Created: Washington, D.C.: Scientific and Technical Information Branch, National Aeronautics and Space Administration: For sale by the Supt. of Docs., U.S. G.P.O., 1984. Description: xix, 385 p.: ill. (some col.); 26 cm. Notes: Includes index. Bibliography: p. 363-376. Subjects: Dryden Flight Research Facility--History. Series: The NASA history series NASA SP; 4303 LC Classification: TL521.312 .H34 1984 Dewey Class No.: 629.1/072079488 19

Hallion, Richard. On the frontier: flight research at Dryden, 1946-1999 / Richard P. Hallion and Michael H. Gorn. Published/Created: Washington, D.C.: National Aeronautics and Space Administration, 2001. Related Authors: Gorn, Michael H. Description: p. cm. Notes: "NASA SP-2001-4315." Includes bibliographical references and index. Subjects: Dryden Flight Research Facility--History. Series: NASA history series LC Classification: TL521.312 .H342 2001 Dewey Class No.: 629.1/07/2079488 21

Hanover, Paul N. The next 100 years [sound recording]: your future on earth & in space / Paul N. Hanover. Published/Created: Tucson, AZ: P.N. Hanover, p1996. Related Authors: United States. National Aeronautics and Space Administration. Description: 1 sound cassette: analog. Summary: Presents a lecture about the future of space exploration and the United States space program. Subjects: Outer space--Exploration--United States. LC Classification: RYG 1476

Hansen, James R. Engineer in charge: a history of the Langley Aeronautical Laboratory, 1917-1958 / James R. Hansen. Published/Created: Washington, D.C.: Scientific and Technical Information Office, National Aeronautics and Space Administration, 1987. Description: xxxviii, 620 p.: ill.; 26 cm. Notes: Includes index. Bibliography: p. 515-565. Subjects: Langley Aeronautical Laboratory--History. Series: The NASA history series NASA SP; 4305 LC Classification: TL568.L34 H36 1987 Dewey Class No.: 629.13/00720755/412 19

Hansen, James R. Spaceflight revolution: NASA Langley Research Center from Sputnik to Apollo / James R. Hansen. Published/Created: Washington, DC: National Aeronautics and Space Administration: For sale by Supt. of Docs., U.S. G.P.O., [1995] Description: xxxi, 542 p.: ill., maps; 25 cm. Notes: Includes bibliographical references (p. 447-518) and index. Subjects: Langley Research Center--History. Astronautics--United States--History. Series: NASA history series NASA SP; 4308 LC Classification: TL521.312 .H36 1995 Dewey Class No.: 629.4/0973 20

Harris, Gordon L. Selling Uncle Sam / Gordon L. Harris. Edition Information: 1st ed. Published/Created: Hicksville, N.Y.: Exposition Press, c1976. Description: xi, 222 p., [4] leaves of plates: ill.; 22 cm. ISBN: 0682484814 : Notes: Includes index. Subjects: United States. National Aeronautics and Space Administration--Public relations. LC Classification: TL789.8.U5 H33 Dewey

Class No.: 353.008/77/8

Hartmann, William K. The new Mars: the discoveries of Mariner 9: prepared for the NASA Office of Space Science / William K. Hartmann and Odell Raper; with the cooperation of the Mariner 9 Science Experiment Team. Published/Created: Washington: Scientific and Technical Information Office, National Aeronautics and Space Administration: for sale by the Supt. of Docs., U.S. Govt. Print. Off., 1974. Related Authors: Raper, Odell, joint author. United States. Office of Space Science. Description: ix, 179 p.: ill.; 30 cm. Notes: Includes bibliographical references and index. Subjects: Project Mariner (U.S.) Mars (Planet) Series: NASA SP; 337. LC Classification: QB641 .H33 Dewey Class No.: 559.9/23

Heppenheimer, T. A., 1947- The space shuttle decision: NASA's search for a reusable space vehicle / T.A. Heppenheimer. Published/Created: Washington, DC: National Aeronautics and Space Administration, NASA History Office, Office of Policy and Plans, 1999. Description: xiv, 470 p.: ill.; 25 cm. Notes: Includes bibliographical references (p. 437-446) Subjects: United States. National Aeronautics and Space Administration. Space shuttles--United States--Design and construction History. Series: NASA SP; 4221 The NASA history series LC Classification: TL795.5 .H47 1999 Dewey Class No.: 629.44/1/0973 21

Herring, Mack R., 1935- Way station to space: a history of the John C. Stennis Space Center / by Mack R. Herring. Published/Created: Washington, DC: National Aeronautics and Space Administration, NASA History Office, Office of Policy and Plans, 1997. Description: xvii, 484 p.: ill.; 25 cm. Notes: Includes bibliographical references and index. Subjects: John C. Stennis Space Center--History. Series: The NASA history series NASA SP; 4310 LC Classification: TL862.J65 H47 1997 Dewey Class No.: 629.4/0720762/14 21

High energy phenomena on the sun; [papers] Edited by R. Ramaty and R. G. Stone. Published/Created: Washington, Scientific and Technical Information Office, National Aeronautics and Space Administration; [for sale by the Supt. of Docs., U.S. Govt. Print. Off.] 1973. Related Authors: Ramaty, R., ed. Stone, Robert G. (Robert Gilbert), 1928- ed. United States. National Aeronautics and Space Administration. Description: viii, 641 p. illus. 26 cm. Notes: Includes bibliographies. Subjects: Solar activity--Congresses. Particles (Nuclear physics)--Congresses. Series: NASA SP; 342. LC Classification: QB524 .S9 1972 Dewey Class No.: 523.7 Govt. Doc. No.: NAS 1.21:342

Highlights of the year ... / NASA, Earth Science and Applications Division. Published/Created: Washington, D.C.: National Aeronautics and Space Administration, [1989- Description: v.: ill.; 28 cm. 1989- Continues: United States. National Aeronautics and Space Administration. Earth Science and Applications Division. Program and plans for FY ... (DLC) 88641449 (OCoLC)17673463 ISSN: 1055-6427 Cancel/Invalid LCCN: sn 90039847 Notes: SERBIB/SERLOC merged record Additional Form Avail.: Issued also on microfiche, Supt. of Docs. class no.: NAS 1.15: Distributed to depository libraries in microfiche, Supt. of Docs. class no.: NAS 1.1/6: Subjects: Earth sciences--Research--United States--Periodicals. Astronautics in earth sciences--Periodicals. LC Classification: QE47.A1 P76 Dewey Class No.: 550/.72073 20 Govt. Doc. No.: NAS 1.15: NAS 1.1/6:

Hirsch, Richard, 1912-1971. The National Aeronautics and Space Administration [by] Richard Hirsch and Joseph John Trento. Published/Created: New York, Praeger [1973] Related Authors: Trento, Joseph John, 1947- Description: x, 245 p. 21 cm. Notes: Bibliography: p. 234-235. Subjects: United States. National Aeronautics and Space Administration. LC Classification: TL789.8.U5 H55 Dewey Class No.: 353.008/242

Historical origins of the National Aeronautics and Space Administration. Published/Created: Washington [For sale by the Superintendent of Documents, U. S. Govt. Print. Off., 1963] Description: 22 p. illus., ports., map. 24 cm. Notes: Cover title. Bibliography: p. 20-22. Subjects: U.S. National Aeronautics and Space Administration. LC Classification: TL521 .A54283 1963b

Historical sketch of NASA.
Published/Created: Washington [For sale by the Superintendent of Documents, U.S. Govt. Print. Off.] 1965. Description: 56 p. illus., map, ports. 23 cm. Notes: Bibliography: p. 51-56. Subjects: United States. National Aeronautics and Space Administration. LC Classification: TL521.312.A5 T7 Dewey Class No.: 629.406173

History at NASA / the NASA History Office. Published/Created: Washington, DC: NASA Headquarters, [1986] Related Authors: United States. National Aeronautics and Space Administration. History Office. Description: ix, 61 p.: ill.; 28 cm. Notes: "June 1986." "NASA HHR-50." "NASA history publications": p. 53-55. Subjects: United States. National Aeronautics and Space Administration--Research. United States. National Aeronautics and Space Administration--History--Sources. LC Classification: TL521.312 .H57 1986 Dewey Class No.: 016.6294/0973 20

Hoban, Francis T. Where do you go after you've been to the moon?: a case study of NASA's pioneer effort at change / by Francis T. Hoban with William M. Lawbaugh and Edward J. Hoffman. Published/Created: Malabar, Fla.: Krieger Pub. Co., 1997. Related Authors: Lawbaugh, William M. Hoffman, Edward J. (Edward Jay), 1959- Description: xvi, 223 p.: ill.; 30 cm. ISBN: 0894640607 (hardback: alk. paper) Notes: Includes index. Subjects: United States. National Aeronautics and Space Administration--Reorganization. Astronautics and state--United States. Series: An Orbit series book LC Classification: TL521.312 .H587 1997 Dewey Class No.: 387.8/0973 20

Holme, Molly. First five years of NASA; a concise chronology. Published/Created: Washington, NASA Historical Staff, National Aeronautics and Space Administration, 1963. Description: 68 l. 27 cm. Subjects: United States. National Aeronautics and Space Administration. Series: United States. National Aeronautics and Space Administration. NASA Historical Staff. NASA historical report. LC Classification:

TL521.312 .H6

Hoyt, Edwin Palmer. The space dealers; a hard look at the role of American business in our space effort [by] Edwin P. Hoyt. Published/Created: New York, John Day Co. [1971] Description: ix, 243 p. 22 cm. Notes: Bibliography: p. [235-236] Subjects: United States. National Aeronautics and Space Administration. Aerospace industries--United States. LC Classification: HD9711.5.U6 H68 Dewey Class No.: 338.4/7/62910973

Human spaceflight: mission analysis and design / edited by Wiley J. Larson, Linda K. Pranke. Published/Created: New York: McGraw-Hill, [2000?] Related Authors: Larson, Wiley J. Pranke, Linda K. Description: xxiii, 1035 p.: ill.; 24 cm. ISBN: 007236811X Notes: "This book is published as part of the Space Technology Series, a cooperative activity of the United States Department of Defense and the National Aeronautics and Space Administration." Includes bibliographical references and index. Subjects: Space flight--Planning. Space vehicles--Design and construction. Manned space flight. Series: Space technology series LC Classification: TL790 .H85 2000 Dewey Class No.: 629.45 21

Improved reporting needed on National Aeronautics and Space Administration projects: report to the Congress / by the Comptroller General of the United States. Published/Created: [Washington: U.S. General Accounting Office, 1977] Description: ii, 44 p.; 27 cm. Notes: Cover title. "PSAD-77-54." "B-182956." Publication date stamped on cover. Subjects: United States. National Aeronautics and Space Administration. Astronautics and state--United States. LC Classification: TL789.8.U5 U53 1977 Dewey Class No.: 353.008/55

In orbit at Jupiter: contributions of the Galileo Imaging Science Team / edited by Michael J.S. Belton. Published/Created: [S.l.: s.n., 1999?] Related Authors: Belton, M. J. S. Galileo Imaging Science Team. Description: 1 v. (various pagings): ill. (some col.); 28 cm. Notes: "This work constitutes the final report of the Galileo Imaging Science Team to the Galileo Project and the National Aeronautics and Space Administration on the orbital phase of the Galileo Nominal Mission." "Galileo Project Document 625-801" Consists mainly of reprints of articles from various scientific journals. Includes bibliographical references. Subjects: Galileo Project. Jupiter probes. Space flight to Jupiter. Jupiter (Planet)--Exploration. LC Classification: QB661 .I52 1999 Dewey Class No.: 523.45 21

"In this decade ...": mission to the moon. Published/Created: [Washington; For sale by Supt. of Docs., U.S. Govt. Print. Off., 1969] Description: 46 p. illus., ports. (both part col.) 20 x 27 cm. Notes: Cover title. "EP 71." Subjects: Project Apollo (U.S.) LC Classification: TL789.8.U6A63 1969 Dewey Class No.: 629.45/4

Index of NACA technical publications / National Advisory Committee for Aeronautics. Published/Created: Washington: NACA, Related Authors: United States. National Aeronautics and Space Administration. Description: 9 v. 26 cm. Cumulation for 1915-1947 superseded by 1915-1949, which is supplemented by subsequent volumes.

Began with: 1915-1947. -July 1957-Sept. 1958. Continued by: United States. National Aeronautics and Space Administration. Index of NASA technical publications (OCoLC)2262218 Notes: Description based on: 1915-1949; title from cover. Vol. for July 1957-Sept. 1958 issued by: National Aeronautics and Space Administration. Indexes: Author index: 1915/49. 1 v. Subjects: United States. National Advisory Committee for Aeronautics Bibliography. Aeronautics--Abstracts. LC Classification: Z5063 .U634

Information security: many NASA mission-critical systems face serious risks: report to the Committee on Governmental Affairs, U.S. Senate / United States General Accounting Office. Published/Created: Washington D.C. (P.O. Box 37050, Washington D.C. 20013): The Office, 1999. Related Authors: United States. Congress. Senate. Committee on Governmental Affairs. Description: 29 p.; 28 cm. Notes: "May 1999." "B-277744"--P. 1. Includes bibliographical references. Subjects: United States. National Aeronautics and Space Administration--Information systems--Evaluation. Computer networks--Security measures--United States. LC Classification: TL521.312 .U52 1999 Dewey Class No.: 005.8 21

Issues concerning the future operation of the space transportation system: study / by the staff of the U.S. General Accounting Office. Published/Created: Washington, D.C.: The Office, [1982] Related Authors: National Aeronautics and Space Administration. Description: v, 26 p.; 28 cm. Notes: Cover title. "December 1982." "GAO/MASAD-83-6." Item 546-D (microfiche) Subjects: Space shuttles. LC Classification: TL795.5 .U55 1982 Dewey Class No.: 353.0087/78 19 Govt. Doc. No.: GA 1.13:MASAD-83-6

Janson, Bette R. Astronautics and aeronautics, 1979-1984: a chronology / by Bette R. Janson and Eleanor H. Ritchie. Published/Created: Washington, DC: National Aeronautics and Space Administration, Office of Management, Scientific and Technical Information Division: For sale by the Supt. of Docs., U.S. G.P.O., 1990. Related Authors: Ritchie, Eleanor H. Description: v, 778 p.; 24 cm. Notes: Includes bibliographical references. Subjects: Astronautics--History--Chronology. Aeronautics--History--Chronology. Series: The NASA history series LC Classification: TL788.5 .J37 1990 Dewey Class No.: 629.4/09 20

Jenkins, Dennis R. Hypersonics before the shuttle: a concise history of the X-15 research airplane / by Dennis R. Jenkins Published/Created: Washington, D.C.: National Aeronautics and Space Administration, 2000. Description: p. cm. Notes: Includes index. Subjects: X-15 (Rocket aircraft)--History. Series: Monographs in aerospace history; no. 18 NASA history series NASA SP; 4518. LC Classification: TL789.8.U6 X553 2000 Dewey Class No.: 629.133/38 21

Jones, Robert Lewis, 1934- Evaluation and comparison of three space suit assemblies, by R. L. Jones. Published/Created: [Washington] National Aeronautics and Space Administration; for sale by the Clearinghouse for Federal Scientific and Technical Information, Springfield, Va.

[1966] Description: ix, 135 p. illus. 27 cm. Subjects: Space suits. LC Classification: TL521 .A3525 no. 3482 Dewey Class No.: 629.47/7

Kaplan, Judith. Space patches: from Mercury to the space shuttle / Judith Kaplan and Robert Muniz. Published/Created: New York: Sterling Pub. Co., c1986. Related Authors: Muniz, Robert. Description: 128 p.: col. ill.; 27 cm. ISBN: 0806962925 0806962941 (pbk.) Notes: Includes index. Subjects: United States. National Aeronautics and Space Administration--Insignia. LC Classification: TL521 .K33 1986 Dewey Class No.: 387.8/027 19

Kay, W. D. Can democracies fly in space?: the challenge of revitalizing the U.S. space program / W.D. Kay. Published/Created: Westport, Conn.: Praeger, 1995. Description: xi, 244 p.; 25 cm. ISBN: 0275952541 (alk. paper) Notes: Includes bibliographical references (p. [197]-236) and index. Subjects: United States. National Aeronautics and Space Administration--Reorganization. Astronautics and state--United States. Astronautics--Research--Government policy--United States. LC Classification: TL789.8.U5 K34 1995 Dewey Class No.: 387.8/0973 20

Kennan, Erlend A. Mission to the moon; a critical examination of NASA and the space program, by Erlend A. Kennan and Edmund H. Harvey, Jr. Published/Created: New York, Morrow, 1969. Related Authors: Harvey, Edmund H., joint author. Description: xviii, 396 p. illus. 24 cm. Notes: Bibliography: p. 353-388. Subjects: United States. National Aeronautics and Space Administration. Project Apollo (U.S.) Astronautics and state--United States. LC Classification: TL789.8.U5 K4 Dewey Class No.: 629.4/0973

Kerrigan, Evans E. The NASA space medals: the medals and awards of the National Aeronautics and Space Administration for achievements in the exploration of outer space / Evans E. Kerrigan. Published/Created: Noroton Heights, Conn.: Medallic Pub. Co., c1989. Description: 68 p.: ill.; 22 cm. ISBN: 0962466336 : Notes: Includes bibliographical references (p. 68). Subjects: United States. National Aeronautics and Space Administration--Medals. United States. National Aeronautics and Space Administration--Awards. Astronautics--Medals--United States. Astronautics--Awards--United States. LC Classification: TL537 .K47 1989 Dewey Class No.: 629.4/079/73 20

Kerrod, Robin. NASA: visions of space: capturing the history of NASA / Robin Kerrod. Published/Created: Philadelphia, Pa.: Courage Books, c1990. Description: 167 p.: col. ill.; 28 x 30 cm. ISBN: 0894718533 Notes: Includes index. Subjects: United States. National Aeronautics and Space Administration--History--Pictorial works. LC Classification: TL521 .K37 1990 Dewey Class No.: 500.5/072073 20

Kibler, James F. An Earth-Mars mission-analysis program, by James F. Kibler. Published/Created: Washington, National Aeronautics and Space Administration; [for sale by the National Technical Information Service, Springfield, Va.] 1971. Description: 73 p. illus. 27 cm. Notes: Cover title. Subjects: Space flight to Mars. Space trajectories--Data processing.

FORTRAN (Computer program language) Series: United States. National Aeronautics and Space Administration. NASA technical note, NASA TN D-5985 LC Classification: TL521 .A3525 no. 5985 Dewey Class No.: 629.1/08 s

Kloman, Erasmus H. Unmanned space project management; surveyor and lunar orbiter, by Erasmus H. Kloman. Published/Created: Washington, Scientific and Technical Information Office, National Aeronautics and Space Administration; [for sale by the Supt. of Docs., U.S. Govt. Print. Off.] 1972. Description: ix, 41 p. illus. 23 cm. Notes: Includes bibliographical references. Subjects: Surveyor Program (U.S.) Lunar Orbiter (Artificial satellite) Project management. Series: NASA SP; 4901. LC Classification: TL789.8.U6 S953 Dewey Class No.: 629.43/53 Govt. Doc. No.: NAS 1.21:4901

Kraemer, Robert S. Beyond the moon: a golden age of planetary exploration, 1971-1978 / Robert S. Kraemer. Published/Created: Washington, DC: Smithsonian Institution Press, c2000. Description: xxix, 270 p.: ill. (some col.); 24 cm. ISBN: 1560989548 (alk. paper) Notes: Includes bibliographical references (p. 253-259) and index. Subjects: Kraemer, Robert S.--Career in space sciences. United States. National Aeronautics and Space Administration. Space probes--History. Space sciences--United States--History. Planets--Exploration--History. LC Classification: TL795.3 .K73 2000 Dewey Class No.: 629.43/54/09 21

Kraft, Christopher C. Flight: my life in mission control / Chris Kraft. Published/Created: New York: Dutton, c2001. Description: x, 371 p.: ill.; 24 cm. ISBN: 0525945717 (alk. paper) Notes: Includes index. Subjects: Kraft, Christopher C. United States. National Aeronautics and Space Administration--History. Aerospace engineers--United States--Biography. Astronautics--United States--History. LC Classification: TL789.85.K7 A3 2001 Dewey Class No.: 629.4/092 B 21

Kraft, Christopher C. Flight: my life in mission control / Chris Kraft. Published/Created: Thorndike, ME: Thorndike Press, 2001. Description: p. cm. ISBN: 0786234261 (lg. print: hc: alk. paper) Notes: Includes index. Originally published: New York: Dutton, 2001. Subjects: Kraft, Christopher C. United States. National Aeronautics and Space Administration--History. Aerospace engineers--United States--Biography. Astronautics--United States--History. LC Classification: TL789.85.K7 A3 2001a Dewey Class No.: 629.4/092 B 21

Kranz, Gene. Failure is not an option: mission control from Mercury to Apollo 13 and beyond / Gene Kranz. Published/Created: New York: Berkley Books, 2001. Description: p. cm. ISBN: 0425179877 Notes: Originally published: New York: Simon & Schuster, 2000. Includes index. Subjects: Kranz, Gene--Career in aerospace engineering. United States National Aeronautics and Space Administration. Manned space flight--Systems engineering--United States History. Ground support systems (Astronautics)--History. Astronautics--United States--History. LC Classification: TL873 .K73 2001 Dewey Class No.: 629.45/3/0973 21

Kranz, Gene. Failure is not an option: mission control from Mercury to Apollo 13 and beyond / Gene Kranz. Published/Created: New York: Simon & Schuster, c2000. Description: 415 p.: ill.; 25 cm. ISBN: 0743200799 Notes: Includes index. Subjects: Kranz, Gene--Career in aerospace engineering. United States National Aeronautics and Space Administration. Manned space flight--Systems engineering--United States History. Ground support systems (Astronautics)--History. Astronautics--United States--History. LC Classification: TL873 .K73 2000 Dewey Class No.: 629.45/3/0973 21

Lambright, W. Henry, 1939- Powering Apollo: James E. Webb of NASA / W. Henry Lambright. Published/Created: Baltimore: The John Hopkins University Press, 1995. Description: xi, 271 p.: ill.; 24 cm. ISBN: 0801849020 (alk. paper) Notes: Includes bibliographical references (p. [219]-253) and index. Subjects: Webb, James E. (James Edwin), 1906- United States. National Aeronautics and Space Administration--Officials and employees--Biography. Project Apollo (U.S.) Government executives--United States--Biography. Series: New series in NASA history LC Classification: TL789.85.W43 L36 1995 Dewey Class No.: 353.0087/78 20

Large space structures & systems in the space station era: a bibliography with indexes. Published/Created: Washington, DC: National Aeronautics and Space Administration, Office of Management, Scientific and Technical Information Division; Springfield, Va.: Available from the National Technical Informtion Service, 1990- Related Authors: United States. National Aeronautics and Space Administration. Scientific and Technical Information Division. United States. National Aeronautics and Space Administration. Scientific and Technical Information Program. Description: v.; 28 cm. 01 (Nov. 1990)- Notes: Title varies slightly. Vols. for 1991- issued by: National Aeronautics and Space Administration, Office of Management, Scientific and Technical Information Program. SERBIB/SERLOC merged record Formed by the union of: Technology for large structures. Supplement; and: Space station systems. Supplement. Subjects: Large space structures (Astronautics)--Bibliography. Space stations--Bibliography--Periodicals. Series: NASA SP LC Classification: Z5064.S8 L37 TL875 Dewey Class No.: 016.62947 20 Govt. Doc. No.: NAS 1.21:7085

Launius, Roger D. An annotated bibliography of the Apollo program / by Roger D. Launius and J.D. Hunley. Published/Created: Washington, DC: NASA History Office, [1994] Related Authors: Hunley, J. D., 1941- United States. National Aeronautics and Space Administration. History Office. Description: vi, 99 p.: ill. (some col.); 28 cm. Notes: "July 1994". Includes index. Subjects: Project Apollo (U.S)--Bibliography. APOLLO FLIGHT BIBLIOGRAPHIES ASTRONAUTS MANNED SPACE FLIGHT HISTORIES MOON UNITED STATES Series: Monographs in aerospace history; no. 2 LC Classification: Z5064.S7 L38 1994 TL789.8.U6

Launius, Roger D. NASA & the exploration of space: with works from the NASA art collection / by Roger D. Launius and

Bertam Ulrich; ; foreword by John Glenn. Published/Created: New York: Stewart, Tabori & Chang, c1998. Related Authors: Ulrich, Bertram. Description: 224 p.: ill. (some col.); 32 cm. ISBN: 1556706960 Notes: Includes bibliographical references (p. 218-220) and index. Subjects: United States. National Aeronautics and Space Administration--History. Astronautics--United States--History. Space vehicles in art. Space ships in art. Outer Space--Exploration--United States--History. Outer space--In art. LC Classification: TL789.8.U5 L38 1998 Dewey Class No.: 629.4/0973 21

Launius, Roger D. NASA: a history of the U.S. civil space program / Roger D. Launius. Edition Information: Original ed. Published/Created: Malabar, Fla.: Krieger Pub. Co., 1994. Description: ix, 276 p.; 20 cm. ISBN: 089464727X Notes: "An Anvil original." Includes bibliographical references (p. 265-269) and index. Subjects: United States. National Aeronautics and Space Administration--History. Space sciences--United States--History. Astronautics--United States--History. LC Classification: TL521.312 .L28 1994 Dewey Class No.: 387.8/0973 20

Legislative history of the Space Law of 1958 establishing the world's first civilian space agency (a detailed documentation of the enactment by Congress of a new law): report to the Committee on Science and Astronautics, U.S. House of Representatives, Eighty-sixth Congress, second session / [compiled by Raymond Wilcove]. Published/Created: Washington: U.S. Govt. Print. Off., 1960. Related Authors: Wilcove, Raymond. United States. Congress. House. Committee on Science and Astronautics. Description: 1465 p. in various pagings; 24 cm. Notes: At head of Committee print. Subjects: United States. Laws, statutes, etc. National aeronautics and space act. United States. National Aeronautics and Space Council. Space law. LC Classification: KF2474.53 .A15 1960 Dewey Class No.: 344/.73/095

Levine, Arnold S. Managing NASA in the Apollo era / Arnold S. Levine. Published/Created: Washington, D.C.: Scientific and Technical Information Branch, National Aeronautics and Space Administration, 1982. Description: xxi, 343 p.: ill.; 25 cm. Notes: "NASA SP-4102." Includes index. Bibliography: p. 301-303. Subjects: United States. National Aeronautics and Space Administration--History. Series: United States. National Aeronautics and Space Administration. LC Classification: TL521.312 .L38 1982 Dewey Class No.: 353.0085/6 19

Levine, Arthur L. The future of the U.S. space program / Arthur L. Levine. Published/Created: New York: Praeger, 1975. Description: xi, 198 p.; 25 cm. ISBN: 027508700X : Notes: Includes index. Bibliography: p. 190-191. Subjects: United States. National Aeronautics and Space Administration. Astronautics--United States. LC Classification: TL789.8.U5 L44 Dewey Class No.: 629.4/0973

Levine, Sol, 1914- Your future in NASA. Edition Information: [1st ed.] Published/Created: New York, R. Rosen Press [1969] Description: 185 p. illus., ports. 22 cm. Summary: Describes the many careers available in the National Aeronautics and Space Administration and explains how to prepare for a future in this agency. Includes a salary scale for NASA and a list of accredited engineering colleges. Notes: Bibliography: p. 182-185. Subjects: United States. National Aeronautics and Space Administration--Vocational guidance. United States. National Aeronautics and Space Administration--Vocational guidance. LC Classification: TL521.312 .L4 Dewey Class No.: 629.1/023

Levine, Sol, 1914- Your future in NASA. Edition Information: [Rev. ed.] Published/Created: New York, Arco [1971] Description: 185 p. illus., ports. 21 cm. ISBN: 0668022558 Summary: Describes the many careers available for both men and women in the National Aeronautics and Space Administration and explains how to prepare for a future in this agency. Includes a salary scale for NASA and a list of accredited engineering colleges. Notes: Bibliography: p. 182-185. Subjects: United States. National Aeronautics and Space Administration--Vocational guidance. United States. National Aeronautics and Space Administration--Vocational guidance. LC Classification: TL521.312 .L4 1971 Dewey Class No.: 629.1/023

Lewis, Richard S., 1916- Appointment on the moon; the full story of Americans in space, from Explorer I to the lunar landing and beyond [by] Richard S. Lewis. Edition Information: [Rev. ed.] Published/Created: New York, Viking Press [1969] Description: xiv, 560 p. illus., ports. 25 cm. ISBN: 0670129690 Notes: "Reference notes": p. 541-546. Subjects: United States. National Aeronautics and Space Administration--History. Astronautics--United States--History. LC Classification: TL789.8 .U8L46 1969b Dewey Class No.: 629.45/4

Lewis, Richard S., 1916- Appointment on the moon; the inside story of America's space venture [by] Richard S. Lewis. Published/Created: New York, Viking Press [1968] Description: xiv, 434 p. illus. 25 cm. Notes: Includes bibliographical references. Subjects: United States. National Aeronautics and Space Administration--History. Astronautics--United States--History. LC Classification: TL789.8.U5 L46 Dewey Class No.: 629.45/4

Life into space: space life sciences experiements, NASA Ames Research Center, 1965-1990 / edited by Kenneth Souza, Robert Hogan, Rodney Ballard. Published/Created: Moffett Field, CA: National Aeronautics and Spcae Administration, Ames Research Center, [1995?] Related Authors: Souza, Kenneth A. Hogan, Robert (Robert P.) Ballard, Rodney, d. 1993. Ames Research Center. Description: vii, 606 p.: ill., maps; 24 x 28 cm. Notes: Includes bibliographical references (p. [391]-450) and indexes. Subjects: Space biology--Research--United States--History. Space biology--Research. Space flight--History. Manned space flight. Series: NASA reference publication; 1372 LC Classification: QH327 .L535 1995 Dewey Class No.: 574.19/19/072 20

List of selected references on NASA programs. Prepared for the National Aeronautics and Space Administration. Published/Created: Washington, National Aeronautics and Space Administration, 1962. Related Authors: United States. National Aeronautics and

Space Administration. Description: iii, 236 p. 26 cm. Subjects: United States. National Aeronautics and Space Administration--Bibliography. Aeronautics--Bibliography. Series: United States. National Aeronautics and Space Administration. NASA; SP-3. LC Classification: TL521 .A333 no. 3

Lord, Douglas R. Spacelab, an international success story / Douglas R. Lord. Published/Created: Washington, DC: Scientific and Technical Information Branch, National Aeronautics and Space Administration, 1987. Related Titles: Spacelab. Description: xv, 554 p.: ill. (some col.); 26 cm. Notes: Bibliography: p. 403-433. Subjects: Spacelab Program. Series: NASA SP; 487 LC Classification: TL797 .L67 1987 Dewey Class No.: 629.44/5 19

Ludwig, George H. The NASA program for particles and fields research in space. Published/Created: Washington, National Aeronautics and Space Administration, 1962. Description: iii, 26 p. illus. 27 cm. Subjects: Astronautics--United States. Series: National Aeronautics and Space Administration. NASA technical note D-1173. LC Classification: TL521 .A3525 no. 1173

Lunar orbit rendezvous; news conference on Apollo plans at NASA headquarters on July 11, 1962. Published/Created: [Washington, 1962] Description: 32 p. illus. 27 cm. Notes: Cover title. Subjects: Project Apollo (U.S.) LC Classification: TL789.8.U6 A6

Lunar orbiter II: photography. Published/Created: [Washington] National Aeronautics and Space Administration; for sale by the Clearinghouse for Federal Scientific and Technical Information, Springfield, Va. [1967] Related Authors: Langley Research Center. Description: viii, 97 p. illus. 27 cm. Notes: Prepared under contract no. NAS 1-3800 for Langley Research Center. Subjects: Lunar Orbiter (Artificial satellite) Lunar photography. Lunar landing sites. Series: NASA contractor report; NASA CR-931. LC Classification: TL521.3.C6 A3 no. 931 Dewey Class No.: 629.43/53

Lunar Orbiter II; photographic mission summary. Published/Created: Washington, National Aeronautics and Space Administration; [for sale by the Clearinghouse for Federal Scientific and Technical Information, Springfield, Va.] 1967. Related Authors: Langley Research Center. Description: vii, 86 p. illus. (1 fold.) 27 cm. Notes: "Prepared under contract no. NASL-3800 by the Boeing Company, Seattle, Washington for Langley Research Center." Subjects: Lunar Orbiter (Artificial satellite) Lunar landing sites. Moon--Photographs from space. Series: NASA contractor report; NASA CR-883. LC Classification: TL521.3.C6 A3 no. 883 Dewey Class No.: 629.43/53

Lunar Orbiter III; photographic mission summary. Published/Created: [Washington] National Aeronautics and Space Administration; for sale by the Clearinghouse for Federal Scientific and Technical Information, Springfield, Va. [1968] Related Authors: Langley Research Center. Description: viii, 108 p. illus. 27 cm. Notes: "Prepared under contract no. NAS 1-3800 ... for Langley Research Center." Subjects: Lunar Orbiter (Artificial satellite) Lunar landing sites. Moon--Photographs from space. Series: NASA contractor report;

NASA CR-1069. LC Classification: TL521.3.C6 A3 no. 1069 Dewey Class No.: 629.43/53

Lunar Orbiter III; photography. Published/Created: [Washington] National Aeronautics and Space Administration; for sale by the Clearinghouse for Federal Scientific and Technical Information, Springfield, Va. [1968] Related Authors: Langley Research Center. Description: viii, 120 p. illus. 27 cm. Notes: Prepared under contract no. NAS 1-3800 for Langley Research Center. Subjects: Lunar Orbiter (Artificial satellite) Lunar photography. Lunar landing sites. Series: NASA contractor report; NASA CR-984. LC Classification: TL521.3.C6 A3 no. 984 Dewey Class No.: 629.43/53

Lunar Orbiter IV; photography. Published/Created: [Washington] National Aeronautics and Space Administration; for sale by the Clearinghouse for Federal Scientific and Technical Information, Springfield, Va. [1968] Related Authors: Langley Research Center. Description: ix, 169 p. illus. 27 cm. Notes: Prepared under contract no. NAS 1-3800 for Langley Research Center. Subjects: Lunar Orbiter (Artificial satellite) Lunar photography. Lunar landing sites. Series: NASA contractor report; NASA CR-1093. LC Classification: TL521.3.C6 A3 no. 1093 Dewey Class No.: 629.43/53

Magellan: revealing the face of Venus. Published/Created: Pasadena: National Aeronautics and Space Administration, Jet Propulsion Laboratory, Calif. Institute of Technology, [1993] Related Authors: Jet Propulsion Laboratory (U.S.) Description: 22 p.: ill. (some col.); 31 cm. ISBN: 0160421977 Notes: Cover title. Subjects: Magellan (Spacecraft) Venus (Planet)--Surface. Venus (Planet)--Exploration. LC Classification: QB621 .M34 1993 Dewey Class No.: 559.9/22 20

Maintaining U.S. leadership in aeronautics: scenario-based strategic planning for NASA's aeronautics enterprise / Steering Committee for a Workshop to Develop Long-Term Global Aeronautics Scenarios ... [et al.]. Published/Created: Washington, D.C.: National Academy Press, 1997. Description: viii, 129 p.: ill.; 28 cm. ISBN: 0309056969 Notes: Includes bibliographical references (p. 123-125). Subjects: United States. National Aeronautics and Space Administration. Aeronautics--United States--Planning. Aeronautics and state--United States. LC Classification: TL521.312 .N394 1997 Dewey Class No.: 387.8/0973 21

Maisel, Martin D. The history of the XV-15 tilt rotor research aircraft: from concept to flight / Martin D. Maisel, Demo J. Giulianetti, and Daniel C. Dugan. Published/Created: Washington, DC: National Aeronautics and Space Administration, Office of Policy and Plans, NASA History Division, 2000. Related Authors: Dugan, Daniel C. Giulianetti, Demo J. United States. National Aeronautics and Space Administration. History Office. Description: xxi, 194 p.: ill.; 28 cm. ISBN: 0160502764 Notes: Includes bibliographical references (p. 164-184) and index. Subjects: Research aircraft--United States. Aeronautics--Research--United States. Tilt rotor aircraft. Series: NASA SP; 2000-4517 The NASA history series Monographs in aerospace history; no. 17. LC Classification:

TL685 .M25 2000 Dewey Class No.: 629.133/35 21

Major activities in the programs of the National Aeronautics and Space Administration, October 1, 1959 - March 31, 1960. Published/Created: [Washington, U.S. Govt. Print. Off., 1960] Description: x, 180 p. illus., diagrs., tables. 24 cm. Subjects: Astronautics--United States. Aeronautics--Research--United States. Outer space--Exploration. LC Classification: TL789.8.U5 A58 1960 Dewey Class No.: 629.40973

Malik, P. W. Project Gemini; a technical summary, by P. W. Malik and G. A. Souris. Published/Created: [Washington] National Aeronautics and Space Administration; for sale by the Clearinghouse for Federal Scientific and Technical Information, Springfield, Va. [1968] Related Authors: Souris, G. A., joint author. McDonnell Douglas Corporation. Manned Spacecraft Center (U.S.) Description: vi, 343 p. illus. 27 cm. Notes: "Prepared under contract no NAS 9-170 by McDonnell Douglas Corporation, St. Louis, Mo. for Manned Spacecraft Center." Subjects: Project Gemini (U.S.) Series: NASA contractor report; NASA CR-1106. LC Classification: TL521.3.C6 A3 no. 1106 Dewey Class No.: 629.47

Managing the moon program: lessons learned from Project Apollo: proceedings of an oral history workshop / conducted July 21, 1989; moderator, John M. Logsdon; participants, Howard W. Tindall ... [et al.].
Published/Created: Washington, DC: National Aeronautics and Space Administration, NASA History Division, Office of Policy and Plans, 1999. Related Authors: Logsdon, John M., 1937- Tindall, Howard W. United States. National Aeronautics and Space Administration. History Office. Description: 52 p.: ill.; 28 cm. Notes: "July 1999." Includes index. Subjects: Project Apollo (U.S.)--History--Congresses. Space flight--Planning--Congresses. Astronauts--United States--Interviews. Series: Monographs in aerospace history; no. 14 LC Classification: TL789.8.U6 A5537 1999 Dewey Class No.: 629.45/4/0973 21

Manned space flight program of the National Aeronautics and Space Administration: Projects Mercury, Gemini, and Apollo; staff report. Published/Created: Washington, U.S. Govt. Print. Off., 1962. Description: vii, 242 p. illus., maps. 24 cm. Notes: At head of 87th Congress, 2d Session, committee print Subjects: Project Apollo (U.S.) Project Mercury (U.S.) Project Gemini (U.S.) LC Classification: TL789.8.U5 A53 1962a

Manned space flight, 1963.
Published/Created: Washington [1963] Related Authors: Holmes, Dyer Brainerd. Description: 87 p. ill. Notes: Based on the testimony of D. Brainerd Holmes, Deputy Associate Administrator and Director of Manned Space Flight, National Aeronautics and Space Administration, before the Subcommittee on Manned Space Flight, Committee on Science and Astronautics, House of Representatives and before the Committee on Aeronautical and Space Sciences, U. S. Senate. Subjects: Project Apollo (U.S.) Space flight. LC Classification: TL789.8.U6 A63

Mariner Mars 1964 Project report: Mission operations. Published/Created: Pasadena, 1966. Related Authors: United States. National Aeronautics and Space Administration. Description: ix, 174 p. illus. 28 cm. Notes: At head of National Aeronautics and Space Administration. Subjects: Project Mariner (U.S.) Series: JPL technical report; no. 32-881. LC Classification: TL789.8.U6 M32 Dewey Class No.: 629.43/54/3

Mariner Mars 1971 television picture catalog. Published/Created: Pasadena, 1972-74 [v. 1, 1974] Related Authors: Koskela, P. E. Cutts, James A. United States. National Aeronautics and Space Administration. Description: 2 v. illus. 28 cm. Contents: v. 1. Experiment design and picture data, by J. A. Cutts.-- v. 2. Sequence design and picture coverage, by P. E. Koskela, and others. Notes: "Prepared under contract no. NAS 7-100, National Aeronautics and Space Administration." Subjects: Project Mariner (U.S.)--Catalogs. Video recording--Catalogs. Mars (Planet)--Photographs from space--Catalogs. Series: California Institute of Technology, Pasadena. Jet Propulsion Laboratory. Technical memorandum 33-585. LC Classification: QB641 .C3 1972a Dewey Class No.: 523.4/3/0222

Mariner mission to Venus. Prepared for the National Aeronautics and Space Administration. Compiled by Harold J. Wheelock; foreword by W.H. Pickering. Published/Created: New York, McGraw-Hill [1963] Related Authors: Wheelock, Harold J., comp. Description: 118 p. illus. 21 cm. Subjects: Project Mariner (U.S.) LC Classification: TL789.8.U6 M34 Dewey Class No.: 629.43542

Mariner-Mars 1969: a preliminary report. Published/Created: Washington; [For sale by the Clearinghouse for Federal Scientific and Technical Information, Springfield, Va.] 1969. Description: vii, 145 p. illus. 27 cm. Notes: Includes bibliographical references. Subjects: Mars (Planet) Series: NASA SP; 225. LC Classification: QB641 .U55 Dewey Class No.: 550

Mariner-Venus, 1962; final project report. Published/Created: Washington, Scientific and Technical Information Division, National Aeronautics and Space Administration; [for sale by the Superintendent of Documents, U.S. Govt. Print. Off.] 1965. Related Authors: United States. National Aeronautics and Space Administration. Description: xi, 344 p. illus. (1 col.) 26 cm. Notes: "Prepared under contract for NASA." Bibliography: p. 342-344. Subjects: Project Mariner (U.S.) Series: United States. National Aeronautics and Space Administration. NASA SP; -59. LC Classification: TL789.8.U6 M343 Dewey Class No.: 629.43542

Mariner-Venus, 1967; final project report. Published/Created: Washington; [For sale by the Supt. of Docs., U.S. Govt. Print. Off.] 1971 [i.e. 1972] Description: x, 301 p. illus. 26 cm. Notes: Includes bibliographical references. Subjects: Project Mariner (U.S.) Series: NASA SP; 190. LC Classification: TL789.8.U6 M383 Dewey Class No.: 629.43/54/2 Govt. Doc. No.: NAS 1.21:190

Mars as viewed by Mariner 9: a pictorial presentation by the Mariner 9 television team and the planetology program principal investigators. Edition

Information: Rev. Published/Created: Washington: Scientific and Technical Information Office, National Aeronautics and Space Administration: for sale by the Supt. of Docs., U.S. Govt. Print. Off., 1976. Related Authors: United States. National Aeronautics and Space Administration. Scientific and Technical Information Office. Description: 225 p.: ill.; 30 cm. Subjects: Project Mariner (U.S.) Landforms. Mars (Planet)--Photographs from space. Series: NASA SP; 329. LC Classification: QB641 .M36 1976 Dewey Class No.: 559.9/23/0222 Govt. Doc. No.: NAS 1.21:329/2.

Mars as viewed by Mariner 9: a pictorial presentation by the Mariner 9 television team and the planetology program principal investigators. Published/Created: Washington: Scientific and Technical Information Office, National Aeronautics and Space Administration: for sale by the Supt. of Docs., U.S. Govt. Print. Off., 1974. Related Authors: United States. National Aeronautics and Space Administration. Scientific and Technical Information Office. Description: 225 p.: ill.; 30 cm. Subjects: Project Mariner (U.S.) Landforms. Mars (Planet)--Photographs from space. Series: NASA SP; 329. LC Classification: QB641 .M36 Dewey Class No.: 523.4/3/0222

Masursky, Harold. Viking site selection and certification / Harold Masursky and Norman L. Crabill; prepared by NASA Langley Research Center. Published/Created: Washington, D.C.: National Aeronautics and Space Administration, Scientific and Technical Information Branch, 1981. Related Authors: Crabill, Norman L. United States. National Aeronautics and Space Administration. Scientific and Technical Information Branch. Langley Research Center. Description: v, 34 p.: ill.; 27 cm. Notes: Includes bibliographical references. Subjects: Viking Mars Program (U.S.) Space vehicles--Landing. Series: NASA SP; 429 LC Classification: TL789.8.U6 V527 1981 Dewey Class No.: 629.43/543 19 Govt. Doc. No.: NAS 1.21:429

McCurdy, Howard E. Faster, better, cheaper: low-cost innovation in the U.S. space program / Howard E. McCurdy. Published/Created: Baltimore: Johns Hopkins University Press, 2001. Description: p. cm. ISBN: 0801867207 (hardcover) Notes: Includes bibliographical references and index. Subjects: United States. National Aeronautics and Space Administration. Astronautics--United States--Cost control. Organizational effectiveness. Astronautics--Technological innovations. Outer space--Exploration--Cost control. Series: New series in NASA history LC Classification: TL796.5.U5 M37 2001 Dewey Class No.: 629.4/068/1 21

McCurdy, Howard E. Inside NASA: high technology and organizational change in the U.S. space program / Howard E. McCurdy. Published/Created: Baltimore: Johns Hopkins University Press, c1993. Description: xiv, 215 p.: ill., map; 24 cm. ISBN: 0801844525 Notes: Includes bibliographical references (p. 193-208) and index. Subjects: United States. National Aeronautics and Space Administration--Management. United States. National Aeronautics and Space Administration--History. Organizational sociology. Corporate culture. Series: New series in NASA history LC Classification:

TL521.312 .M33 1993 Dewey Class No.: 353.0087/78 20

McRoberts, Joseph J. Space telescope / Joseph J. McRoberts. Published/Created: Washington, D.C.: National Aeronautics and Space Administration, Division of Public Affairs: For sale by the Supt. of Docs., U.S. G.P.O., [1982] Related Authors: United States. National Aeronautics and Space Administration. Public Affairs Division. Description: 64 p.: ill. (some col.); 28 cm. Notes: "February 1982"-- P. 2 of cover. S/N 033-000-00862-7 Item 830-G Subjects: Hubble Space Telescope (Spacecraft) Series: NASA EP; 166 LC Classification: QB88 .M395 1982 Dewey Class No.: 522/.2 19 Govt. Doc. No.: NAS 1.19:166

Meteor orbits and dust; the proceedings of a symposium. Published/Created: Washington, Scientific and Technical Information Division, National Aeronautics and Space administration; [for sale by the Supt. Of Docs., U.S. Govt. Print. Off.] 1967. Related Authors: Hawkins, Gerald S., ed. Description: 412 p. Notes: Includes bibliographies. Subjects: Meteors. Interstellar matter. Series: Smithsonian contributions to astrophysics, v. 11 NASA SP; -135. LC Classification: QB461 .S6 vol. 11 Govt. Doc. No.: NAS 1.21:135

Mirabito, Michael M., 1956- The exploration of outer space with cameras: a history of the NASA unmanned spacecraft missions / by Michael M. Mirabito. Published/Created: Jefferson, N.C.: McFarland, 1983. Description: vi, 170 p., [32] p. of plates: ill.; 24 cm. ISBN: 0899500617 : Notes: Includes index. Bibliography: p. 155-164.

Subjects: United States. National Aeronautics and Space Administration--History. Space probes. Outer space--Exploration--History--United States. LC Classification: TL789.8.U5 M57 1983 Dewey Class No.: 629.43/54/09 19

Mission description and in-flight operations of ERBE instruments on ERBS and NOAA 9 spacecraft: November 1984 through January 1986 / William L. Weaver ... [et al.]. Published/Created: Washington, D.C.: National Aeronautics and Space Administration, Office of Management, Scientific and Technical Information Program; Springfield, Va.: For sale by the National Technical Information Service, 1991. Related Authors: Weaver, William L. Description: ix, 272 p.: ill.; 28 cm. Notes: Includes bibliographical references (p. 34). Subjects: Earth Radiation Budget Experiment. Earth Radiation Budget Satellite. NOAA (Weather satellite) Terrestrial radiation--Measurement. Terrestrial radiation--Measurement--Equipment and supplies. Energy budget (Geophysics)--Measurement. Series: NASA reference publication; 1256 LC Classification: QC809.T4 M58 1991 Dewey Class No.: 551.5/272 20

Mission to earth: Landsat views the world / Nicholas M. Short ... [et al.]. Published/Created: Washington: Scientific and Technical Office, National Aeronautics and Space Administration: for sale by the Supt. of Docs., U.S. Govt. Print. Off., 1976. Related Authors: Short, Nicholas M. Description: ix, 459 p.: ill.; 36 cm. Notes: Includes index. Subjects: Astronautics in earth sciences. Earth--Photographs from space. Series: NASA SP; 360. LC Classification: QB637

.M57 Dewey Class No.: 910/.02/0222
Govt. Doc. No.: NAS 1.21:360

Mission to the solar system: exploration and discovery: a mission and technology roadmap / The Roadmap Development Team; edited by S. Gulkis, D.S. Stetson, E.R. Stofan. Published/Created: Pasadena, Calif: National Aeronautics and Space Administration, Jet Propulsion Laboratory, California Institute of Technology, [1998] Related Authors: Gulkis, S. (Samuel) Stetson, D. S. Stofan, Ellen Renee, 1961- Roadmap Development Team. Description: 1 v. (various pagings): col. ill.; 28 cm. Notes: "March 1998." Subjects: Outer space--Exploration. Planets--Exploration. Astronautics in astronomy. LC Classification: QB501.5 .M57 1998 Dewey Class No.: 629.43/5 21

Models of Mars atmosphere, 1967. Published/Created: [Washington, For sale by the Clearinghouse for Federal Scientific and Technical Information, Springfield, Va.] 1968. Related Titles: NASA space vehicle design criteria: environment. Description: iii, 21 p. illus. 27 cm. Notes: Cover title. At head of NASA space vehicle design criteria: environment. Bibliography: p. 15-16. Subjects: Mars (Planet)--Atmosphere. Series: NASA SP; 8010. LC Classification: QB641 .U5 Dewey Class No.: 551.5/1

Morgan, Clay, 1950- Shuttle-Mir = [Mir-shattl]: the U.S. and Russia share history's highest stage / by Clay Morgan. Published/Created: Washington, D.C.: National Aeronautics and Space Administration, 2001. Description: p. cm. Notes: Parallel title romanized. Includes bibliographical references and index. In English. Subjects: Mir (Space station) Astronautics--United States--International cooperation. Astronautics--Russia--International cooperation. Space shuttles. Series: NASA SP; 4225 The NASA history series LC Classification: TL788.4 .M67 2001 Dewey Class No.: 629.44 21

Morgan, Norman Evan, 1925- Development of a hydrogen-oxygen internal combustion engine space power system, by N.E. Morgan and W.D. Morath. Published/Created: Springfield, Va., For sale by the Clearinghouse for Federal Scientific and Technical Information [1965] Related Authors: Morath, W. D., joint author. Vickers Incorporated, Detroit. United States. National Aeronautics and Space Administration. Description: iv, 200 p. illus. 27 cm. Notes: "Prepared under contract no. NAS 3-2787 by Vickers, inc., Detroit, Mich., for National Aeronautics and Space Administration." Bibliography: p. 178. Subjects: Internal combustion engines. Space vehicles--Auxiliary power supply. Series: United States. National Aeronautics and Space Administration. NASA contractor report; CR-255. LC Classification: TL521.3.C6 A3 no. 255 Dewey Class No.: 629.47/4

Morrison, David, 1940- Voyage to Jupiter / David Morrison and Jane Samz. Published/Created: Washington, D.C.: Scientific and Technical Information Branch National Aeronautics and Space Administration: For sale by the Supt. of Docs., U.S. G.P.O., 1980. Related Authors: Samz, Jane. United States. National Aeronautics and Space Administration. Scientific and Technical Information Branch. Description: xi,

199 p.: ill. (some col.), charts; 26 cm. Notes: S/N 033-000-00797-3 Item 830-I Bibliography: p. 199. Subjects: Voyager Project. Jupiter probes. Series: NASA SP; 439 LC Classification: QB661 .M67 1980 Dewey Class No.: 559.9/25 19 Govt. Doc. No.: NASA 1.21:439

Morrison, David, 1940- Voyages to Saturn / David Morrison. Published/Created: Washington, D.C.: Scientific and Technical Information Branch, National Aeronautics and Space Administration: For sale by the Supt. of Docs., U.S. G.P.O., 1982. Related Authors: United States. National Aeronautics and Space Administration. Scientific and Technical Information Branch. Description: ix, 227 p.: ill., (some color); 26 cm. Notes: Includes index. S/N 033-000-00842-2 Item 830-I Bibliography: p. 217-218. Subjects: Voyager Project. Saturn probes. Pioneer (Space probes) Saturn (Planet)--Exploration. Series: NASA SP; 451 LC Classification: QB671 .M67 1982 Dewey Class No.: 919.9/26 19 Govt. Doc. No.: NAS 1.21:451

Mudgway, Douglas J., 1923- Uplink-downlink: a history of the NASA Deep Space Network, 1957-1997 / Douglas J. Mudgway. Published/Created: Washington, D.C.: National Aeronautics and Space Administration, 2000. Description: p. cm. Notes: Includes bibliographical references and index. Subjects: Deep Space network--History. Series: NASA SP; 2001-4227 The NASA history series LC Classification: TL3026 .M84 2000 Dewey Class No.: 629.47/43/0973 21

Muenger, Elizabeth A. Searching the horizon: a history of Ames Research Center, 1940-1976 / Elizabeth A. Muenger. Published/Created: Washington, DC: Scientific and Technical Information Branch, National Aeronautics and Space Administration: For sale by the Supt. of Docs., U.S. G.P.O., 1985. Related Authors: United States. National Aeronautics and Space Administration. Scientific and Technical Information Branch. Description: xiii, 299 p.: ill. (some col.); 23 cm. Notes: Includes index. Bibliography: p. 285-288. Subjects: Ames Research Center--History. Series: The NASA history series NASA SP; 4304 LC Classification: TL565 .M84 1985 Dewey Class No.: 629.1/072079473 19

NASA 1958-1983: remembered images. Published/Created: Washington, D.C.: NASA: On sale by the Supt. of Docs., U.S. G.P.O., 1983. Related Titles: Remembered images. Description: 133, [3] p.: chiefly col. ill.; 30 cm. Notes: "25th anniversary, 1958-1983." S/N 033-000-00901-1 Item 830-G Subjects: United States. National Aeronautics and Space Administration--Addresses, essays, lectures. Space sciences--United States--Addresses, essays, lectures. Series: NASA EP; 200 LC Classification: TL521.312 .U54 1983 Dewey Class No.: 629.4/0973 19 Govt. Doc. No.: NAS 1.19:200

NASA accident and injury data. Published/Created: Washington. Description: ill. 27 cm. Continued by: United States. National Aeronautics and Space Administration. NASA mishap and injury data ISSN: 0095-6066 Notes: SERBIB/SERLOC merged record Subjects: United States. National Aeronautics and Space Administration--Officials and employees--Accidents. Aerospace industries--United States--Accidents. LC Classification: TL521.3.A2 U54a Dewey Class No.:

312/.4/3

NASA astronauts. Published/Created: [Washington; For sale by the Supt. of Docs., U.S. Govt. Print. Off., 1967] Description: 32 p. illus., ports. 27 cm. Notes: Cover title. Subjects: Astronauts--United States. LC Classification: TL789.85.A1 U53 Dewey Class No.: 629.45/0092/2 B

NASA day, April 27, 1962. Published/Created: Washington, National Aeronautics and Space Administration [1962] Related Authors: United States. National Aeronautics and Space Administration. Description: i l., 66p. illus.,ports.,maps. 26cm. Notes: Cover title. Sponsored by National Aeronautics and Space Administration. Subjects: United States. National Aeronautics and Space Administration Astronautics--United States LC Classification: TL521 .A333 no. 4

NASA factbook: guide to National Aeronautics and Space Administration programs and activities. Edition Information: 2d ed. Published/Created: Chicago: Marquis Academic Media, c1975. Related Authors: Renetzky, Alvin. NASA factbook. Related Titles: National Aeronautics and Space Administration factbook. Description: 613 p.; 29 cm. ISBN: 0837918014 : Notes: Cover National Aeronautics and Space Administration factbook. Previous ed. by A. Renetzky. Includes indexes. Subjects: United States. National Aeronautics and Space Administration. LC Classification: TL521.312 .R45 1975 Dewey Class No.: 629.4/072/073

NASA historical data book. Published/Created: Washington, D.C.: Scientific and Technical Information Division, National Aeronautics and Space Administration: For sale by Supt. of Docs., U.S. G.P.O., 1988-<2000> Related Authors: Van Nimmen, Jane, 1937- Bruno, Leonard C. Ezell, Linda Neuman. Description: v. <1-6 >: ill.; 26 cm. ISBN: 0160502667 (v. 6) Contents: v. 1. NASA resources, 1958-1968 / Jane Van Nimmen and Leonard C. Bruno -- v. 2. Programs and projects, 1958-1968 / Linda Neuman Ezell -- v. 3. Programs and projects, 1969-1978 / Linda Neuman Ezell -- v. 4. NASA resources, 1969-1978 / Ihor Gawadiak with Helen Fedor -- v. 5. NASA launch systems, space transportation, human spaceflight, and space science, 1979-1988 / Judy A. Rumerman -- v. 6. NASA space applications, aeronautics and space research and technology, tracking and data aquisition/support operations, commercial programs, and resources 1979-1988 / Judy A. Rumerman. Notes: Vol. 1 is a republication of: NASA historical data book, 1958-1968 / Jane Van Nimmen and Leonard C. Bruno. Vol. 4 published by NASA History Office. Vol. 4 in series: The NASA history series. Includes bibliographical references and indexes. Subjects: United States. National Aeronautics and Space Administration--History. Series: The NASA historical series NASA SP; 4012 NASA history series. LC Classification: TL521.312 .N37 1988 Dewey Class No.: 353.0087/78 19

NASA magazine [microform]. Published/Created: Washington, DC: National Aeronautics and Space Administration: For sale by the Supt. of Docs., U.S. G.P.O., Description: v.: ill.; 28 cm. Began with fall 1991. Continues: NASA activities 0190-3292 (OCoLC)13902180 (DLC)sf 86091814

Notes: Description based on: Spring 1992; title from cover. Microfilm. Ann Arbor, Mich: University Microfilms International. microfilm reels; 35 mm. SERBIB/SERLOC merged record Indexed by: Index to U.S. government periodicals 0098-4604 Additional Form Avail.: Distributed to depository libraries in microfiche. Subjects: United States. National Aeronautics and Space Administration--Periodicals. Aeronautics--United States--Periodicals. Astronautics--United States--Periodicals. LC Classification: Microfilm (o) 85/142

NASA magazine. Published/Created: Washington, DC: National Aeronautics and Space Administration: For sale by the Supt. of Docs., U.S. G.P.O., Description: 4 v.: ill.; 28 cm. Began with fall 1991; ceased with winter 1994 issue. Continues: NASA activities 0190-3292 (DLC)sf 77000364 (OCoLC)1348188 Cancel/Invalid LCCN: sn 92023694 Notes: Description based on: Spring 1992; title from cover. SERBIB/SERLOC merged record Indexed by: Index to U.S. government periodicals 0098-4604 Additional Form Avail.: Distributed to depository libraries in microfiche. Subjects: United States. National Aeronautics and Space Administration--Periodicals. Aeronautics--United States--Periodicals. Astronautics--United States--Periodicals. LC Classification: TL521.3 N16 Dewey Class No.: 353 11 Govt. Doc. No.: NAS 1.87:

NASA mishap and injury data. Published/Created: [Washington] National Aeronautics and Space Administration. Description: v. ill. 27 cm. Continues: United States. National Aeronautics and Space Administration. NASA accident and injury data ISSN: 0360-5086 Notes: SERBIB/SERLOC merged record Subjects: United States. National Aeronautics and Space Administration--Officials and employees--Accidents. Aerospace industries--United States--Accidents. LC Classification: TL521.3.A2 U54a Dewey Class No.: 312/.4/3

NASA patent abstracts bibliography. Published/Created: Washington. Description: 27 cm. Vols. for Jan. 1972- issued in two sections (e.g., section 1: Abstracts; section 2: Indexes) Jan. 1972- Continued by: United States. National Aeronautics and Space Administration. Scientific and Technical Information Branch. NASA patent abstracts bibliography 0091-0384 (DLC)sn 87042257 (OCoLC)4820173 ISSN: 0091-0384 Notes: SERBIB/SERLOC merged record Subjects: Astronautics--United States--Patents--Abstracts Periodicals. Astronautics--United States--Patents--Indexes Periodicals. Series: NASA SP LC Classification: TL788.35 .U55a Dewey Class No.: 629.13/0027/2 Govt. Doc. No.: NAS 1.21:

NASA photography from five years of space. Published/Created: [Washington, For sale by the Superintendent of Documents, U. S. Govt. Print Off., 1964] Description: 1 v. (chiefly illus., ports) 23 x 11 cm. Notes: Cover title. Subjects: Astronautics--United States--Pictorial works. Photography--Exhibitions. LC Classification: TL793.5 .U55

NASA pocket statistics. Published/Created: Washington DC: NASA, Description: v.: ill.; 10 x 17-20 cm. Cancel/Invalid LCCN: sn 96034076 Notes:

Description based on: 1995 ed.; title from cover. SERBIB/SERLOC merged record Subjects: United States. National Aeronautics and Space Administration--Statistics--Periodicals. Astronautics--United States--Registers. LC Classification: TL789.8.U5 U56a TL789.8.U5 U55 Dewey Class No.: 629.4/0973 21 Govt. Doc. No.: NAS 1.2:ST 2/13/

NASA PPMI lessons learned [computer file] / National Aeronautics and Space Administration. Published/Created: [Washington, DC]: NASA, [1998?] Description: 1 computer laser optical disc; 4 3/4 in. + 1 booklet ([3] p.; 12 cm.) Computer File Info.: Computer data and program. Notes: Title from p. [4] of cover: "Lessons learned: program & project management initiatives." ISO 9660 format. Booklet includes table of contents. "Gamma Ray Observatory." Shipping list no. 98-65-E. Disc characteristics: CD-ROM. System requirements: for personal computers only. Subjects: United States. National Aeronautics and Space Administration--Management. LC Classification: IN PROCESS (COPIED) Dewey Class No.: ----- 13 Govt. Doc. No.: NAS 1.86:G 14/CD

NASA reactor facility hazards summary. Published/Created: [Washington, National Aeronautics and Space Administration, 1959- Related Authors: Hallman, Theodore M., ed. Lubarsky, Bernard, ed. Description: 1 v. illus., maps (part fold.) diagrs., tables. 27 cm. Subjects: Airplanes--Nuclear power plants. Nuclear engineering--Safety measures. Series: United States. National Aeronautics and Space Administration. NASA memorandum.

LC Classification: TL783.5 .U57

NASA reports required by Congress: report prepared by the Subcommittee on Space Science and Applications, transmitted to the Committee on Science, Space, and Technology. Published/Created: Washington: U.S. G.P.O.: For sale by the Supt. of Docs., Congressional Sales Office, U.S. G.P.O., -1990. Description: v.: ill.; 24 cm. Vol. for 1989 covers 101st Congress, 1st session. -1989. Former Frequency: Biennial, 19 -1987-1988 Continued by: United States. Congress. House. Committee on Science, Space, and Technology. Subcommittee on Space. NASA reports required by Congress (DLC)sn 93028751 (OCoLC)25685432 Cancel/Invalid LCCN: sn 89023378 Notes: At head of title, <1987-1988>-1989: Committee print. Description based on: 1987-1988. SERBIB/SERLOC merged record Additional Form Avail.: Vols. for <1987-1988>-1989 distributed to some depository libraries in microfiche. Subjects: United States. National Aeronautics and Space Administration--Periodicals. Astronautics--United States--Periodicals. Aeronautics--United States--Periodicals. LC Classification: TL521 .U62a Dewey Class No.: 353.0087/78 20 Govt. Doc. No.: Y 4.SCI 2:

NASA space communications program: report / prepared by the Subcommittee on Space Science and Applications, transmitted to the Committee on Science and Technology, U.S. House of Representatives, Ninety-seventh Congress, second session. Published/Created: Washington: U.S. G.P.O., 1982. Related Titles: N.A.S.A. space communications program.

Description: v, 23 p.; 24 cm. Notes: "Serial S." Item 1025-A-1, 1025-A-2 (microfiche) Subjects: United States. National Aeronautics and Space Administration. Artificial satellites in telecommunication--United States. LC Classification: TK5104 .U48 1982 Dewey Class No.: 621.38/0423/072073 19 Govt. Doc. No.: Y 4.Sci 2:97/S

NASA space plans and scenarios to 2000 and beyond / edited by National Aeronautics and Space Administration, Office of External Relations, Management Support Division. Published/Created: Park Ridge, N.J., U.S.A.: Noyes Publications, c1986. Related Authors: United States. National Aeronautics and Space Administration. Office of External Relations. Management Support Division. Description: xv, 239 p.: ill.; 25 cm. ISBN: 0815510713 : Notes: Includes index. Subjects: United States. National Aeronautics and Space Administration. Astronautics--United States. LC Classification: TL789.8.U5 N28 1986 Dewey Class No.: 387.8/0973 19

NASA specifications and standards. Published/Created: Washington [For sale by the Clearinghouse for Federal Scientific and Technical Information, Springfield, Va.] 1967. Description: vii, 380 p. 27 cm. Subjects: Space vehicles--Specifications--United States--Indexes. Series: United States. National Aeronautics and Space Administration. NASA SP-9000. LC Classification: TL869 .U53

NASA thesaurus. Published/Created: [Washington, D.C.?]: National Aeronautics and Space Administration, Office of Technology Utilization, Scientific and Technical Information Division, Related Authors: United States. National Aeronautics and Space Administration. Scientific and Technical Information Division. United States. National Aeronautics and Space Administration. Scientific and Technical Information Branch. United States. National Aeronautics and Space Administration. Scientific and Technical Information Program. Related Titles: [NASA thesaurus. Supplement. Description: v.; 28 cm. Vols. for <1967-> issued in 2 or more vols. ISSN: 0899-5257 Cancel/Invalid LCCN: sn 88006124 sn 88001971 Notes: Description based on: 1967 ed., v. 1. Updated by cumulative supplement, issued semiannually: NASA thesaurus. Supplement. Vols. for <1982>-1985 issued by: U.S. National Aeronautics and Space Administration, Scientific and Technical Information Branch; <1967>-19 , 1988-19 by: National Aeronautics and Space Administration, Scientific and Technical Information Division; <1994-> by: National Aeronautics and Space Administration, Scientific and Technical Information Program. SERBIB/SERLOC merged record Additional Form Avail.: Some issues also available via Internet from the NASA web site. Address as of 12/10/1999: http://www.sti.NASA.gov/thesfrm1.htm ; current access available via PURL. Subjects: Subject headings--Aeronautics--Periodicals. Subject headings--Astronautics--Periodicals. Subject headings--Technology--Periodicals. Aeronautics--Abstracting and indexing--Periodicals. Astronautics--Abstracting and indexing--Periodicals. Technology--Abstracting and indexing--Periodicals. Series: NASA SP LC Classification: Z695.1.A25 N36 Dewey

Class No.: 025.4/962913 20 Govt. Doc. No.: NAS 1.21:

NASA visitor information facilities; report. Published/Created: Washington, U.S. Govt. Print. Off., 1971. Description: v, 17 p. 24 cm. Notes: At head of Committee print. "Serial H." Subjects: United States. National Aeronautics and Space Administration--Information services. LC Classification: TL521.312 .A517 Dewey Class No.: 629.4/07

NASA, the first 25 years, 1958-1983: a resource for teachers: a curriculum project. Published/Created: Washington, D.C.: NASA: For sale by the Supt. of Docs., U.S. G.P.O., 1983. Related Authors: United States. National Aeronautics and Space Administration. Description: 132 p.: ill. (some col.); 28 cm. Notes: "25th anniversary, 1958-1983"--P. [4] of cover. S/N 033-000-00909-7 Item 830-G Subjects: United States. National Aeronautics and Space Administration--Study and teaching. Series: NASA EP; 182. LC Classification: TL521.312 .N38 1983 Dewey Class No.: 353.0087/78 19 Govt. Doc. No.: NAS 1.19:182

NASA's Solar Maximum Mission: a look at a new sun / prepared by the SMM principal investigator teams; edited by Joseph B. Gurman. Published/Created: [Greenbelt, Md.]: National Aeronautics and Space Administration, Goddard Space Flight Center, [1987] Related Authors: Gurman, Joseph Bearak. Goddard Space Flight Center. Description: xiii, 34 p.: ill. (some col.); 28 cm. Notes: "June 1987." Subjects: Solar Maximum Mission (Project) Solar flares. Solar atmosphere. LC Classification: QB526.F6 N37 1987 Dewey Class No.: 523.7/5 19

NASA--the 25th year [computer file]. Published/Created: Arlington, VA: Troika Multimedia, Inc., c1991. Related Authors: Troika Multimedia, Inc. Description: 1 computer laser optical disc; 4 3/4 in. Computer File Info.: Computer program. Commodore XL CDTV player Summary: Digital motion picture (50 minutes in length) about America's accomplishments in space from the first Mercury flights to Neil Armstron's walk on the moon. Highlights include the Viking missions to MARS, in-depth coverage of the first Space Shuttle mission, extensive footage from NASA's flight testing and research programs, and a look to the future in space. Adapted from a NASA film of the same name. Includes stereo sound. Notes: Title from disc label. System requirements: Commodore XL computer or other Commodore with the capability of playing CDTV format optical discs. Subjects: United States. National Aeronautics and Space Administration--History--Software. Space flight--History. Series: Heroic age of spaceflight LC Classification: TL521.312 Dewey Class No.: 629.4 12

National aeronautics and space act of 1958 as amended through October 6, 1961 (with footnotes) Published/Created: Washington, U. S. Govt. print. off., 1961. Related Authors: United States. Congress. Senate. Committee on Aeronautical and Space Sciences. Description: v, 21 p. 24 cm. Subjects: United States. National Aeronautics and Space Administration. Space law. Astronautics--Patents. LC Classification: KF4280.S7 A3 1961

National Aeronautics and Space Act of 1958, as amended, and related legislation / prepared at the request of Hon. Don Fuqua, chairman, Committee on Science and Technology, U.S. House of Representatives, Ninety-eighth Congress, first session. Published/Created: Washington: U.S. G.P.O., 1983. Related Authors: United States. Congress. House. Committee on Science and Technology. Description: ix, 212 p.; 24 cm. Notes: At head of Committee print. Distributed to some depository libraries in microfiche. "October 1983." "Serial O." Item 1025-A-1, 1025-A-2 (microfiche) Includes bibliographical references. Subjects: United States. National Aeronautics and Space Administration. Astronautics--Research--Law and legislation--United States. Aeronautics--Research--Law and legislation--United States. LC Classification: KF4280.S7 A3 1983 Dewey Class No.: 346.7304/694 347.3064694 19 Govt. Doc. No.: Y 4.Sci 2:98/0

National aeronautics and space act of 1958, as amended, and related legislation; staff report. Published/Created: Washington, U.S. Govt. Print. Off., 1965. Related Authors: United States. Congress. Senate. Committee on Aeronautical and Space Sciences. Description: v, 38 p. 24 cm. Notes: At head of 89th Congress, 1st session. Committee print. "Prepared for the Committee on Aeronautical and Space Sciences, United States Senate." Subjects: United States. National Aeronautics and Space Administration Astronautics--United States. Aeronautics--Research--United States. LC Classification: KF4280.S7 A3 1965 Dewey Class No.: 340

National aeronautics and space act of 1958, as amended, and related legislation: staff report prepared for the Committee on Aeronautical and Space Sciences, United States Senate. Published/Created: Washington: U.S. Govt. Print Off., 1975. Related Authors: United States. Congress. Senate. Committee on Aeronautical and Space Sciences. Description: iv, 101 p.; 24 cm. Notes: At head of 94th Congress, 1st session. Committee print. Subjects: United States. National Aeronautics and Space Administration. Astronautics--Research--Law and legislation--United States. Aeronautics--Research--Law and legislation--United States. LC Classification: KF4280.S7 A3 1975 Dewey Class No.: 343/.73/09702632

National aeronautics and space act of 1958, as amended, and related legislation; staff report prepared for the Committee on Aeronautical and Space Sciences, United States Senate. Published/Created: Washington, U.S. Govt. Print. Off., 1969. Related Authors: United States. Congress. Senate. Committee on Aeronautical and Space Sciences. Description: iii, 43 p. 24 cm. Notes: At head of 90th Congress, 2d session. Committee print. Subjects: United States. National Aeronautics and Space Administration. Astronautics--Research--Law and legislation--United States. Aeronautics--Research--Law and legislation--United States. LC Classification: KF4280.S7 A3 1969 Dewey Class No.: 340

Naugle, John E. First among equals: the selection of NASA space science experiments / John E. Naugle. Published/Created: Washington, DC: National Aeronautics and Space Administration, Office of Management,

Scientific and Technical Information Program: For sale by the Sup. of Doc., U.S.G.P.O., 1991. Description: xi, 134 p.; 23 cm. Notes: Includes bibliographical references (p. [125]-134). Subjects: National Academy of Sciences (U.S.). Space Science Board History. United States. National Aeronautical and Space Administration--History. Space sciences--Research--United States--History. Series: NASA SP; 4215 The NASA history series LC Classification: QB500.266.U6 N38 1991 Dewey Class No.: 500.5/072073 20 Govt. Doc. No.: NAS 1.21:4215

Naumann, Robert J. Materials processing in space: early experiments / by Robert J. Naumann and Harvey W. Herring. Published/Created: Washington, D.C.: Scientific and Technical Information Branch, National Aeronautics and Space Administration, 1980. Related Authors: Herring, Harvey W., joint author. Description: ix, 114 p.: ill.; 29 cm. Notes: Includes index. Subjects: Skylab Program. Materials. Manufacturing processes. Series: NASA SP; 443. LC Classification: TA410 .N28 Dewey Class No.: 670 Govt. Doc. No.: NAS 1.21:443

Neal, Valerie. Exploring the universe with the Hubble Space Telescope. Published/Created: [Washington, DC]: National Aeronautics and Space Administration: For sale by the Supt. of Docs., U.S. G.P.O., [1990?] Related Authors: United States. National Aeronautics and Space Administration. Related Titles: Hubble Space Telescope. Description: 71 p.: ill. (some col.); 30 cm. Notes: Shipping list no.: 90-111-P. S/N 033-000-01066-4. Item 830-I. Subjects: Hubble Space Telescope (Spacecraft) Orbiting astronomical observatories. Outer space--Exploration. Series: NP (Series); 126. LC Classification: QB500.268 .N43 1990 Dewey Class No.: 522/.2919 20 Govt. Doc. No.: NAS 1.83:126

Near-earth phase operations plan for Mariner Mars 1969 project. Published/Created: Cape Kennedy, Fla., California Institute of Technology, Jet Propulsion Laboratory, ETR Field Station, 1969- Description: v. illus., maps. 29 cm. Incomplete Contents: v. 1. Near-earth phase launch operations plan.--v. 4. Near-earth phase expected coverage capabilities. Notes: "Prepared under contract no. NAS 7-100, National Aeronautics & Space Administration." Subjects: Project Mariner (U.S.) LC Classification: TL789.8.U6M344 Dewey Class No.: 629.45/5/3

New horizons / National Aeronautics and Space Administration. Published/Created: [Washington: Office of Public Affairs, NASA: for sale by the Supt. of Docs., U.S. Govt. Print. Off., 1975] Description: 40 p.: ill.; 26 cm. Notes: Cover title. "Stock number 033-000-00631-4." Subjects: United States. National Aeronautics and Space Administration. Aeronautics--United States. Astronautics--United States. LC Classification: TL521.312 .U54 1975 Dewey Class No.: 629.1/0973

Newkirk, Roland W. Skylab: a chronology / Roland W. Newkirk and Ivan D. Ertel, with Courtney G. Brooks. Published/Created: Washington: Scientific and Technical Information Office, National Aeronautics and Space Administration: for sale by the Supt. of Docs., U.S. Govt. Print. Off., 1977. Related Authors: Ertel, Ivan D., 1914- joint author. Brooks, Courtney G., joint author. Historical Services and Consultants Company. Description: xvii, 458 p.; 26 cm. Notes: "Prepared by the Historical Services and Consultants Company, Houston, TX, under contract NASW-2590." Includes bibliographical references and index. Subjects: Skylab Program. Series: NASA history series. NASA SP; 4011. LC Classification: TL789.8.U6 S56 Dewey Class No.: 629.44/5

Nicks, Oran W. A review of the Mariner IV results [by] Oran W. Nicks. Published/Created: Washington, Scientific and Technical Information Division, Office of Technology Utilization, National Aeronautics and Space Administration; [for sale by the Supt. of Docs., U.S. Govt. Print. Off.] 1967. Description: iii, 39 p. illus. 23 cm. Notes: Bibliography: p. 38-39. Subjects: Project Mariner (U.S.)--History. Series: NASA SP; 130. LC Classification: TL789.8.U6 M352 Dewey Class No.: 629.43/54/3

Nimbus 6 Random access measurement system applications experiments / editors, Charles Cote, Ralph Taylor, and Eugene Gilbert. Published/Created: Washington, D.C.: Scientific and Technical Information Br., National Aeronautics and Space Administration, 1982. Related Authors: Cote, Charles (Charles E.) Taylor, Ralph E., 1923- Gilbert, Eugene (Eugene L.) Goddard Space Flight Center. Related Titles: Nimbus 6 Random access measuring system applications experiments. Description: vi, 99 p.: ill.; 28 cm. Subjects: Nimbus (Artificial satellite) Random Access Measurement System. Atmosphere--Research. Oceanography--Research. Polar regions--Research. Series: NASA SP; 457 LC Classification: QC808.5 .N55 1982 Dewey Class No.: 551.5/072 19

Oakley, Eric Gilbert, 1916- Project Telstar; the amazing history of the world's first communications satellite, by Gilbert Oakley. With the cooperation of Bell Telephone Laboratories, New York, and National Aeronautics and Space Administration, Washington, D. C. Published/Created: London, New York, W. Foulsham [1963] Description: 95 p. illus., maps (on lining papers) 20 cm. Subjects: Project Telstar. LC Classification: TK6649 .T4015

Orloff, Richard W., 1948- Apollo by the numbers: a statistical reference / by Richard W. Orloff. Published/Created: Washington, D.C.: National Aeronautics and Space Administration, 2000. Description: p. cm. Notes: "NASA SP-2000-4029." Includes bibliographical references. Subjects: Project Apollo (U.S.) Project Apollo (U.S.)--Statistics. Space flight to the moon. Moon--Exploration. Series: NASA history series LC Classification: TL789.8.U6 A564 2000 Dewey Class No.: 629.45/4/0973 21

Other worlds from earth: the future of planetary astronomy: a report of the Planetary Astronomy Committee of the Solar System Exploration Division. Published/Created: Washington, D.C.: National Aeronautics and Space Administration: For sale by the Supt. of Docs., U.S. G.P.O., 1989. Related Authors: United States. National Aeronautics and Space Administration. Solar System Exploration Division. Planetary Astronomy Committee. Related Titles: Future of planetary astronomy. Description: 97 p.: col. ill.; 26 cm. Notes: Shipping list no.: 89-402-

P. S/N 033-000-01052-4. Item 830-C. Subjects: Planets--Exploration. Outer space--Exploration. LC Classification: QB601 .O79 1989 Dewey Class No.: 523.4 20 Govt. Doc. No.: NAS 1.2:P 69/4

Outlook for space / report to the NASA administrator by the Outlook for Space Study Group; prepared by a task group consisting of participants from Ames Research Center ... [et al]. Published/Created: Washington: National Aeronautics and Space Administration, Scientific and Technical Information Office, 1976. Description: 383 p. in various pagings: ill.; 27 cm. Notes: Includes bibliography. Subjects: Astronautics--United States. Series: NASA SP; 386. LC Classification: TL789.8.U5 U56 1976 Dewey Class No.: 629.4/0973

Oversight reports during the 95th Congress: report / prepared by the Subcommittee on Space Science and Applications of the Committee on Science and Technology, U.S. House of Representatives, Ninety-sixth Congress, first session. Published/Created: Washington: Govt. Print. Off., 1979. Description: v, 167 p.: ill.; 24 cm. Notes: At head of Committee print. "Serial M." Subjects: United States. National Aeronautics and Space Administration. Astronautics--United States. LC Classification: TL789.8.U5 U52 1979 Dewey Class No.: 629.4/0973

Pace, George. Mariner Venus 67 guidance and control system [by] G. Pace. Published/Created: Pasadena, Jet Propulsion Laboratory, California Institute of Technology, 1968. Related Authors: United States. National Aeronautics and Space Administration. Description: vi, 43 p. illus. 28 cm. Notes: At head of National Aeronautics and Space Administration. Bibliography: p. 43. Subjects: Project Mariner (U.S.) Series: California Institute of Technology, Pasadena. Jet Propulsion Laboratory. JPL technical report no. 32-1258. LC Classification: TL789.8.U6M356 Dewey Class No.: 629.46

Pace, Scott. Data policy issues and barriers to using commercial resources for Mission to Planet Earth / Scott Pace, Brant Sponberg, Molly Macauley. Published/Created: Santa Monica, CA: RAND, c1999. Related Authors: Macauley, Molly K. Sponberg, Brant. United States. National Aeronautics and Space Administration. United States. Office of Science and Technology Policy. Science and Technology Policy Institute (Rand Corporation) Description: xxiii, 248 p.: ill.; 28 cm. ISBN: 0833027018 Notes: "Science and Technology Policy Institute." "Prepared for the National Aeronautics and Space Administration and Office of Science and Technology Policy." Includes bibliographical references (p. 237-248). Subjects: Mission to Planet Earth (Program) Earth sciences--Research--Government policy--United States. Earth sciences--United States--Remote sensing--Finance. Science and industry--United States. Industry and state--United States. Series: Documented briefing (Rand Corporation) LC Classification: QE33.2.R4 P33 1999 Dewey Class No.: 550/.72073 21

Page, Lou Williams. Apollo-Soyuz pamphlet[s] / prepared by Lou Williams Page and Thornton Page from investigators' reports of experimental results and with the help of advising teachers. Published/Created: Washington: National Aeronautics and Space Administration: for sale by the Supt. of Docs., U.S. Govt. Print. Off., 1977. Related Authors: Page, Thornton, joint author. United States. National Aeronautics and Space Administration. Description: 9 v.: ill.; 27 cm. Contents: no. 1 The flight.--no. 2. X-rays, gamma-rays.--no. 3. Sun, stars, in between.--no. 4. Gravitational field.--no. 5. The earth from orbit.--no. 6. Cosmic ray dosage.--no. 7. Biology in zero-G.--no. 8. Zero-G technology.--no. 9. General science. Notes: Includes bibliographies. Subjects: Apollo Soyuz Test Project. LC Classification: TL788.4 .P34 Dewey Class No.: 500.5 19

Paine, Thomas O., 1921- Papers of Thomas O. Paine, 1931-1992 (bulk 1960-1982) Description: 64,050 items. 183 containers plus 1 CL. 1 microfilm reel. 73.2 linear feet. Access Advisory: CLASSIFIED, in part. Biog./History Note: Engineer, corporate executive, and National Aeronautics and Space Administration official. Full name: Thomas Otten Paine; died 1992. Summary: Correspondence, memoranda, reports, minutes of meetings, appointment books, family and genealogical papers, and printed matter chiefly relating to Paine's engineering career with General Electric Company and Northrop Corporation and as deputy and acting administrator at NASA, where he directed seven Apollo missions, including the first to the moon. Also includes a journal (1945) kept by Paine while serving in the U.S. Navy describing the demilitarization of Japanese submarines during the early days of the Allied occupation of Japan; and material relating to Paine's service as chairman of the National Commission on Space and as a member of the Advisory Committee on the Future of the United States Space Program and Engineers Joint Council. Paine's interest in interplanetary exploration and colonization is documented by papers relating to the Case for Mars conferences and drafts of books and screenplays by others on outer space exploration. Correspondents include Buzz Aldrin, Ray Bradbury, John Glenn, J. Herbert Holloman, Thomas V. Jones, and Robert C. Seamans. Notes: Sound recordings transferred to Library of Congress Motion Picture, Broadcasting, and Recorded Sound Division. Some photographs transferred to Library of Congress Prints and Photographs Division. MSS49866 Source of Acquisition: Gift, Thomas O. Paine, 1972, 1983-1992. Additional Form Avail.: Microfilm edition available of a petition (6 Feb. 1969) concerning the use of religious references by the Apollo astronauts, no. 21,385. Subjects: Aldrin, Buzz. Bradbury, Ray, 1920- Glenn, John, 1921- Hollomon, J. Herbert (John Herbert) Jones, Thomas V. Seamans, Robert C. Engineers Joint Council. General Electric Company. Northrop Corporation. Project Apollo (U.S.) United States. Advisory Committee on the Future of the U.S. Space Program United States. National Aeronautics and Space Administration. United States. National Commission on Space. United States. Navy--Foreign service--Japan. Case for Mars Conference. Planets--Exploration. Space colonies. Space flight to the moon. Submarine boats--Japan. Japan--History--Allied occupation, 1945-1952. Mars (Planet) Outer space--Exploration. Engineers. Local Call/Shelving: 0813T VAULT

Microfilm 21,385-1P

Passage to a ringed world: the Cassini-Huygens mission to Saturn and Titan / Linda J. Spilker, editor. Published/Created: Washington, D.C.: National Aeronautics and Space Administration, [1997] Related Authors: Spilker, Linda J. United States. National Aeronautics and Space Administration. Description: 157 p.: ill. (some col.); 28 cm. Notes: Shipping list no.: 98-0021-P. "October 1997." Includes index. Includes bibliographical references (p. 152-157). Subjects: Saturn probes. Saturn (Planet)--Exploration. Saturn (Planet)--Exploration--Equipment and supplies. Titan (Satellite)--Exploration. Series: NASA SP; 533 LC Classification: QB671 .P37 1997 Dewey Class No.: 629.43/546 21 Govt. Doc. No.: NAS 1.21:533

Pearcy, Arthur. Flying the frontiers: NACA and NASA experimental aircraft / Arthur Pearcy. Published/Created: Annapolis, Md.: Naval Institute Press, 1993. Description: 200 p.: ill. (some col.); 30 cm. ISBN: 1557502587 Notes: Includes bibliographical references (p. 200). Subjects: United States. National Aeronautics and Space Administration--History. Research aircraft--United States--History. LC Classification: TL567.R47 P43 1993 Dewey Class No.: 629.13/0072073 20

Physical properties of the surface materials at the Viking landing sites on Mars / by H.J. Moore ... [et al.]. Published/Created: Washington: U.S. G.P.O.; Denver, CO: For sale by the Books and Open-File Reports Section, U.S. Geological Survey, 1987. Related Authors: Moore, H. J. (Henry J.), 1928- United States. National Aeronautics and Space Administration. Description: xiii, 222 p.: ill.; 29 cm. Notes: "A summary of the results of the Physical Properties Investigation of Viking Lander 1 and 2 during the Viking 1975 project." "Prepared on behalf of the National Aeronautics and Space Administration." Maps on 2 folded leaves in pocket. Bibliography: p. 131-135. Subjects: Mars (Planet)--Geology. Mars (Planet)--Surface. Mars (Planet)--Exploration. Series: Geological Survey professional paper; 1389. LC Classification: QB641 .P47 1987 Dewey Class No.: 559.9/23 19 Govt. Doc. No.: I 19.16:1389

Piccard family papers, ca. 1470-1983 (bulk 1926-1983) Related Authors: Piccard, Don, 1926- Papers. Piccard, Jean Felix, 1884-1963. Papers. Piccard, Jeannette, 1895-1981. Papers. Description: 73,000 items. 204 containers plus 1 OV. 82 linear feet. Access Advisory: Restrictions apply. Biog./History Note: Family members represented include Jean Felix Piccard (1884-1963), chemist, balloonist, and aeronautical engineer; his wife, Jeannette Ridlon Piccard (1895-1981), balloonist, aerospace consultant, and clergywoman; and their son, Don Piccard (1926-), balloonist and entrepreneur. Summary: Correspondence, memoranda, diaries, journals, logbooks, drafts of writings and speeches, reports, notes, financial papers, biographical and genealogical material, scrapbooks, newspaper clippings, printed material, blueprints, patent specifications and descriptions, drawings, photographs, and other papers documenting the careers of Piccard family members in the fields of aeronautics, ballooning, bathyscaphe exploration, chemistry, education, religion, and stratospheric exploration. Includes Ridlon (Ridlen) family papers and Swiss legal and genealogical documents concerning the Piccard (Picard) family dating from the 1470s to the late 18th century. The papers of Jean Piccard document his career as explorer, inventor, and scientist collaborating with his brother Auguste on balloon flights for stratospheric exploration and the study of cosmic rays and on other mutual scientific interests; as organic chemist and consultant with Calco Chemical Company and Hercules Powder Company; and as professor of aeronautical engineering. Includes material pertaining to the Piccards's stratospheric balloon flight of 1934 and to Jean Piccard's 1937 test flight in the aerostat Pleiades. Also includes records Piccard kept as treasurer of the Aéro-Club Suisse and transcript of interrogations which he conducted for the U.S. Army Air Forces in 1945. Papers of Jeannette Piccard pertain chiefly to her role as member and priest in the Episcopal Church; her work as aerospace consultant with the National Aeronautics and Space Administration; and her career as an educator. Don Piccard's papers include his files as newsletter editor for the Balloon Federation of America and material pertaining to his sport balloon manufacturing company, Don Piccard Balloons. Family correspondents include Auguste Piccard, Jean and Auguste Piccard's father Jules, Jeannette Piccard's father John Ridlon, and Jean and Jeannette Piccard's son Jacques, granddaughter Kathryn Ann Piccard, and several foster children. Other correspondents include Denzil A. Carty, Daniel Corrigan, Albert Einstein, Robert R. Gilruth, Suzanne R. Hiatt, Philip F. McNairy, Robert Andrews Millikan, A. W. Stevens, W. F. G. Swann, and Francis Zielinski. Includes correspondence between Jean Piccard and the National Research Council concerning trinitrotoluene (TNT). Notes: Some blueprints, drawings, and photographs transferred to Library of Congress Prints and Photographs Division. Some maps transferred to Library of Congress Geography and Maps Division. Motion picture films and sound recordings transferred to Library of Congress Motion Picture, Broadcasting and Recorded Sound Division. In English, French, and German. MSS36145 Source of Acquisition: Gift, Jeannette Piccard, 1969. Gift, Don Piccard, 1985. Subjects: Carty, Denzil A. Corrigan, Daniel, b. 1897. Einstein, Albert, 1879-1955. Gilruth, Robert R. (Robert Rowe), 1913- Hiatt, Suzanne R. McNairy, Philip F. Millikan, Robert Andrews, 1868-1953. Piccard, Auguste, 1884-1962. Piccard, Jacques. Ridlon, John, 1852-1936. Stevens, A. W. (Albert William), 1886-1949. Swann, W. F. G. (William Francis Gray), 1884-1962. Zielinski, Francis. Picard family. Ridlen family. Piccard, Jules, b. 1840. United States. Army Air Forces. United States. National Aeronautics and Space Administration. Aéro-Club Suisse. Balloon Federation of America. Don Piccard Balloons. Pleiades (Aerostat) Calco Chemical Company. Episcopal Church. Hercules Powder Company. National Research Council (U.S.) Aeronautics. Aerospace engineering. Airships. Ballooning. Bathyscaphe. Chemistry. Chemistry, Organic. Cosmic rays. Education. Explosives. Legal documents--Switzerland. Religion. Stratosphere. Women in the Episcopal Church. World War, 1939-1945. Switzerland--Genealogy. Local Call/Shelving: 0629AA OVSD 3:7

Pocket statistics history. Published/Created: [Washington, Program and Special Reports Division, Executive Secretariat, National Aeronautics and Space Administration] Description: 20 x 27 cm. ISSN: 0361-0306 Notes: SERBIB/SERLOC merged record Subjects: United States. National Aeronautics and Space Administration. LC Classification: TL521.3.P6 U5a Dewey Class No.: 353.008/55

Poole, Lynn. Scientists who work with astronauts, by Lynn and Gray Poole. With a foreword by Hugh L. Dryden. Published/Created: New York, Dodd, Mead [1964] Related Authors: Poole, Gray Johnson, joint author. Description: xvi, 172 p. ports. 21 cm. Subjects: United States. National Aeronautics and Space Administration--Officials and employees. Scientists--United States. Series: Makers of our modern world books LC Classification: Q141 .P66 Dewey Class No.: 925

Portable life support systems. Prepared by Ames Research Center. Published/Created: Washington, Scientific and Technical Information Division, National Aeronautics and Space Administration; [for sale by the Clearinghouse for Federal Scientific and Technical Information, Springfield, Va.] 1970. Related Authors: Ames Research Center. Description: vi, 389 p. illus. 27 cm. Notes: Includes bibliographical references. Subjects: Life support systems (Space environment) Series: NASA SP; 234. LC Classification: TL1500 .A64 1969m Dewey Class No.: 629.47/7

Portree, David S. F. Humans to Mars: fifty years of mission planning, 1950-2000 / by David S.F. Poretree. Published/Created: Washington, D.C.: National Aeronautics and Space Administration, 2000. Description: p. cm. Notes: Includes bibliographical references and index. Subjects: United States. National Aeronautics and Space Administration. Space flight to Mars--Planning. Series: Monographs in aerospace history; no. 20 NASA SP; no. 4521. LC Classification: TL799.M3 P67 2000 Dewey Class No.: 629.45/53 21

Preliminary inventory of the records of the National Aeronautics and Space Council, record group 220 / compiled by Jarritus Wolfinger. Published/Created: Washington: National Archives & Records Service, General Services Administration, 1977. Related Authors: Wolfinger, Jarritus. United States. National Aeronautics and Space Council. Description: vii, 12 p.; 27 cm. Subjects: United States. National Aeronautics and Space Council Archives--Catalogs. United States. National Archives and Records Service Catalogs. Series: United States. National Archives and Records Service. Preliminary inventory - National Archives and Records Service; 190. LC Classification: CD3026 .A32 no. 190 CD3041.N25 Dewey Class No.: 016.973 s 016.387/0973 Govt. Doc. No.: GS 4.10:190

Proceedings of the Sixth Annual Meeting, [held at] Aerospace Medical Division, Brooks Air Force Base, Tex., February 19-21, 1968. Published/Created: Washington, Scientific and Technical Information Division, National Aeronautics and Space Administration; [for sale by the Supt. of Docs., U.S. Govt. Print. Off.] 1968 [i.e., 1969] Description: v, 273 p. illus. 27 cm. Notes: Includes bibliographies.

Subjects: Lunar geology--Congresses. Moon--Exploration--Congresses. Series: NASA SP; 177. LC Classification: QB592 .W67 1969 Dewey Class No.: 622/.0999/1

Proceedings. Prepared by Goddard Space Flight Center. Published/Created: Washington, Scientific and Technical Information Division, National Aeronautics and Space Administration; [for sale by the Clearinghouse for Federal Scientific Information, Springfield, Va.] 1965. Related Authors: Goddard Space Flight Center Description: ix, 302 p. illus. 27 cm. Subjects: Project Apollo (U.S.) Astronautics--Communication systems. Space vehicles--Tracking. Series: NASA SP; 87. LC Classification: TL4030 .A58 1965 Dewey Class No.: 629.457

Project Gemini. Published/Created: [Washington, 1966] Description: 52 p. illus. 27 cm. Subjects: Project Gemini (U.S.) LC Classification: TL789.8.U6 G78 Dewey Class No.: 629.45

Project Mercury: man-in-space program of the National Aeronautics and Space Administration. Report of the Committee on Aeronautical and Space Sciences, United States Senate. [Compiled by Earl W. Lindveit of the staff] Published/Created: Washington, U.S. Govt. Print. Off., 1959. Description: vii, 97 p. illus., map. 24 cm. Subjects: Project Mercury (U.S.) Series: United States. 86th Cong. 1st sess., 1959. Senate. Report; no. 1014. LC Classification: TL789.8.U5 A53 1959b Dewey Class No.: 629.1388

Project Orion: a design study of a system for detecting extrasolar planets / David C. Black, editor. Published/Created: [Washington, D.C.]: National Aeronautics and Space Administration, Scientific and Technical Information Branch: [for sale by the Supt. of Docs., U.S. Govt. Print. Off.], 1980. Related Authors: Black, David C. Description: xii, 204 p.: ill.; 23 cm. Notes: Includes bibliographical references. Subjects: Project Orion. Extrasolar planets. Series: NASA SP; 436. LC Classification: QB602.9 .P76 Dewey Class No.: 523.1/13

Quimby, Freeman Henry, 1915- ed. Concepts for detection of extraterrestrial life. Edited by Freeman H. Quimby. Published/Created: Washington, Scientific and Technical Information Division, National Aeronautics and Space Administration; [for sale by the Supt. of Docs., U.S.Govt.Print.Off.] 1964. Description: 53p. illus.(part col.) 26cm. Notes: Bibliography: p.51-53. Subjects: Life on other planets Outer space--Exploration. Series: United States. National Aeronautics and Space Administration. NASA SP- 56 LC Classification: QB54 .Q5

Ragsdale, George C. Mariner Venus 67 spacecraft environmental test results [by] George C. Ragsdale [and] Donald C. Mesnard. Published/Created: Pasadena, Jet Propulsion Laboratory, California Institute of Technology, 1968. Related Authors: Mesnard, Donald C., joint author. Description: xviii, 245 p. illus. 28 cm. Notes: "Prepared under contract no. NAS 7-100, National Aeronautics & Space Administration." Bibliography: p. 245. Subjects: Project Mariner (U.S.) Series: California Institute of Technology, Pasadena. Jet Propulsion Laboratory. JPL technical report 32-1249. LC

Classification: TL796.5.U6M37 Dewey Class No.: 629.43/54/2

Ranger IX lunar chart[s] compiled and published for National Aeronautics and Space Administration by Aeronautical Chart and Information Center, United States Air Force. Edition Information: 1st ed. Published/Created: St. Louis, 1966. Related Authors: United States. Aeronautics and Space Administration. Description: 5 col. maps on sheets 56 x 74 cm. Subjects: Moon--Maps, Topographic. Series: United States. Aeronautical Chart and Information Center, St. Louis. Ranger lunar charts, RLC 13-17. LC Classification: G3195 svar..U5 RLC13

Ranger VII lunar charts. Edition Information: 1st ed. Published/Created: St. Louis, 1964. Related Authors: United States. National Aeronautics and Space Administration. Description: 5 col. maps on sheets 56 x 74 cm. Subjects: Moon--Maps, Topographic. Series: United States. Aeronautical Chart and Information Center, St. Louis. Ranger lunar charts, RLC 1-5. LC Classification: G3195 svar. .U5

Ranger VII; a special report, August 5, 1964. Published/Created: Washington, National Aeronautics and Space Administration; Pasadena, Calif., Jet Propulsion Laboratory; for sale by the Superintendent of Documents, U.S. Govt. Print. Off. [1964] Related Authors: Jet Propulsion Laboratory (U.S.) Description: 35 p. illus. 20 cm. Subjects: Project Ranger. Moon--Photographs. Outer space--Exploration. LC Classification: TL789.8.U6 R35 1964

Ranger VII; special report to Congress, August 4, 1964. [Prepared by the Office of Public Information, National Aeronautics and Space Administration. Published/Created: Washington, 1964] Description: 40 p. illus. 26 cm. Notes: Cover title. Subjects: Project Ranger. Moon--Photographs. Outer space--Exploration. LC Classification: TL789.8.U6 R35

Renetzky, Alvin. NASA factbook; guide to National Aeronautics and Space Administration programs and activities. Alvin Renetzky, editor-in-chief; Barbara J. Flynn, editor; and the staff of Academic media. Published/Created: Orange, N.J., Academic Media [1971] Related Authors: Flynn, Barbara J., joint author. Description: 456 p. 29 cm. ISBN: 0878760121 Subjects: United States. National Aeronautics and Space Administration. LC Classification: TL521.312 .R45 Dewey Class No.: 629.4/072/073

Report from Mars: Mariner IV, 1964-1965. Published/Created: [Washington, National Aeronautics and Space Administration; for sale by the Superintendent of Documents, U.S. Govt. Print. Off., 1966] Related Authors: United States. National Aeronautics and Space Administration. Description: 45 p. illus. (part col.) 27 cm. Notes: "Prepared under contract no. NAS 7-100, National Aeronautics and Space Administration." Subjects: Project Mariner (U.S.) LC Classification: TL789.8.U6 M345 Dewey Class No.: 629.45/5/3

Report on the coordination of proposed NASA/DOD FY 1967 facilities / [National Aeronautics and Space Administration]. Published/Created:

Washington: [NASA], 1966. Description: 5, 74 leaves; 27 cm. Notes: Cover title. Subjects: United States. National Aeronautics and Space Administration--Facilities. United States--Armed Forces--Facilities. LC Classification: TL521.312 .U54 1966 Dewey Class No.: 353.008/77/8

Report to the Administrator, National Aeronautics and Space Administration. Published/Created: [Washington, For sale by the Supt. of Docs., U.S. Govt. Print. Off., 1967] Description: 1 v. (various pagings) illus. 27 cm. Notes: Supplemented by: "Appendix" (7 v. in 15. illus. 27 cm.) published: [Washington, For sale by the Supt. of Docs., U.S. Govt. Print. Off., 1967] Call number: TL789.8.U6A587 Suppl. Contents.--A. Board minutes.--B. Witness statements and releases.--C. Apollo operations handbook, command and service module, Spacecraft 012 (12 November 1966) prepared by North American Aviation, inc., Space and Information Systems Division, Training and Support Documentation, Dept. 671. 2 v.--D. Report[s] of Panels. 7 v.--E. Management and organization.--F. Schedule of physical evidence.--G. Addenda and corrigenda. 2 v. Subjects: Project Apollo (U.S.) Space vehicle accidents. LC Classification: TL789.8.U6 A587

Report to the President: actions to implement the recommendations of the Presidential Commission on the Space Shuttle Challenger Accident / National Aeronautics and Space Administration. Published/Created: Washington, D.C.: NASA: [Supt. of Docs., U.S. G.P.O., distributor, 1986] Related Authors: United States. Presidential Commission on the Space Shuttle Challenger Accident. Related Titles: Actions to implement the recommendations of the Presidential Commission on the Space Shuttle Challenger Accident. Description: 49 p.: ill., forms; 28 cm. Notes: Shipping list no.: 86-621-P. "July 14, 1986." S/N 033-000-00990-9 Item 830-C Subjects: Challenger (Spacecraft) United States. Presidential Commission on the Space Shuttle Challenger Accident. Space vehicle accidents-- United States. LC Classification: TL867 .U54 1986 Dewey Class No.: 629.45/4 19 Govt. Doc. No.: NAS 1.2:R 24/2

Research program on pyroelectric detection techniques and materials, phase II. Published/Created: [Washington] National Aeronautics and Space Administration; for sale by the Clearinghouse for Federal Scientific and Technical Information, Springfield, Va. [1966] Description: vi, 66 p. illus. 27 cm. Subjects: Pyrometry. Series: NASA contractor report; NASA CR-629. LC Classification: TL521.3.C6 A3 no. 629 Dewey Class No.: 537.6/5

Results of the Project Mercury ballistic and orbital chimpanzee flights. Edited by James P. Henry and John D. Mosely. Published/Created: Washington, Office of Scientific and Technical Information, National Aeronautics and Space Administration; [for sale by the Supt. of Docs., U.S. Govt. Print. Off.] 1963. Related Authors: Henry, James Paget, 1914- ed. Mosely, John D., ed. Description: vi, 71 p. illus. 27 cm. Notes: Includes bibliographies. Subjects: Project Mercury (U.S.) Space medicine LC Classification: TL521 .A333 no. 39

Results of the second U.S. Manned orbital space flight, May 24, 1962.

Published/Created: [Washington] National Aeronautics and Space Administration, Manned Spacecraft Center, Project Mercury [1962] Related Authors: Carpenter, M. Scott (Malcolm Scott), 1925- Description: 107 p. Notes: Includes bibliographies. Subjects: Project Mercury (U.S.) Series: United States. National Aeronautics and Space Administration. NASA SP -6 LC Classification: TL521 .A333 no. 6

Results of the third U.S. manned orbital space flight, October 3, 1962. Published/Created: [Washington] National Aeronautics and Space Administration, Manned Spacecraft Center, Project Mercury; [for sale by the Superintendent of Documents, U.S. Govt. Print. Off., 1962] Related Authors: Schirra, Wally. Description: v, 120 p. illus., ports., maps, diagrs. Notes: Includes bibliographies. Subjects: Project Mercury (U.S.) Series: United States. National Aeronautics and Space Administration. NASA SP-12. LC Classification: TL521 .A333 no. 12

Review of recent launch failures; report, Ninety-second Congress, first session. Published/Created: Washington, U.S. Govt. Print. Off., 1971. Description: v, 19 p. 24 cm. Notes: At head of Committee print. "Serial G." Subjects: United States. National Aeronautics and Space Administration. Project Mariner (U.S.) Orbiting astronomical observatories. Centaur rocket. LC Classification: TL789.8.U5 A52 1971 Dewey Class No.: 629.46

Review of the National Aeronautics and Space Act of 1958, as amended: report together with additional views / prepared by the Subcommittee on Space Science and Applications, transmitted to the Committee on Science and Technology, House of Representatives, Ninety-eighth Congress, second session. Published/Created: Washington: U.S. G.P.O., 1984. Description: v, 56 p.; 24 cm. Notes: At head of Committee print. Distributed to some depository libraries in microfiche. "Serial EE." Item 1025-A-1, 1025-A-2 (microfiche) Includes bibliographical references. Subjects: United States. National Aeronautics and Space Administration. Astronautics and state--United States. Space law--United States. LC Classification: TL521 .U62 1984 Dewey Class No.: 387.8/0973 19 Govt. Doc. No.: Y 4.Sci 2:98/EE

Ride, Sally. Leadership and America's future in space: a report to the Administrator / by Sally K. Ride. Published/Created: [Washington, D.C.?]: NASA, [1987] Related Authors: United States. National Aeronautics and Space Administration. Description: 63 p.: ill. (some col.); 28 cm. Notes: Shipping list no.: 87-636-P. Item 830-C Bibliography: p. 62. Subjects: United States. National Aeronautics and Space Administration. Astronautics--United States. Space flight. LC Classification: TL521.312 .R53 1987 Dewey Class No.: 353.0087/78 19 Govt. Doc. No.: NAS 1.2:L 46

Roland, Alex, 1944- Model research: the National Advisory Committee for Aeronautics, 1915-1958 / Alex Roland. Published/Created: Washington, D.C.: Scientific and Technical Information Branch, National Aeronautics and Space Administration: For sale by the Supt. of Docs., U.S. G.P.O., 1985. Description: 2 v. (xx, 769, 25 p.): ill.; 26 cm. Notes: Includes bibliographies and indexes. Subjects: United States. National Advisory Committee for Aeronautics

History. Series: The NASA history series NASA SP; 4103 LC Classification: TL521.312 .R58 1985 Dewey Class No.: 353.0085/6 19

Rosenthal, Alfred. Venture into space: early years of Goddard Space Flight Center. Published/Created: Washington, Scientific and Technical Information Division, National Aeronautics and Space Administration; [for sale by the Supt. of Docs., U.S. Govt. Print. Off.] 1968. Description: xv, 354 p. illus., ports. 25 cm. Notes: Bibliography: p. 333-339. Subjects: Goddard Space Flight Center. Astronautics--United States. Series: NASA Center history series NASA SP; 4301. LC Classification: TL862.G6 R6 Dewey Class No.: 629.4/061/73

Rosholt, Robert L. An administrative history of NASA, Published/Created: Washington, Scientific and Technical Information Division, National Aeronautics and Space Administration; 1966. Description: p. cm. Subjects: United States. National Aeronautics and Space Administration. Series: United States. National Aeronautics and Space Administration. NASA SP-4101. LC Classification: TL521.312 .R6

Samonski, Frank H. Technical history of the environmental control system for Project Mercury, by Frank H. Samonski, Jr. Published/Created: [Washington] National Aeronautics and Space Administration; for sale by the Clearinghouse for Federal Scientific and Technical Information, Springfield, Va. [1967] Description: viii, 110 p. illus. 27 cm. Notes: Bibliography: p. 110. Subjects: Project Mercury (U.S.) Life support systems (Space environment) Series: United States. National Aeronautics and Space Administration. NASA technical note D-4126 LC Classification: TL521 .A3525 no. 4126 Dewey Class No.: 629.47/7

Saturn SA-1. Published/Created: Washington: Office of Congressional Relations, National Aeronautics and Space Administration, [1964?] Description: 22 leaves in various foliations, [22] leaves of plates: ill.; 27 cm. Notes: Cover title. Subjects: Saturn launch vehicles. LC Classification: TL789.8.U6 S374 1964

Sayles, Leonard R. Managing large systems: organizations for the future / Leonard R. Sayles, Margaret K. Chandler; with a new introduction by Leonard R. Sayles. Published/Created: New Brunswick, N.J.: Transaction Publishers, [1993] Related Authors: Chandler, Margaret K. Description: xxx, 332 p.: ill.; 23 cm. ISBN: 1560006420 (paper) Notes: Originally published: New York: Harper & Row, 1971. Includes bibliographical references and index. Subjects: United States. National Aeronautics and Space Administration--Management--History. Aeronautics and state--United States--History. Series: Classics in organization and management series. LC Classification: TL521.312 .S38 1993 Dewey Class No.: 658.4/04 20

Schmitz, Bruce W. Development of the post-injection propulsion system for the Mariner C spacecraft [by] Bruce W. Schmitz, Thomas A. Groudle [and] James H. Kelley. Published/Created: Pasadena, Calif., Jet Propulsion Laboratory, California Institute of Technology, 1966. Related Authors: Groudle, Thomas A., joint author. Kelley, James Howard, 1935- joint author. United States. National

Aeronautics and Space Administration. Description: v, 43 p. illus. 28 cm. Notes: At head of National Aeronautics and Space Administration. Bibliography: p. 43. Subjects: Project Mariner (U.S.) Space vehicles--Propulsion systems. Series: California Institute of Technology, Pasadena. Jet Propulsion Laboratory. JPL technical report no. 32-830. LC Classification: TL789.8.U6M358 Dewey Class No.: 629.42/22

Schultz, James. Winds of change: expanding the frontiers of flight: Langley Research Center's 75 years of accomplishment 1917-1992 / by James Schultz. Published/Created: Washington, DC: National Aeronautics and Space Administration: For sale by the U.S. G.P.O., Supt. of Docs., [1992?] Related Authors: United States. National Aeronautics and Space Administration. Description: ix, 133 p.: ill. (some col.); 31 cm. ISBN: 0160379245 : Notes: Shipping list no.: 92-0415-P. S/N 033-000-01117-2 (GPO) Includes bibliographical references (p. 132). Subjects: Langley Research Center--History. Aeronautics--Research--United States--History. Astronautics--Research--United States--History. Series: NP (Series); 130. LC Classification: T521.312 .S387 1992 Govt. Doc. No.: NAS 1.83:130

Science in orbit: the shuttle & spacelab experience: 1981-1986 / prepared by Marshall Space Flight Center, Huntsville, Alabama, 1988. Published/Created: [Washington, D.C.]: National Aeronautics and Space Administration; Supt. of Docs., U.S. G.P.O., distributor, 1988. Related Authors: George C. Marshall Space Flight Center. Description: vii, 124 p.: col. ill.; 29 cm. Notes: Shipping list no.: 89-199-P. Includes index. S/N 033-000-01039-7 Item 830-C Subjects: Space shuttles--Scientific applications. LC Classification: TL795.5 .S35 1988 Dewey Class No.: 500.5/072073 20 Govt. Doc. No.: NAS 1.2:Or 1/3

Seamans, Robert C. Aiming at targets: the autobiography of Robert C. Seamans, Jr. / Robert C. Seamans, Jr. Published/Created: Washington, DC: NASA History Office, Office of Policy and Plans, NASA Headquarters: For sale by the U.S. G.P.O., Superintendent of Documents 1996. Description: ix, 291 p.: ill.; 24 cm. Notes: Includes index. Subjects: Seamans, Robert C. United States. National Aeronautics and Space Administration--Officials and employees--Biography. United States. Air Force--Officials and employees Biography. Astronautics--United States--Biography. Series: The NASA history series NASA SP; 4106 LC Classification: TL789.85.S43 A3 1996 Dewey Class No.: 629.4/092 B 20

Search for the universal ancestors / editors, H. Hartman, J.G. Lawless, P. Morrison. Published/Created: Washington, D.C.: Scientific and Technical Information Branch, National Aeronautics and Space Administration: For sale by the Supt. of Docs., U.S. G.P.O., 1985. Related Authors: Hartman, H. Lawless, J. G. (James G.) Morrison, Philip. Description: xiv, 129 p.: ill., port.; 23 cm. Subjects: Life--Origin--History. Life--Origin. Series: NASA SP; 477 LC Classification: QH325 .S42 1985 Dewey Class No.: 577 19

Selling to NASA. Published/Created: [Washington] 1963. Description: 38 p. illus. 27 cm. Subjects: Research and development contracts--United States. Government purchasing--United States. LC Classification: TL724.1.G6 U5

Sharpless, Jack. The earthbound observer: a personal look at the people of NASA and the space shuttle effort / Jack Sharpless. Published/Created: Michigan City [Ind.]: Sharpless Corp., 1983. Related Authors: United States. National Aeronautics and Space Administration. Description: v, 89 p.: ill.; 25 cm. ISBN: 0961069600 Subjects: Space shuttles. LC Classification: TL795.5 .S48 1983 Dewey Class No.: 387.8 19

Short, Nicholas M. The Heat Capacity Mapping Mission (HCMM) anthology / Nicholas M. Short and Locke M. Stuart, Jr. Published/Created: Washington, DC: Scientific and Technical Information Branch, National Aeronautics and Space Administration: For sale by the Supt. of Docs., U.S. G.P.O., 1982. Related Authors: Stuart, Locke M. United States. National Aeronautics and Space Administration. Scientific and Technical Information Branch. Related Titles: Heat Capacity Mapping Mission (H.C.M.M.) anthology. Description: vii, 264 p.: ill. (some col.); 36 cm. Notes: Includes bibliographical references and index. Subjects: Heat Capacity Mapping Mission Program. Series: NASA SP; 465 LC Classification: QE33.2.A7 S48 1982 Dewey Class No.: 551.1/4 19 Govt. Doc. No.: NAS 1.21:465

Shuttle atmospheric lidar research program: final report of Atmospheric Lidar Working Group. Published/Created: Washington, D.C.: Scientific and Technical Information Branch, National Aeronautics and Space Administration; Springfield, Va.: for sale by the National Technical Information Service, 1979. Description: v, 220 p.: ill.; 27 cm. Notes: Includes bibliographies. Subjects: Shuttle Atmospheric Lidar Research Program. Series: NASA SP; 433. LC Classification: QC869.4.U5 U53 1979 Dewey Class No.: 551.5/1/072

Siddiqi, Asif A., 1966- Challenge to Apollo: the Soviet Union and the space race, 1945-1974 / by Asif A. Siddiqi. Published/Created: Washington, D.C.: National Aeronautics and Space Administration, NASA History Div., Office of Policy and Plans, 2000. Description: xvi, 1011 p.: ill.; 25 cm. Cancel/Invalid LCCN: 00031047 Notes: Includes bibliographical references and index. Subjects: Astronautics--Soviet Union--History. Space race--History. Series: NASA history series NASA SP; 4408. LC Classification: TL789.8.S65 S47 2000 Dewey Class No.: 629.4/0947 21

Sightseeing: a space panorama: 84 photographs from the NASA archives / selected by Barbara Hitchcock; with a foreword by Arthur C. Clarke. Edition Information: 1st ed. Published/Created: New York, N.Y.: Knopf, 1985. Related Authors: Hitchcock, Barbara. Riva, Peter, 1950- United States. National Aeronautics and Space Administration. Description: x, 105 p.: chiefly col. ill.; 29 cm. ISBN: 0394542436 : Notes: Introductory text written by Barbara Hitchcock and Peter Riva. Subjects: Space photography--United States--Exhibitions. LC Classification: TR713 .S54 1985 Dewey Class No.: 779/.96294 19

Significant NASA inventions. Published/Created: Washington, D.C.:

National Aeronautics and Space Administration: [For sale by the Supt. of Docs., U.S. G.P.O., Description: v.: ill.; 23 cm. ISSN: 0360-5264 Notes: Title from cover. "Available for licensing in foreign countries." Description based on: 1981. SERBIB/SERLOC merged record Subjects: Inventions--United States. Series: NASA SP LC Classification: T223.C2 U55a Dewey Class No.: 608/.7/73 Govt. Doc. No.: NAS 1.21:7038

Simmons, Gene, 1929- On the moon with Apollo 15; a guidebook to Hadley Rille and the Apennine Mountains. Published/Created: [Washington] National Aeronautics and Space Administration; [for sale by the Supt. of Docs., U.S. Govt. Print. Off.] 1971. Description: v, 46 p. illus. 26 cm. Notes: Bibliography: p. 35. Subjects: Project Apollo (U.S.) Moon--Surface. LC Classification: QB591 .S62 Dewey Class No.: 559.91 Govt. Doc. No.: NAS 1.18:M77

Simmons, Gene, 1929- On the moon with Apollo 16; a guidebook to the Descartes region. Published/Created: [Washington] National Aeronautics and Space Administration; [for sale by the Supt. of Docs., U.S. Govt. Print. Off.] 1972. Description: vi, 90 p. illus. 27 cm. Notes: "EP-95." Bibliography: p. 71-72. Subjects: Project Apollo (U.S.) Space flight to the moon. Moon--Exploration. LC Classification: TL789.8.U6 A58178 Dewey Class No.: 508.99/1 Govt. Doc. No.: NAS 1.19:95

Simmons, Gene, 1929- On the moon with Apollo 17; a guidebook to Taurus-Littrow. Published/Created: [Washington] National Aeronautics and Space Administration; [for sale by the Supt. of Docs., U.S. Govt. Print. Off.] 1972. Description: viii, 111 p. illus. 26 cm. Notes: "EP-101." Bibliography: p. 87-88. Subjects: Project Apollo (U.S.) Lunar landing sites. Moon--Exploration. LC Classification: TL789.8.U6 A58179 Dewey Class No.: 559.9/1 Govt. Doc. No.: NAS 1.19:101

Sjogren, W. L. The Ranger III flight path and its determination from tracking data Published/Created: Padasena, Jet Propulsion Laboratory, California Institute of Technology, 1965. Related Authors: United States. National Aeronautics and Space Administration. Description: iv, 128 p. illus. 28 cm. Subjects: Project Ranger. Space vehicles--Tracking. Series: California Institute of Technology, Pasadena. Jet Propulsion Laboratory. JPL technical report no. 32-563. LC Classification: TL789.8.U6 R354

Skylab 2 photography catalog: multispectral, earth terrain, Hasselblad, Nikon cameras. Edition Information: [1st ed.]. Published/Created: Albuquerque: Technology Application Center, University of New Mexico, [1974] Related Authors: United States. National Aeronautics and Space Administration. Related Titles: Catalog of NASA's Skylab 2 photography. Description: ii, 106 p.: maps (4 fold. in pocket); 28 cm. Notes: Half Catalog of NASA's skylab 2 photography. "Based on 'Skylab 2, Photographic Index and Scene Identification' prepared by Richard W. Underwood and John W. Holland for NASA." Subjects: Skylab Program--Catalogs. Earth--Photographs from space--Catalogs. LC Classification: QB637 .N48 1974a Dewey Class No.: 016.55/022/2

Skylab EREP investigations summary / prepared by NASA Lyndon B. Johnson Space Center. Published/Created: Washington: Scientific and Technical Information Office, National Aeronautics and Space Administration: for sale by Supt. of Docs, U.S. Govt. Print. Off., 1978. Related Authors: Lyndon B. Johnson Space Center. Description: ix, 386 p.: ill.; 30 cm. Notes: Includes bibliographies. Subjects: Skylab Program. Remote sensing. Series: NASA SP; 399. LC Classification: G70.4 .S57 Dewey Class No.: 621.36/7 Govt. Doc. No.: NAS 1.21:399

Skylab experiments. Produced by the Skylab Program and NASA's Education Programs Division in cooperation with the University of Colorado. Published/Created: Washington, National Aeronautics and Space Administration; for sale by the Supt. of Docs., U.S. Govt. Print. Off., 1973- Related Authors: United States. National Aeronautics and Space Administration. Educational Programs Division. University of Colorado (Boulder campus) Description: v. illus. 27 cm. Incomplete Contents: v. 1. Physical science solar astronomy.--v. 2. Remote sensing of earth resources.--v. 3. Materials science.--v. 4. Life sciences.--v. 5. Astronomy and space physics.--v. 6. Mechanics.--v. 7. Living and working in space. Notes: Includes bibliographies. Subjects: Skylab Program. Astronomy--Experiments. LC Classification: QB61 .S52 1973 Dewey Class No.: 522

Skylab explores the Earth / prepared by NASA Lyndon B. Johnson Space Center. Published/Created: Washington: Scientific and Technical Information Office, National Aeronautics and Space Administration, 1977. Related Authors: Lyndon B. Johnson Space Center. Description: xii, 517 p.: ill.; 30 cm. Notes: Includes bibliographical references. Subjects: Geophysics-- Remote sensing. Skylab Program. Series: NASA SP; 380. LC Classification: QC808.5 .S48 Dewey Class No.: 550

Skylab, classroom in space / edited by Lee B. [i.e. R.] Summerlin; prepared by George C. Marshall Space Flight Center. Published/Created: Washington: Scientific & Technical Information Office, National Aeronautics and Space Administration: for sale by the Supt. of Docs., U.S. Govt. Print. Off., 1977. Related Authors: Summerlin, Lee R. George C. Marshall Space Flight Center. Description: 182 p.: ill.; 30 cm. Notes: Includes index. Subjects: Skylab Program. Series: NASA SP; 401. LC Classification: TL789.8.U6 S5675 Dewey Class No.: 500.5

Skylab, our first space station / edited by Leland F. Belew; prepared by George C. Marshall Space Flight Center. Published/Created: Washington: Scientific Technical Information Office, National Aeronautics and Space Administration: for sale by the Supt. of Docs., U.S. Govt. Print. Off., 1977. Related Authors: Belew, Leland F. Description: viii, 164 p.: ill.; 30 cm. Notes: "Stock no. 033-000-00670-5." Includes index. Subjects: Skylab Program. Series: NASA SP; 400 LC Classification: TL789.8.U6 S572 1977 Dewey Class No.: 629.44/5

Skylab's astronomy and space sciences / edited by Charles A. Lundquist. Published/Created: Washington:

Scientific and Technical Information Branch, National Aeronautics and Space Administration: for sale by the Supt. of Docs., U.S. Govt. Print. Off., 1979. Related Authors: Lundquist, Charles A. Description: vi, 122 p.: ill.; 30 cm. Notes: Includes index. Subjects: Astronomy. Atmosphere. Skylab Project. Series: NASA SP; 404. LC Classification: QB4 .U47 1979 Dewey Class No.: 520 Govt. Doc. No.: NAS 1.21:404

Small business participation in the NASA research and development programs; staff study of the Committee on Science and Astronautics, U.S. House of Representatives, Eighty-seventh Congress, first session. Published/Created: Washington, U.S. Govt. Print. Off., 1961. Description: v, 15 p. diagrs., tables. 24 cm. Notes: At head of Committee print. Subjects: United States. National Aeronautics and Space Administration. Research and development contracts--United States. Small business. LC Classification: TL521 .A54147 1961b

Smetana, Frederick O., 1928- A study of NACA and NASA published information of pertinence in the design of light aircraft, by Frederick O. Smetana. Published/Created: [Washington] National Aeronautics and Space Administration; for sale by the Clearinghouse for Federal Scientific and Technical Information, Springfield, Va. [1970- Related Authors: North Carolina State University. Dept. of Mechanical and Aerospace Engineering. Langley Research Center. Description: v. illus. 27 cm. Incomplete Contents: v. 1. Structures. Notes: Prepared under contract no. 1-7265 by Dept. of Mechanical and Aerospace Engineering, North Carolina State University, Raleigh, N.C., for Langley Research Center. Subjects: Airplanes--Design and construction--Abstracts. Series: NASA contractor report; NASA CR-1484. LC Classification: TL521.3.C6 A3 no. 1484, etc. Dewey Class No.: 629.1/08 s 629.134/1

Smith, Howard Everett, 1927- Daring the unknown: a history of NASA / Howard E. Smith. Edition Information: 1st ed. Published/Created: San Diego: Harcourt Brace Jovanovich, c1987. Description: xi, 178 p., [8] p. of plates: ill. (some col.), ports.; 25 cm. ISBN: 0152004351 Summary: A chronicle of the technological and political challenges, the people, and the discoveries of the American space program during the last thirty years. Notes: "Gulliver books." Includes index. Bibliography: p. 171-172. Subjects: United States. National Aeronautics and Space Administration--Juvenile literature. United States. National Aeronautics and Space Administration--History. LC Classification: TL793 .S586 1987 Dewey Class No.: 353.0087/78 19

Smull, T L K The nature and scope of the NASA university program [by] T. L. K. Smull. Published/Created: Washington, Scientific and Technical Information Division, National Aeronautics and Space Administration; [for sale by the Superintendent of Documents, U.S. Govt. Print. Off.] 1965. Description: 39 p. illus., maps. 20 cm. Notes: Bibliography: p. 39. Subjects: United States. National Aeronautics and Space Administration. Astronautics--Study and teaching. Space sciences--Study and teaching. Series: United States. National Aeronautics and Space Administration. NASA SP-; 73. LC Classification:

TL846 .S6

Space benefits: the secondary application of aerospace technology in other sectors of the economy: prepared for the Committee on Aeronautical and Space Sciences, United States Senate / [by the NASA Technology Utilization staff]. Published/Created: Washington: U.S. Govt. Print. Off., 1975. Related Authors: United States. Congress. Senate. Committee on Aeronautical and Space Sciences. Description: v, 54 p.; 24 cm. Notes: At head of 94th Congress, 1st session. Committee print. Includes index. Subjects: United States. National Aeronautics and Space Administration. Technology Utilization Office. Technology. LC Classification: T49.5 .U55 1975 Dewey Class No.: 301.24/3

Space exploration: opposing viewpoints / Charles P. Cozic, book editor. Published/Created: San Diego, CA: Greenhaven Press, c1992. Related Authors: Cozic, Charles P., 1957- Description: 192 p.: ill.; 22 cm. ISBN: 0899081975 (alk. paper): 089908172X (pbk.: alk. paper) : Summary: Debates the space exploration issues of proper goals of space exploration, which programs should be pursued, the elimination of NASA, and the appropriateness of using space for warfare. Notes: Includes bibliographical references (p. 185-187) and index. Subjects: United States. National Aeronautics and Space Administration--Juvenile literature. Astronautics--United States--Juvenile literature. Astronautics, Military--United States--Juvenile literature. Astronautics. Outer space--Exploration--United States--Juvenile literature. Outer space--Exploration. Series: Opposing viewpoints series

(Unnumbered) LC Classification: TL789.8.U5 S58 1992 Dewey Class No.: 333.9/4 20

Space flight: the first 30 years. Published/Created: Washington, DC (400 Maryland Ave., S.W., Washington 20546): National Aeronautics and Space Administration, Office of Space Flight, [1991] Related Authors: United States. Office of Space Flight. Description: iii, 36 p.: ill.; 28 cm. Notes: Item 830-I Shipping list no.: 92-170-P. "December 1991." Subjects: United States. National Aeronautics and Space Administration--History. Space flight--History--Chronology. Series: NP (Series); 150. LC Classification: TL790 .S72 1991 Dewey Class No.: 629.4/0973 20 Govt. Doc. No.: NAS 1.83:150

Space missions to comets: a conference / sponsored by NASA Office of Space Science and held at the Goddard Space Flight Center, Greenbelt, Maryland, October 1977; editors, M. Neugebauer ... [et al.]. Published/Created: [Washington]: National Aeronautics and Space Administration, Scientific and Technical Information Branch, 1979. Related Authors: Neugebauer, Marcia. United States. Office of Space Science. Description: v, 226 p.: ill.; 27 cm. Notes: Includes bibliographies. Subjects: Comets--Congresses. Halley's comet--Congresses. Outer space--Exploration--Congresses. Series: NASA conference publication; 2089. LC Classification: QB721 .S97 1977 Dewey Class No.: 523.6 19

Space scientists and engineers: selected biographical and bibliographical listing, 1957-1961. Published/Created: Washington, For sale by the Superintendent of Documents, U.S.

Govt. Print. Off., 1962. Description: iii, 332 p. 26 cm. Subjects: Astronautics--Bio-bibliography. Series: United States. National Aeronautics and Space Administration. NASA SP- 5. LC Classification: TL521 .A333 no. 5

Space shuttle facility program: more definitive cost information needed, National Aeronautics and Space Administration: report to the Congress / by the Comptroller General of the United States. Published/Created: Washington: U.S. General Accounting Office, 1977. Description: ii, 29 p.: ill.; 27 cm. Notes: Cover title. "PSAD-77-17." "B-183134." Subjects: United States. National Aeronautics and Space Administration. Space shuttles--Costs. Ground support systems (Astronautics)--Costs. LC Classification: TL795.5 .U55 1977a Dewey Class No.: 353.008/77/8 Govt. Doc. No.: GA 1.13:PSAD-77-17

Space shuttle: human capital and safety upgrade challenges require continued attention: report to the Chairman, Committee on Commerce, Science, and Transportation, U.S. Senate / United States General Accounting Office. Published/Created: Washington, D.C.: The Office, [2000] Related Authors: United States. Congress. Senate. Committee on Commerce, Science, and Transportation. Description: 20 p.: ill.; 28 cm. Notes: Cover title. "August 2000." "GAO/NSIAD-00-186." "B-285463"--P. 3. Includes bibliographical references. Subjects: United States. National Aeronautics and Space Administration--Officials and employees--Planning. Space shuttles--United States--Safety appliances Planning. LC Classification: TL795.5 .U55 2000 Dewey Class No.: 629.44/1/0973 21

Space Station Program: description, applications, and opportunities / by Space Station Task Force, National Aeronautics and Space Administration. Published/Created: Park Ridge, N.J., U.S.A.: Noyes Publications, c1985. Related Authors: United States. National Aeronautics and Space Administration. Space Station Task Force. Description: xxix, 754 p.: ill.; 25 cm. ISBN: 0815510241 : Notes: Includes index. Subjects: Space Station Program (U.S.) Space stations. LC Classification: TL797 .S6454 1985 Dewey Class No.: 387.8 19

Space station: key to the future. Published/Created: [Washington; For sale by the Supt. of Docs., U.S. Govt. Print. Off., 1970] Description: 40 p. illus. 28 cm. Notes: Cover title. Subjects: Space stations. LC Classification: TL797 .U5 Dewey Class No.: 629.44

Space transportation system: past, present, future, National Aeronautics and Space Administration, Department of Defense: report to the Congress / by the Comptroller General of the United States. Published/Created: [Washington: U.S. General Accounting Office, 1977] Description: iv, 75 p.: ill.; 28 cm. Notes: Cover title. Publication date stamped on cover. "PSAD-77-113." "B-183134." Subjects: United States. National Aeronautics and Space Administration. United States. Dept. of Defense. Reusable space vehicles. LC Classification: TL795.5 .U55 1977 Dewey Class No.: 387.8 Govt. Doc. No.: GA 1.13:PSAD-77-113

Spaceborne radar observations: a guide for Magellan radar-image analysis / J.P. Ford ... [et al.]. Published/Created: Pasadena, Calif.: National Aeronautics and Space Administration, Jet Propulsion Laboratory, California Institute of Technology, [1989] Related Authors: Ford, J. P. Jet Propulsion Laboratory (U.S.) Description: xi, 126 p.: ill., col. maps; 28 cm. Notes: "JPL publication 89-41"--T.p. verso. "December 15, 1989." Includes bibliographical references (p.). Subjects: Megallan (Spacecraft)--Observations. Venus (Planet)--Surface--Observations. Venus (Planet)--Geology--Observations. Earth--Observations. Remote sensing. LC Classification: QB621 .S63 1989 Dewey Class No.: 559.9/22 20

Spacelab 1. Published/Created: Washington, D.C.: National Aeronautics and Space Administration: [Supt. of Docs., U.S. G.P.O., distributor], [1982] Related Authors: United States. National Aeronautics and Space Administration. George C. Marshall Space Flight Center. Description: 30 p.: ill. (some col.), ports.; 28 cm. Notes: Cover title. "George C. Marshall Space Flight Center"--P. 4 Cover. Includes index. S/N 033-000-00846-5 Item 830-C Subjects: Spacelab Program. LC Classification: TL797 .S648 1982 Dewey Class No.: 629.44/5 Govt. Doc. No.: NAS 1.2:SP 1/46

Spangenburg, Ray, 1939- The history of NASA / Ray Spangenburg and Kit Moser. Published/Created: New York: Franklin Watts, 2000. Related Authors: Moser, Diane, 1944- Description: p. cm. ISBN: 0531117189 Summary: Surveys the history of the National Aeronautics and Space Administration, describing the major space craft and missions launched. Notes: Includes bibliographical references and index. Subjects: United States. National Aeronautics and Space Administration--History--Juvenile literature. United States. National Aeronautics and Space Administration--History. Astronautics--United States--History--Juvenile literature. Astronautics--History. Outer space--Exploration. Series: Out of this world (Franklin Watts, Inc.) LC Classification: TL789.8.U5 S66 2000 Dewey Class No.: 629.4/0973 21

Spehalski, Richard J. Mariner IV mechanical operations [by] Richard J. Spehalski. Published/Created: Pasadena, Jet Propulsion Laboratory, California Institute of Technology, 1966. Related Authors: United States. National Aeronautics and Space Administration. Description: iii, 13 p. illus. 28 cm. Notes: At head of National Aeronautics and Space Administration. Subjects: Project Mariner (U.S.) Series: California Institute of Technology. Jet Propulsion Laboratory. JPL technical report no. 32-954. LC Classification: TL789.8.U6M359

Spehalski, Richard J. Mariner Mars 1964 mechanical configuration [by] R. J. Spehalski. Published/Created: Pasadena, Jet Propulsion Laboratory, California Institute of Technology, 1966. Related Authors: United States. National Aeronautics and Space Administration. Description: iv, 28 p. illus. 28 cm. Notes: "Prepared under Contract no. NAS 7-100, National Aeronautics & Space Administration." Bibliography: p. 27-28. Subjects: Project Mariner (U.S.) Series: California Institute of Technology, Pasadena. Jet Propulsion Laboratory. JPL technical report no. 32-

933. LC Classification: TL789.8.U6M36 Dewey Class No.: 629.43/54

Sperling, F. A treatise on the Surveyor lunar landing dynamics and an evaluation of pertinent telemetry data returned by Surveyor I [by] F. Sperling [and] J. Garba. Published/Created: Pasadena, Jet Propulsion Laboratory, California Institute of Technology, 1967. Related Authors: Garba, J. A., joint author. United States. National Aeronautics and Space Administration. Description: vii, 25 p. illus. 28 cm. Notes: "Prepared under contract no. NAS 7-100, National Aeronautics & Space Administration." Bibliography: p. 24-25. Subjects: Surveyor Program (U.S.) Space vehicles--Landing. Series: California Institute of Technology, Pasadena. Jet Propulsion Laboratory. JPL technical report no. 32-1035 LC Classification: TL789.8.U6 S955 Dewey Class No.: 629.43/53

Status and issues pertaining to the proposed development of the space telescope project, National Aeronautics and Space Administration: report to the Congress / by the Comptroller General of the United States. Published/Created: [Washington: U.S. General Accounting Office, 1977] Related Titles: Status and issues pertaining to the proposed development of the space telescope project ... Description: v, 45 p.: ill.; 27 cm. Notes: Cover title. "PSAD-77-98." Publication date stamped on cover. Subjects: Hubble Space Telescope (Spacecraft) Orbiting astronomical observatories. LC Classification: QB82.U6 U54 1977 Dewey Class No.: 522/.109/19 Govt. Doc. No.: GA 1.13:PSAD-77-98

Status of actions taken on recommendations of the Apollo 204 Accident Review Board. Published/Created: Washington, U.S. Govt. Print. Off., 1968. Related Authors: United States. Apollo 204 Review Board. Report to the Administrator of the National Aeronautics and Space Administration. United States. Congress. House. Committee on Science and Astronautics. Subcommittee on NASA Oversight. Description: v, 50 p. 23 cm. Notes: At head of Committee print. "Serial L." Report prepared by the Apollo Program Office in conjunction with NASA centers and headquarters offices, and based largely on the Report of the Apollo 204 Review Board to the Administrator, National Aeronautics and Space Administration, dated April 5, 1967. Prepared for the Subcommittee on NASA Oversight of the Committee on Science and Astronautics, U.S. House of Representatives, Ninetieth Congress, second session. Subjects: Project Apollo (U.S.) Space vehicle accidents. LC Classification: TL789.8.U6A5855 Dewey Class No.: 629.45/4

Status of the Mariner Jupiter/Saturn 1977 Project, National Aeronautics and Space Administration: report to the Congress / by the Comptroller General of the United States. Published/Created: [Washington: U.S. General Accounting Office], 1977. Description: ii, 27 p.; 27 cm. Notes: Cover title. "B-183134." Subjects: Mariner Jupiter/Saturn 1977 Project. LC Classification: TL789.8.U6 M382 Dewey Class No.: 353.008/5 Govt. Doc. No.: GA1.13:PSAD-77-103

Stein, John P. Pricing policies for the space shuttle / John P. Stein and Charles Wolf, Jr. Published/Created: Santa Monica,

CA: Rand Corp., 1977. Related Authors: Wolf, Charles, 1924- joint author. Description: 30 p.; 28 cm. Notes: Cover title. Subjects: United States. National Aeronautics and Space Administration--Pricing. Reusable space vehicles. Series: The Rand paper series; P-5971 LC Classification: AS36 .R28 no. 5971 TL795.5 Dewey Class No.: 081 s 387.8 19

Steinberg, Florence S. Aboard the space shuttle / Florence S. Steinberg. Published/Created: Washington, D.C.: National Aeronautics and Space Administration, Division of Public Affairs: [Supt. of Docs., U.S. G.P.O., distributor], 1980. Related Authors: United States. National Aeronautics and Space Administration. Public Affairs Division. Description: 32 p.: col. ill.; 23 cm. Notes: Item 830-G Subjects: Reusable space vehicles. Space simulators. Series: NASA EP; 169 LC Classification: TL795.5 .S73 Dewey Class No.: 629.45/4 19 Govt. Doc. No.: NAS 1.19:169

Stillwell, Wendell H X-15 research results. By Wendell H. Stillwell. Published/Created: Washington, Scientific and Technical Information Division, National Aeronautics and Space Administration; [for sale by the Superintendent of Documents, U.S. Govt. Print. Off.] 1965. Description: vi, 128 p. illus. 24 cm. Notes: Bibliography: p. 103-116. Subjects: X-15 (Rocket aircraft). Series: United States. National Aeronautics and Space Administration. NASA SP-; 60. LC Classification: TL789.8.U6 X574

Strasser, Bruce E. Sounds from space. [Sound recording] Published/Created: [New York] Bell Telephone Laboratories [1960] Matrix no. LO8L-5128-5129. Related Authors: Bell Telephone Laboratories, inc. Description: p. 2 s. 10 in. 33 1/3 rpm. microgroove. Contents: Prelude to space communications.--Project Echo. Notes: At head of Bell Telephone Laboratories, inc. Commemorates the communication experiment of Project Echo, Aug. 12, 1960. Produced with the cooperation of the National Aeronautics and Space Administration, Jet Propulsion Laboratory, and U. S. Naval Research Laboratory. Descriptive notes on slipcase. Subjects: Project Echo. Artificial satellites in telecommunication.

Structural systems and program decisions. Published/Created: Washington, National Aeronautics and Space Administration; [for sale by the Clearinghouse for Federal Scientific and Technical Information, Springfield, Va.] 1966. Description: 2 v. illus. 27 cm. Notes: Includes bibliography. Subjects: Project Apollo (U.S.) Space vehicles--Design and construction--Data processing. Series: NASA SP; 6008. LC Classification: TL875 .U5 Dewey Class No.: 629.42/22

Sturms, F. M. Trajectory analysis of an Earth-Venus-Mercury mission in 1973 [by] F. M. Sturms, Jr. Published/Created: Pasadena, Jet Propulsion Laboratory, California Institute of Technology, 1967. Related Authors: United States. National Aeronautics and Space Administration. Description: x, 52 p. illus. 28 cm. Notes: "Prepared under contract no. NAS 7-100, National Aeronautics & Space Administration." Bibliography: p. 52. Subjects: Space trajectories. Space flight to Mercury. Series: California

Institute of Technology, Pasadena. Jet Propulsion Laboratory. JPL technical report, no. 32-1062 LC Classification: TL1075 .S78 Dewey Class No.: 629.4/11

Summary of Gemini extravehicular activity. Edited by Reginald M. Machell. Published/Created: Washington, Scientific and Technical Information Division, Office of Technology Utilization, National Aeronautics and Space Administration; [for sale by the Clearinghouse for Federal Scientific and Technical Information, Springfield, Va.] 1967. Related Authors: Machell, Reginald M., ed. United States. National Aeronautics and Space Administration. Scientific and Technical Information Division. Related Titles: Gemini extravehicular activity. Description: 1 v. (various pagings) illus. 27 cm. Notes: Includes bibliography. Subjects: Project Gemini (U.S.) Extravehicular activity (Manned space flight) Series: NASA SP; 149. LC Classification: TL789.8.U6 G77 Dewey Class No.: 629.45/4

Surveyor I mission report. Published/Created: Pasadena, Jet Propulsion Laboratory, California Institute of Technology, 1966- Description: v. illus. (part fold.) 28 cm. Incomplete Contents: pt. 1. Mission description and performance. Notes: "Prepared under contract no. NAS 7-100, National Aeronautics & Space Administration." Bibliography: v. 1, p. 165-168. Subjects: Surveyor Program (U.S.) Series: California Institute of Technology, Jet Propulsion Laboratory. JPL technical report no. 32-1023 LC Classification: TL789.8.U6 S965 Dewey Class No.: 629.43/53

Surveyor I; a preliminary report. Published/Created: Washington, Scientific and Technical Information Division, National Aeronautics and Space Administration; [available from the Clearinghouse for Federal Scientific and Technical Information, Springfield, Va.] 1966. Description: iii, 39 p. illus. 27 cm. Notes: Bibliography: p. 37. Subjects: Surveyor Program (U.S.) Series: NASA SP; 126. LC Classification: TL789.8 .U6S973 Dewey Class No.: 629.43/53

Surveyor III, a preliminary report. Published/Created: Washington, Scientific and Technical Information Division, National Aeronautics and Space Administration; [available from the Clearinghouse for Federal Scientific and Technical Information, Springfield, Va.] 1967. Description: viii, 159 p. illus. 27 cm. Notes: Includes bibliographical references. Subjects: Surveyor Program (U.S.) Series: NASA SP; 146. LC Classification: TL789.8.U6 S974 Dewey Class No.: 629.45/4

Surveyor program results. Published/Created: Washington, Scientific and Technical Information Division, National Aeronautics and Space Administration; [for sale by the Supt. of Docs., U.S. Govt. Print. Off.] 1969. Description: ix, 423 p. illus. 27 cm. Notes: Includes bibliographical references. Subjects: Surveyor Program (U.S.) Series: NASA SP; 184. LC Classification: TL789.8.U6S98 Dewey Class No.: 629.45/4

Surveyor v; a preliminary report. Published/Created: Washington, Scientific and Technical Information Division, National Aeronautics and Space Administration; [available from the Clearinghouse for Federal Scientific and Technical Information, Springfield, Va.] 1967. Description: viii, 161 p. illus., maps. 27 cm. Subjects: Surveyor Program (U.S.) Series: NASA SP; 163. LC Classification: TL789.8.U6 S975 Dewey Class No.: 629.43/53

Surveyor VI: a preliminary report. Published/Created: Washington, Scientific and Technical Information Division, National Aeronautics and Space Administration; [available from the Clearinghouse for Federal Scientific and Technical Information, Springfield, Va.] 1968. Description: viii, 165 p. illus. 27 cm. Notes: Includes bibliographies. Subjects: Surveyor Program (U.S.) Series: NASA SP; 166. LC Classification: TL789.8.U6 S976 Dewey Class No.: 629.43/53

Surveyor VII; a preliminary report. Published/Created: Washington, Scientific and Technical Information Division, National Aeronautics and Space Administration; [available from the Clearinghouse for Federal Scientific and Technical Information, Springfield, Va.] 1968. Description: viii, 303 p. illus., maps. 27 cm. Notes: Includes bibliographies. Subjects: Surveyor Program (U.S.) Series: NASA SP; 173. LC Classification: TL789.8.U6 S977 Dewey Class No.: 629.43/53

Sutton, Felix. Conquest of the moon. Written by Felix Sutton and Alvin Maurer. Pictures by Raul Mina Mora and National Aeronautics and Space Administration. Published/Created: New York, Grosset & Dunlap [1969] Related Authors: Maurer, Alvin, joint author. Mora, Raul Mina, illus. United States. National Aeronautics and Space Administration. Description: 63 p. illus. (part col.), ports. 29 cm. Summary: Sixty questions and answers about the moon and various steps in the program to reach the moon. Subjects: Project Apollo (U.S.)--Juvenile literature. Project Apollo (U.S.) Space flight to the moon--Juvenile literature. Space flight to the moon. LC Classification: TL789.8.U6A5823 Dewey Class No.: 629.45/4

Sutton, Felix. The how and why wonder book of the moon / written by Felix Sutton; with additional material by Alvin Maurer and Oscar Weigle; illustrated by Raul Mina Mora; with additional pictures by National Aeronautics and Space Administration. The how and why wonder book of stars / by Norman Hoss; illustrated by James Ponter. The how and why wonder book of planets and interplanetary travel / written by Harold J. Highland; illustrated by Denny McMains. Published/Created: Chicago: J.G. Ferguson Pub. Co., c1987. Related Authors: Maurer, Alvin. Weigle, Oscar. Mora, Raul Mina, ill. Hoss, Norman. How and why wonder book of stars. 1987. Highland, Harold Joseph. How and why wonder book of planets and interplanetary travel. 1987. Description: 48, 47, 47 p.: ill. (some col.); 29 cm. ISBN: 089434076X (set) Summary: Discusses the characteristics and exploration of the moon, stars, and planets. Notes: No collective t.p. Titles transcribed from individual title pages. "Edited under the supervision of Dr. Paul E. Blackwood ... text and illustrations approved by Oakes A. White." Subjects: Moon--Juvenile literature. Stars--Juvenile literature. Space flight--Juvenile literature. Moon. Stars. Planets. Space flight. Series: Science library (Chicago, Ill.); vol. 5.

LC Classification: Q163 .S467 1987 vol. 5 QB582 Dewey Class No.: 500 s 523 19

Sutton, Felix. The how and why wonder book of the moon. Written by Felix Sutton, with additional material by Alvin Maurer and Oscar Weigle. Illustrated by Raul Mina Mora, with additional pictures by National Aeronautics and Space Administration. Edited under the supervision of Paul E. Blackwood. Text and illus. approved by Oakes A. White. Published/Created: New York, Grosset & Dunlap [c1970] Related Authors: Mora, Raul Mina, illus. Description: 48 p. illus. (part col.) 29 cm. Summary: Describes the characteristics and traces the history of man's exploration of earth's nearest neighbor. Subjects: Moon--Juvenile literature. Moon. LC Classification: QB582 .S95 1970 Dewey Class No.: 523.3

Swenson, Loyd S. This new ocean: a history of Project Mercury / by Loyd S. Swenson, Jr., James M. Grimwood, Charles C. Alexander. Published/Created: Washington, DC: National Aeronautics and Space Administration, NASA History Office, Office of Policy and Plans, 1998. Related Authors: Grimwood, James M., joint author. Alexander, Charles C., joint author. Description: xv, 681 p.: ill.; 26 cm. Notes: Bibliographical footnotes: p. 515-604. "Note on source and selected bibliography": p. 609-630. Subjects: Project Mercury (U.S.)--History. Series: NASA SP; 4201 NASA historical series. LC Classification: TL789.8.U6 M547 1998 Dewey Class No.: 629.45/4/0973 21

Swenson, Loyd S. This new ocean; a history of Project Mercury, by Loyd S. Swenson, Jr., James M. Grimwood [and] Charles C. Alexander. Published/Created: [Washington, Scientific and Technical Information Division, Office of Technology Utilization, National Aeronautics and Space Administration; [for sale by the Supt. of Docs., U.S. Govt. Print. Off.] 1966. Related Authors: Grimwood, James M., joint author. Alexander, Charles C., joint author. Description: xv, 681 p. illus. 26 cm. Notes: Bibliographical footnotes: p. 515-604. "Note on source and selected bibliography": p. 609-630. Subjects: Project Mercury (U.S.)--History. Series: United States. National Aeronautics and Space Administration. NASA historical series. NASA SP; 4201. LC Classification: TL789.8.U6 M484 Dewey Class No.: 629.45

Tang, Wen, 1921- Some aspects of the atmospheric circulation on Mars, Published/Created: Springfield, Va., For sale by the Clearinghouse for Federal Scientific and Technical Information [1965] Related Authors: GCA Corporation. United States. National Aeronautics and Space Administration. Description: v, 43 p. illus. 27 cm. Subjects: Mars (Planet)--Atmosphere. Series: United States. National Aeronautics and Space Administration. NASA contractor report CR-262. LC Classification: TL521.3.C6 A3 no. 262

Taylor, Robert, 1948- The space shuttle / by Robert Taylor. Published/Created: San Diego, Calif.: Lucent Books, 2002. Description: p. cm. ISBN: 1560067225 Summary: Discusses the history and development, technological and political challenges, and future of the world's first reusable space vehicle, including the shuttle program's effects on NASA. Notes: Includes bibliographical references and index. Subjects: United States. National Aeronautics and Space Administration. Space shuttles--Juvenile literature. Space shuttles. Series: Building history series LC Classification: TL795.515 .T39 2002 Dewey Class No.: 629.44/1 21

Technical facilities catalog. Edition Information: October 1974 ed. Published/Created: [Washington]: National Aeronautics and Space Administration: [available from the Supt. of Docs., U.S. Govt. Print. Off., 1974] Description: 3 v.: ill.; 26 cm. Notes: Cover title. Subjects: United States. National Aeronautics and Space Administration--Facilities--Directories. LC Classification: TL521.312 .U62 1974 Dewey Class No.: 629.4/07/24

Telstar I. Published/Created: [Washington] National Aeronautics and Space Administration, 1963-66. Related Authors: American Telephone and Telegraph Commpany. Description: 4 v. (739-2373 p.) illus. (part col.) 23 cm. Notes: "Published in accordance with the cooperative agreement executed on July 27, 1961, between the National Aeronautics and Space Administration and the American Telephone and Telegraph Company ... The report consists of four volumes. The first three are composed of papers originally published together as a regular issue of the Bell System technical journal. The fourth volume ... pertinent papers contributed by various authors." Includes bibliographies. Subjects: Project Telstar. Series: United States. National Aeronautics and Space Administration. NASA SP32. LC Classification: TL521 .A333 no. 32 TK5104.2.T4

The Apollo spacecraft: a chronology. Published/Created: Washington: Scientific and Technical Information Division, Office of Technology Utilization, National Aeronautics and Space Administration: for sale by the Supt. of Docs., U.S. Govt. Print. Off., 1969-1978. Related Authors: Ertel, Ivan D., 1914- Description: 4 v.: ill.; 26 cm. Contents: v. 1. Ertel, I. D. and Morse, M. L. Through November 7, 1962.--v. 2. Morse, M.L. and Bays, J. K. November 8, 1962-September 30, 1964.--v. 3. Brooks, C. G. and Ertel, I. D. October 1, 1964-January 20, 1966.--v. 4. Ertel, I. D. and Newkirk, R. W. with Brooks, C. G. January 21, 1966-July 13, 1974. Notes: Vol. 2-4 issued by the Scientific and Technical Information Office. Vol. 3-4 not issued as part of the NASA historical series. Includes bibliographical references and index. Subjects: Project Apollo (U.S.) Series: United States. National Aeronautics and Space Administration. NASA historical series. NASA SP; 4009. LC Classification: TL789.8.U6 A5135 Dewey Class No.: 629.47 19

The Apollo-Soyuz Test Project medical report / compiled by Arnauld E. Nicogossian. Published/Created: Washington: Scientific and Technical Information Office, National Aeronautics and Space Administration; Springfield, Va.: for sale by the National Technical Information Service, 1977. Related Authors: Nicogossian, Arnauld E. Description: x, 129 p.: ill.; 27 cm. Notes: Includes index. Bibliography: p. 123. Subjects: Apollo Soyuz Test Project. Space medicine. Space flight--Physiological effect.

Series: NASA SP; 411. LC Classification: RC1135 .A66 1977 Dewey Class No.: 616.9/80214

The atmosphere of Titan; the proceedings. Edited by Donald M. Hunten. Published/Created: Washington, Scientific and Technical Information Office, National Aeronautics and Space Administration; [for sale by the National Technical Information Service, Springfield, Va.] 1974. Related Authors: Hunten, Donald M., ed. Ames Research Center. Description: x, 177 p. illus. 25 cm. Notes: "Prepared by NASA Ames Research Center." Bibliography: p. 171-177. Subjects: Titan (Satellite)--Atmosphere--Congresses. Series: NASA SP; 340. LC Classification: QB671 .T57 1973 Dewey Class No.: 551.5/0999/26 Govt. Doc. No.: NAS 1.21:340

The Comet Halley archive summary volume / Zdenek Sekanina, editor; Lori Fry, production editor. Published/Created: [Washington, D.C.]: National Aeronautics and Space Administration, [1991] Related Authors: Sekanina, Zdenek. United States. National Aeronautics and Space Administration Jet Propulsion Laboratory (U.S.) Description: vii, 332 p.: ill.; 28 cm. Notes: "Prepared by the Jet Propulsion Laboratory, California Institute of Technology, under a contract with the National Aeronautics and Space Administration"--T.p. verso. "International Halley Watch." "August 1991." Item 830-C Includes bibliographical references. Subjects: Halley's comet--Observations--International cooperation Handbooks, manuals, etc. Giacobini-Zinner comet--Observations--International cooperation--Handbooks, manuals, etc.

LC Classification: QB723.H2 C6 1991 Dewey Class No.: 523.6/42 20

The early years, Published/Created: [Washington] National Aeronautics and Space Administration [1964] Description: ix, 273 p. illus., ports., maps. 26 cm. Subjects: Goddard, Robert Hutchings, 1882-1945. Astronautics--United States. LC Classification: TL862.G6 U5

The earth below: purchasing science data and the role of public-private partnerships / Scott Pace ... [et al.]. Published/Created: Santa Monica, CA: RAND Science and Technology Policy Institute, 2000. Related Authors: Pace, Scott. United States. National Aeronautics and Space Administration. United States. Office of Science and Technology Policy. Science and Technology Policy Institute (Rand Corporation) Description: xxi, 147 p.; 28 cm. ISBN: 0833029436 Notes: "DB-316-NASA/OSTP" -- back cover. "Prepared for the National Aeronautics and Space Administration and Office of Science and Technology Policy." Includes bibliographical references (p. 139-147). Subjects: Earth Science Enterprise Program (U.S.) Earth sciences--Research--Government policy--United States. Public-private sector cooperation--United States. Global environmental change--Remote sensing--Information resources. Earth sciences--United States--Remote sensing--Finance. LC Classification: QE47.A1 E33 2000

The evolving universe: structure and evolution of the universe: roadmap 2000-2020. Published/Created: [Washington, D.C.?]: NASA, [1997?] Related Authors: United States.

National Aeronautics and Space Administration. Description: 76 p.: col. ill.; 22 x 28 cm. Notes: "To understand the growth of structure and the evolution of the component parts of the universe." "Science roadmap"--Cover. Shipping list no.: 98-0018-P. "NP-1997(06)-019-GSFC"--Cover. Additional Form Avail.: Also available via Internet from the World Wide Web. Subjects: Cosmology. Large scale structure (Astronomy)--Research--United States. Galaxies--Evolution--Research--United States. Astronomy--Research--United States. LC Classification: QB981 .E85 1997 Dewey Class No.: 523.1/072073 21 Govt. Doc. No.: NAS 1.83:1997(06)019-GSFC

The Experiments of Biosatellite II. Edited by Joseph F. Saunders. Published/Created: Washington, Scientific and Technical Information Office, National Aeronautics and Space Administration; [for sale by the Supt. of Docs., U.S. Govt. Print. Off.] 1971 [i.e. 1972] Related Authors: Saunders, Joseph F., 1927- ed. Description: vi, 352 p. illus. 26 cm. Notes: Includes bibliographies. Subjects: Space biology. Series: NASA SP; 204. LC Classification: QH327 .E96 Dewey Class No.: 574.1/919 Govt. Doc. No.: NAS 1.21:204

The face of Venus: the Magellan radar-mapping mission / Ladislav E. Roth, Stephen D. Wall, editors. Published/Created: Washington, D.C.: National Aeronautics and Space Administration, 1995. Related Authors: Roth, Ladislav E. Wall, Stephen D. United States. National Aeronautics and Space Administration. Description: 135 p.: ill. (some col.), maps; 31 cm. Notes: "June 1995"--T.p. Includes bibliographical references (p. 131-135). Subjects: Magellan (Spacecraft) Space flight to Venus. Venus (Planet)--Exploration. Venus (Planet)--Geology. Series: NASA SP; 520. LC Classification: QB621 .F33 1995 Dewey Class No.: 559.9/22 20

The fall of NASA's Skylab--the communications problem: fourteenth report / by the Committee on Government Operations. Published/Created: Washington: U.S. Govt. Print. Off., 1980. Description: v, 31 p.; 24 cm. Notes: At head of Union calendar no. 409. "March 11, 1980." Subjects: Skylab Program. United States. National Aeronautics and Space Administration. Space vehicles--Atmospheric entry. Series: United States. Congress. House. Report; no. 96-818. LC Classification: TL796.5.U6 S588 Dewey Class No.: 363.3/497 19

The Final Skylab mission: man at home and at work in space. Published/Created: [Washington, D.C.]: National Aeronautics and Space Administration, [1974] Related Authors: United States. National Aeronautics and Space Administration. Description: [8] p.: ill.; 27 cm. Subjects: Skylab Program--History. Space stations--History. Space sciences. Series: Mission report; MR-15 LC Classification: TL789.8.U6 S5553 1974 Dewey Class No.: 629.44/5 20

The first "A" in NASA: report / prepared by the Subcomittee on Transportation, Aviation, and Weather of the Committee on Science and Technology, U.S. House of Representatives, Ninety-fifth Congress, second session. Published/Created: Washington: U.S. Govt. Print Off., 1978. Description: v,

12 p.; 24 cm. Notes: "Serial TT." At head of Committee print. Subjects: United States. National Aeronautics and Space Administration. LC Classification: TL521 .U62 1978 Dewey Class No.: 387.7/0973

The Long duration exposure facility (LDEF): mission 1 experiments / edited by Lenwood G. Clark ... [et al.]. Published/Created: Washington, DC: Scientific and Technical Information Branch, National Aeronautics and Space Administration; Springfield, Va.: For sale by the National Technical Information Service, 1984. Related Authors: Clark, Lenwood G. Description: vii, 189 p.: ill.; 24 cm. Subjects: Space sciences--Experiments. LDEF (Artificial satellite) Series: NASA SP; 473 LC Classification: QB500.264 .L66 1984 Dewey Class No.: 500.5 19

The Magellan Venus explorer's guide / Carolynn Young, editor. Published/Created: [Washington, D.C.]: National Aeronautics and Space Administration; Pasadena, Calif.: Jet Propulsion Laboratory, California Institute of Technology, 1990. Related Authors: Young, Carolynn. Description: xi, 197 p.: ill., col. map; 23 cm. Notes: One col. map on folded leaf in pocket. "August 1, 1990." "JPL publication 90-24." Subjects: Venus probes. Space flight to Venus. LC Classification: TL799.V45 M34 1990 Dewey Class No.: 523.4/2 20

The Martian landscape / by the Viking Lander Imaging Team. Published/Created: Washington: Scientific and Technical Information Office, National Aeronautics and Space Administration: for sale by the Supt. of Docs., U.S. Govt. Print. Off., 1978. Related Authors: Mutch, Thomas A., 1931- Description: vii, 160 p.: ill; 37 cm. & viewer in pocket. Subjects: Viking Mars Program (U.S.) Mars (Planet)--Surface--Photographs from space. Series: NASA SP; 425. LC Classification: QB641 .V54 1978 Spec Format Dewey Class No.: 559.9/23/028

The NASA scientific and technical information system and how to use it. Published/Created: [Washington] National Aeronautics and Space Administration [1969] Description: 25 p. illus. 23 cm. Notes: Cover title. Subjects: United States. National Aeronautics and Space Administration--Information services. LC Classification: TL521 .312.A583 Dewey Class No.: 629.1/07

The NASA scientific and technical information system: its scope and coverage. Published/Created: [Washington]: NASA Scientific and Technical Information Branch, 1978. Description: ix, 279 p.; 27 cm. Notes: Cover title. At head of NASA, National Aeronautics and Space Administration. Includes index. Subjects: Aeronautics--United States--Information services. Astronautics--United States--Information services. Space sciences--United States--Information services. LC Classification: TL521.312 .U54 1978 Dewey Class No.: 629.1/07

The NASA scientific and technical information system: its scope and coverage. Published/Created: [Washington] National Aeronautics and Space Administration, 1970. Description: vi, 78 p. 23 cm. Notes: Cover title. Subjects: United States. National Aeronautics and Space

Administration--Information services. LC Classification: TL521.312 .A584 Dewey Class No.: 629.1/07

The National space program: its values and benefits; staff study for the Subcommittee on NASA Oversight, Ninetieth Congress, first session. Published/Created: Washington, U.S. Govt. Print. Off., 1967. Related Authors: United States. Congress. House. Committee on Science and Astronautics. Subcommittee on NASA Oversight. Description: viii, 66 p. maps. 24 cm. Notes: At head of Committee print. "Serial D." Subjects: United States. National Aeronautics and Space Administration. Astronautics--United States. LC Classification: TL521.312 .A485 Dewey Class No.: 629.4/0973

The pioneer mission to Jupiter. Prepared for the Office of Space Science and Applications. Published/Created: Washington; [For sale by the Supt. of Docs., U.S. Govt. Print. Off.] 1971. Related Authors: United States. Office of Space Science and Applications. Description: 46 p. illus. 24 cm. Notes: Includes bibliographical references. Subjects: Space flight to Jupiter. Series: NASA SP; 268. LC Classification: TL799.J8 U5 Dewey Class No.: 629.43/54/5 Govt. Doc. No.: NAS 1.21:268

The Search for extraterrestrial intelligence, SETI / edited by Philip Morrison, John Billingham, and John Wolfe; prepared at Ames Research Center. Published/Created: [Washington]: National Aeronautics and Space Administration, Scientific and Technical Information Office: for sale by the Supt. of Docs., U.S. Govt. Print. Off., 1977. Related Authors: Morrison, Philip.
Billingham, John. Wolfe, John, 1933- Ames Research Center. Description: xv, 276 p.: ill.; 25 cm. Notes: Bibliography: p. 239-241. Subjects: Life on other planets--Congresses. Interstellar communication--Congresses. Series: NASA SP; 419. LC Classification: QB54 .S4 Dewey Class No.: 574.999

The SEASAT-A project: where it stands today, National Aeronautics and Space Administration, National Oceanic and Atmospheric Administration: report to the Congress / by the Comptroller General of the United States. Published/Created: [Washington: U.S. General Accounting Office], 1977. Description: iv, 42 p.; 27 cm. Notes: Cover title. "PSAD-77-126." "B-183134." Subjects: United States. National Aeronautics and Space Administration. United States. National Oceanic and Atmospheric Administration. Astronautics in earth sciences. Earth sciences--Research--United States. LC Classification: QE33 .U498 1977 Dewey Class No.: 353.0085/55042 19

The six orbits of Sigma 7: Walter M. Schirra's space flight, October 3, 1962. Published/Created: [Washington] National Aeronautics and Space Administration, Manned Spacecraft Center [1963] Related Titles: Walter M. Schirra's space flight, October 3, 1962. Description: unpaged. illus. 21 x 28 cm. Subjects: Schirra, Wally. Project Mercury (U.S.) LC Classification: TL789.8.U6 M54 1963

The space telescope. Published/Created: Washington: Scientific and Technical Information Office, National Aeronautics and Space Administration: for sale by the Supt. of Docs., U.S.

Govt. Print. 1976. Description: xi, 231 p.: ill.; 24 cm. Notes: "This volume contains [the papers and] the author's summaries of their papers on the space telescope presented at the 21st annual meeting of the American Astronautical Society at Denver, Colorado, August 26-28, 1975." Includes bibliographies and index. Subjects: Hubble Space Telescope (Spacecraft)--Congresses. Orbiting astronomical observatories--Congresses. Series: NASA SP; 392. LC Classification: QB88 .A64 1976 Dewey Class No.: 522/.29/19 Govt. Doc. No.: NAS 1.21:392

The view from Ranger. Published/Created: Washington, National Aeronautics and Space Administration [and] Jet Propulsion Laboratory, California Institute of Technology [Pasadena]; for sale by the Superintendent of Documents, U.S. Govt. Print. Off. [1966] Related Authors: Jet Propulsion Laboratory (U.S.) Description: 58 p. illus., maps. 28 cm. Notes: "NASA EP-38." "Contract no. NAS 7-100." Bibliography: p. 58. Subjects: Project Ranger. Lunar photography. LC Classification: TL789.8.U6 R352 Dewey Class No.: 629.43/53

The Voyager flights to Jupiter and Saturn. Published/Created: Pasadena, Calif.: National Aeronautics and Space Administration, Jet Propulsion Laboratory, California Institute of Technology: For sale by the Supt. of Docs., U.S. G.P.O., 1982. Related Authors: United States. National Aeronautics and Space Administration. Jet Propulsion Laboratory (U.S.) Description: 60 p.: chiefly ill.; 28 cm. Notes: Cover title. S/N 033-000-00854-6 Subjects: Voyager Project. Jupiter probes. Saturn probes. Series: NASA EP; v. 191 LC Classification: QB661 .V69 1982 Dewey Class No.: 523.4/5 19 Govt. Doc. No.: NAS 1.19:191

Thermodynamic equilibrium in prebiological atmospheres of C, H, O, N, P, S, and Cl, by M. O. Dayhoff [and others] Published/Created: Washington, Scientific and Technical Information Division, National Aeronautics and Space Administration; [for sale by the Clearinghouse for Federal Scientific and Technical Information, Springfield, Va.] 1967. Related Authors: Dayhoff, Margaret O. Description: xi, 259 p. illus. 27 cm. Notes: Prepared for NASA under contract no. NSR 21-003-002 by the National Biomedical Research Foundation with the aid of the computer services at the University of Maryland. Includes bibliographies. Subjects: Thermodynamic equilibrium. Gases. Planets--Atmospheres. Series: NASA SP; 3040. LC Classification: QC319 .N37

This is NASA. Published/Created: Washington, D.C.: Public Affairs Division, NASA Headquarters, 1979. Description: 48 p.: ill.; 28 cm. Notes: "EP-155." Includes index. Subjects: United States. National Aeronautics and Space Administration. LC Classification: TL521.312 .U54 1979 Dewey Class No.: 629.4/0973 19

Thomas, Richard Nelson, 1921- Stellar atmospheric structural patterns / Richard N. Thomas. Published/Created: Paris, France: Centre National de la recherche scientifique; Washington, D.C.: National Aeronautics and Space Administration, Scientific and Technical Information Branch, 1983. Description: xxxiv, 369 p.: ill.; 28 cm. Notes: Resume in French. Includes index.

Bibliography: p. 351-364. Subjects: Stars--Atmospheres. Series: Monograph series on nonthermal phenomena in stellar atmospheres NASA SP; 471 LC Classification: QB809 .T46 1983 Dewey Class No.: 523.8 19

Thompson, Milton O. Flight research: problems encountered and what they should teach us / Milton O. Thompson; with a background section by J.D. Hunley. Published/Created: Washington, D.C.: National Aeronautics and Space Administration, 2000. Description: p. cm. Notes: Includes bibliographical references and index. Subjects: Aeronautics--Research--United States. Airplanes--Flight testing. High-speed aeronautics. Series: Monographs in aerospace history; NASA history series NASA SP; 4522. LC Classification: TL565 .T46 2000 Dewey Class No.: 629.13/07/2073 21

Thornton, T. H. Injection accuracy characteristics for lunar missions / T. H. Thornton. Published/Created: Pasadena, Calif.: Jet Propulsion Laboratory, California Institute of Technology, 1965. Related Authors: United States. National Aeronautics and Space Administration. Description: iv, 13 p.; 28 cm. Notes: At head of National Aeronautics and Space Administration. Includes bibliographical references. Subjects: Space flight to the moon. Orbital transfer (Space flight) Series: California Institute of Technology, Pasadena. Jet Propulsion Laboratory. JPL technical report; no. 32-839. LC Classification: TL799.M6 T52

Thostesen, T. O. Mariner Mars absorptivity standard [by] T. O. Thostesen [and] D. W. Lewis. Published/Created: Pasadena, Jet Propulsion Laboratory, California Institute of Technology, 1967. Related Authors: Lewis, D. W., joint author. United States. National Aeronautics and Space Administration. Description: v, 40 p. illus. 28 cm. Notes: "Prepared under contract no. NAS 7-100, National Aeronautics & Space Administration." Bibliography: p. 40. Subjects: Project Mariner (U.S.) Series: California Institute of Technology, Pasadena. Jet Propulsion Laboratory. JPL technical report no. 32-734. LC Classification: TL789.8.U6 M38 Dewey Class No.: 629.46

Tomayko, J. E. (James E.), 1949- Computers in space: journeys with NASA / James E. Tomayko. Published/Created: Indianapolis, Ind.: Alpha Books, c1994. Description: xi, 197 p.: ill. (some col.); 28 cm. ISBN: 1567614639 Notes: Includes index. Subjects: United States. National Aeronautics and Space Administration--History. Astronautics--Data processing--History. Space flight--Data processing--History. Space vehicles--Piloting--Data processing--History. LC Classification: TL1078 .T66 1994 Dewey Class No.: 629.4/0285 20

Toward a better tomorrow with aeronautical and space technology. Prepared at the request of Frank E. Moss, chairman. Published/Created: Washington; for sale by the Supt. of Docs., U.S. Print. Off. [1973] Description: viii, 199 p. illus. 24 cm. Notes: At head of 93d Congress, 1st session. Committee print. Includes bibliographical references. Subjects: United States. National Aeronautics and Space Administration. Technology Utilization Division. Technology transfer. LC Classification: T174.3 .U52 1973 Dewey Class No.: 600 Govt. Doc.

No.: Y 4.Ae8:Ae8

Toward Mach 2: the Douglas D-558 program / edited by J.D. Hunley; featuring comments by Stanley P. Butchart ... [et al.]. Published/Created: Washington, D.C.: National Aeronautics and Space Administration, NASA Office of Policy and Plans, NASA History Office, 1999. Related Authors: Hunley, J. D., 1941- Description: xiii, 161 p.: ill.; 25 cm. Notes: Papers of the NASA Dryden Flight Resarch Center Symposium on the D-55 Program, Feb. 4, 1998. "NASA SP-4222." Includes bibliographical references and index. Subjects: High-speed aeronautics--United States--History Congresses. Skystreak (Supersonic planes)--History--Congresses. Series: NASA history series LC Classification: TL551.5 .T69 1999 Dewey Class No.: 629.132/305/0973 21

Transfer of the Development Operations Division of Army Ballistic Missile Agency to the National Aeronautics and Space Administration. Published/Created: Washington, U. S. Govt. Print. Off., 1960. Description: iii, 41 p. illus. 24 cm. Subjects: United States. National Aeronautics and Space Administration. United States. Army Ballistic Missile Agency. Development Operatons Division. LC Classification: TL789.8.U5 A52 1960a

Trask, N. J. (Newell Jefferson) The contributions of Ranger photographs to understanding the geology of the Moon, by N. J. Trask. Published/Created: Washington, U.S. Govt. Print. Off., 1972. Related Authors: United States. National Aeronautics and Space Administration. Description: iii, 16 p. illus. 29 cm. Notes: Prepared on behalf of the National Aeronautics and Space Administration. Bibliography: p. 15-16. Subjects: Lunar geology. Project Ranger. Series: Contributions to astrogeology Geological Survey professional paper; 599-J. LC Classification: QE75 .P9 no. 599-J QB592 Dewey Class No.: 557.3/08 s 559.9/1 Govt. Doc. No.: I 19.16:599-J

Trento, Joseph John, 1947- Prescription for disaster / Joseph J. Trento with reporting and editing by Susan B. Trento. Edition Information: 1st ed. Published/Created: New York: Crown, c1987. Description: 312 p., [16] p. of plates: ill., ports.; 24 cm. ISBN: 0517564157 Notes: Includes index. Bibliography: p. 296-304. Subjects: United States. National Aeronautics and Space Administration--History. Challenger (Spacecraft) Space vehicle accidents--United States. LC Classification: TL789.8.U5 T74 1987 Dewey Class No.: 363.1/24 19

Tucker, Wallace H. The star splitters: the High Energy Astronomy Observatories / Wallace H. Tucker. Published/Created: Washington, DC: Scientific and Technical Information Branch, National Aeronautics and Space Administration: For sale by the Supt. of Docs., U.S. G.P.O., 1984. Description: ix, 182 p.: ill. (some col.); 24 cm. Notes: Includes bibliographical references. Subjects: High Energy Astronomy Observatories. Astrophysics--Observations. Series: NASA SP; 466 LC Classification: QB81 .T8 1984 Dewey Class No.: 522/.19 19

Twenty-five years of NASA international programs / prepared by the Staff of the International Affairs Division. Published/Created: Washington, D.C.:

National Aeronautics and Space Administration, [1983] Related Authors: United States. National Aeronautics and Space Administration. International Affairs Division. Related Titles: Twenty-five years of N.A.S.A. international programs. 25 years of NASA international programs, January 1, 1983. Description: iv, 135 p., [1] folded leaf of plates: ill.; 21 x 27 cm. Notes: Spine 25 years of NASA international programs, January 1, 1983. Subjects: United States. National Aeronautics and Space Administration. Aeronautics--International cooperation. LC Classification: TL521.312 .T84 1983 Dewey Class No.: 629.4 19

Tyner, Richard L. A catalog of selected Viking Orbiter images / Richard L. Tyner and Roger D. Carroll. Published/Created: Washington, DC: National Aeronautics and Space Administration, Scientific and Technical Information Branch; [Springfield, Va.: For sale by the National Technical Information Service], 1983. Related Authors: Carroll, Roger D. United States. National Aeronautics and Space Administration. Scientific and Technical Information Branch. Geological Survey (U.S.) Description: 399 p.: ill., maps; 37 x 28 cm. Notes: "U.S. Geological Survey." STAR category 91. Item 830-H-11 Includes bibliographical references. Subjects: Viking Mars Program (U.S.)--Catalogs. Mars (Planet)--Exploration--Catalogs. Series: NASA reference publication; 1093 LC Classification: QB641 .T95 1983 Dewey Class No.: 523.4/3 19 Govt. Doc. No.: NAS 1.61:1093

Van Nimmen, Jane, 1937- NASA historical data book, 1958-1968 / Jane Van Nimmen and Leonard C. Bruno, with Robert L. Rosholt. Published/Created: Washington: Scientific and Technical Information Office, National Aeronautics and Space Administration; Springfield, Va.: for sale by the National Technical Information Service, 1976. Related Authors: Bruno, Leonard C., joint author. Rosholt, Robert L., joint author. Description: viii, 543 p.: ill.; 21 x 27 cm. Notes: "Vol. 1: NASA resources." No more published in this format. Includes bibliographical references. Subjects: United States. National Aeronautics and Space Administration. Series: The NASA historical series NASA SP; 4012 LC Classification: TL521.312 .V36 1976 Dewey Class No.: 629.4/0973

Vaughan, Diane. The Challenger launch decision: risky technology, culture, and deviance at NASA / Diane Vaughan. Published/Created: Chicago: University of Chicago Press, 1996. Description: xv, 575 p.: ill.; 24 cm. ISBN: 0226851753 (cloth: alk. paper) Notes: Includes bibliographical references (p. 533-550) and index. Subjects: Challenger (Spacecraft)--Accidents. United States. National Aeronautics and Space Administration--Management. Aerospace industries--United States. Organizational behavior--Case studies. Decision making--Case studies. LC Classification: TL867 .C467 1996 Dewey Class No.: 363.12/465 20

Venus cusp observations during 1969; synopsis of results [by] G. F. Schilling [and others] Published/Created: Santa Monica, Calif., Rand Corp., 1970. Related Authors: Schilling, G. F. United States. Office of Space Science and Applications. Description: ix, 42 p. illus. 28 cm. Notes: Research prepared for the Office of Space Science and

Applications of the National Aeronautics and Space Administration under contract NASw-1762. Includes bibliographical references. Subjects: Venus (Planet)--Atmosphere. Series: Rand Corporation. Research memorandum RM-6261-NASA LC Classification: Q180.A1 R36 no. 6261 Dewey Class No.: 523.4/2

Viking 1: early results. Published/Created: Washington: Scientific and Technical Information Office, National Aeronautics and Space Administration; Springfield, Va.: for sale by the National Technical Information Service, 1976. Description: vii, 67 p.: ill.; 27 cm. Subjects: Mars (Planet)--Observations. Series: NASA SP; 408. LC Classification: QB641 .U55 1976 Dewey Class No.: 559.9/23

Vogt, Gregory. A twenty-fifth anniversary album of NASA / by Gregory Vogt. Published/Created: New York: F. Watts, 1983. Description: 90 p.: ill.; 29 cm. ISBN: 0531046559 (lib. bdg.) 0531035913 Summary: A brief history of NASA's development, from its early stages of research, through the space shuttle programs, to conjectures about future projects in space research and travel. Notes: Includes index. Bibliography: p. 86-87. Subjects: United States. National Aeronautics and Space Administration--Juvenile literature. United States. National Aeronautics and Space Administration. Astronautics--United States--History--Juvenile literature. Astronautics--History. Space sciences. Space flight. LC Classification: TL521 .V64 1983 Dewey Class No.: 629.4/0973 19

Vogt, Gregory. Space mission patches / by Gregory L. Vogt. Published/Created: Brookfield, Conn.: Millbrook Press, c2001. Description: 78 p.: ill. (mostly col.); 26 cm. ISBN: 0761316132 (lib. bdg.) Notes: Includes bibliographical references (p. 73) and index. Subjects: United States. National Aeronautics and Space Administration--Insignia--Juvenile literature. United States. National Aeronautics and Space Administration--Insignia. Astronautics--United States--Collectibles--Juvenile literature. Manned space flight--Collectibles--Juvenile literature. Astronautics--United States--Collectibles--Juvenile literature. Badges--United States--Collectibles--Juvenile literature. Space flights. Astronautics. LC Classification: TL789.8.U5 V54 2001 Dewey Class No.: 629.45/002/75 21

Von Braun, Wernher, 1912-1977. Papers of Wernher Von Braun, 1796-1970 (bulk 1950-1970) Description: 20,000 items. 56 containers plus 5 OVSD. 22.8 linear feet. Biog./History Note: German-born aerospace engineer; emigrated to the United States in 1945; served as director and administrator of U.S. space programs and projects. Summary: Correspondence, fan mail, speeches and writings, public relations material, subject files, scrapbooks, and printed matter, chiefly 1950-1970. The papers relate to Von Braun's career in rocketry and aerospace engineering from his early work on the V-2 rocket in Germany to his service at the U.S. Dept. of Defense, the Redstone Arsenal, and the NASA George C. Marshall Space Flight Center in Huntsville, Ala. Includes material relating to guided missiles, rocketry, and space exploration in general; papers relating to the Mercury, Gemini, and Apollo space programs; and material reflecting Von Braun's associations with the American

Rocket Society and Gesellschaft für Weltraumsforschung, and General Dynamics, General Motors, IBM, Lockheed, Sperry-Rand, and other corporations involved in aerospace work. Correspondents include Otto Wolfgang Bechtle, Wilber Marion Brucker, Frederick C. Durant, Heinz Gartmann, Juppe Gerhards, Aristid von Grosse, Heinz Hermann Kölle, Willy Ley, G. Loeser, John B. Medaris, Cornelius Ryan, Igor Ivan Sikorsky, John Sparkman, Ernst Stublinger, Robert Collins Truax, Armitage Watkins, and Fred Lawrence Whipple. Notes: Photographs and tapes transferred to appropriate custodial divisions of the Library of Congress. In English and German. MSS44172 Source of Acquisition: Gift and deposit, Wernher Von Braun, 1962-1971. Deposit converted to gift, Maria Von Braun, 1997. Subjects: Bechtle, Otto Wolfgang--Correspondence. Brucker, Wilber Marion, 1894-1968 --Correspondence. Durant, Frederick C., 1916- --Correspondence. Gartmann, Heinz, 1917- --Correspondence. Gerhards, Jupps--Correspondence. Grosse, Aristid von, 1905- --Correspondence. Kölle, Heinz Hermann, 1917- --Correspondence. Ley, Willy, 1906-1969 --Correspondence. Loeser, G.--Correspondence. Medaris, John B. (John Bruce), 1902- --Correspondence. Ryan, Cornelius--Correspondence. Sikorsky, Igor Ivan, 1889-1972 --Correspondence. Sparkman, John, 1899-1985 --Correspondence. Stuhlinger, Ernst, 1913- --Correspondence. Truax, Robert Collins, 1917- --Correspondence. Watkins, Armitage, 1906- --Correspondence. Whipple, Fred Lawrence, 1906- --Correspondence. United States. Dept. of Defense. United States. National Aeronautics and Space Administration. American Rocket Society. General Dynamics Corporation. General Motors Corporation. George C. Marshall Space Flight Center. Gesellschaft für Weltraumsforschung. International Business Machines Corporation. Lockheed Aircraft Corporation. Project Apollo (U.S.) Project Gemini (U.S.) Project Mercury (U.S.) Sperry Rand Corporation. Aeronautics. Guided missiles. Rocketry--Germany. Rocketry--United States. Rockets (Aeronautics) Space flight. V-2 rocket. World War, 1939-1945--Aerial operations, German. Redstone Arsenal (Ala.) Aerospace engineers. Public officials. Local Call/Shelving: 0808V OVSD 0114K

Voyager 1 encounters Saturn. Published/Created: Pasadena: National Aeronautics and Space Administration, Jet Propulsion Laboratory, Calif. Institute of Technology; Washington, D.C.: For sale by the Supt. of Docs., U.S. G.P.O., 1980. Related Authors: Jet Propulsion Laboratory (U.S.) Description: 40 p.: col. ill.; 28 cm. Notes: Cover title. Dec. 1980. "JPL 400-100"--P. 4 of cover. S/N 033-000-008171 Item 830-H-9 Subjects: Voyager Project. Saturn probes. Saturn (Planet)--Photographs from space. LC Classification: QB671 .V69 Dewey Class No.: 523.4/6/0222 19 Govt. Doc. No.: NAS 1.12/7:400-100

Voyager 1, encounter with Jupiter. Published/Created: [Washington]: National Aeronautics and Space Administration, 1979. Description: 43 p.: ill.; 28 cm. Subjects: Voyager Project. Jupiter (Planet)--Photographs from space. LC Classification: QB661 .U54 1979a Dewey Class No.:

523.4/5/0222 19

Voyager encounters Jupiter. Published/Created: [Pasadena, Calif.]: NASA, [Jet Propulsion Laboratory, California Institute of Technology, 1979] Description: 40 p.: ill.; 28 cm. Notes: "July 1979." Subjects: Voyager Project. Jupiter (Planet)--Pictorial works. LC Classification: QB661 .U54 1979b Dewey Class No.: 523.4/5/0222 19

Voyager to Jupiter and Saturn. Published/Created: Washington: Scientific and Technical Information Office, National Aeronautics and Space Administration; Springfield, Va.: for sale by National Technical Information Service, 1977. Description: vi, 58 p.: ill.; 23 cm. Subjects: Voyager Project. Jupiter probes. Saturn probes. Series: NASA SP; 420. LC Classification: TL789.8.U6 V69 1977 Dewey Class No.: 629.43/54/5 Govt. Doc. No.: NAS 1.21:420

Voyager: journey to the outer planets. Published/Created: Pasadena, Calif.: Jet Propulsion Laboratory, [1977] Related Authors: Jet Propulsion Laboratory (U.S.) United States. National Aeronautics and Space Administration. Description: 15 p.: ill.; 28 cm. Notes: Cover title. "JPL-SP-43-39." "Prepared under contract no. NAS7-100, National Aeronautics & Space Administration." Subjects: Voyager Project. Outer planets. LC Classification: TL789.8.U6 V695 Dewey Class No.: 629.43/54

Wagner, Richard. Designs on space: blueprints for 21st century space exploration / Richard Wagner, with illustrations by Howard Cook. Published/Created: New York: Simon & Schuster, 2000. Related Authors: Cook, Howard. Description: v, 138 p.: ill.; 20 x 27 cm. ISBN: 068485676X Subjects: United States. National Aeronautics and Space Administration. Space probes-- Design and construction. Outer space-- Exploration. LC Classification: TL795.3 .W34 2000 Dewey Class No.: 629.47 21

Wall, Stephen D. Conclusion of Viking Lander Imaging Investigation: picture catalog of experiment data record / Stephen D. Wall and Teresa C. Ashmore. Published/Created: Washington, D.C.: National Aeronautics and Space Administration, Scientific and Technical Information Branch; Springfield, Va.: For sale by National Technical Information Service, 1985. Related Authors: Ashmore, Teresa C. Description: x, 202 p.: ill.; 28 cm. Notes: "March 1985." Bibliography: p. 173. Subjects: Viking Mars Program (U.S.) Mars (Planet)--Exploration-- Charts, diagrams, etc. Mars (Planet)-- Surface--Photographs from space. Imaging systems in astronomy--Charts, diagrams, etc. Series: NASA reference publication; 1137 LC Classification: QB641 .W24 1985 Dewey Class No.: 559.9/23 19

Wallace, Harold D., 1960- Wallops Station and the creation of an American space program / Harold D. Wallace, Jr. Published/Created: Washington, D.C.: National Aeronautics and Space Administration, NASA History Office, Office of Policy and Plans, 1997. Description: xiii, 167 p.: ill.; 23 cm. Notes: Includes bibliographical references (p. 145-148) and index. Subjects: Wallops Flight Facility-- History. Astronautics--United States-- History. Series: The NASA history series NASA SP; 4311 LC

Classification: TL862.W35 W35 1997 Dewey Class No.: 629.4/09755/16 21

Wallace, Lane E., 1961- Airborne trailblazer / by Lane E. Wallace. Published/Created: Washington, D.C.: National Aeronautics and Space Administration, NASA History Office: For sale by the Supt. of Docs., U.S. G.P.O., 1994. Description: ix, 188 p.: ill. (some col.), col. maps; 31 cm. Notes: "Two decades with NASA Langley's 737 flying laboratory." Includes bibliographical references (p. 125-141) and index. Subjects: Langley Research Center. Aeronautics, Commercial--Research--United States--History. Boeing 737 (Jet transport) Research aircraft--United States. Series: The NASA history series NASA SP; 4216 LC Classification: TL521.312 .W35 1994 Dewey Class No.: 629.1/072073 20

Wallace, Lane E., 1961- Dreams, hopes, realities: NASA's Goddard space flight center: the first forty years / by Lane E. Wallace. Published/Created: Washington, D.C.: National Aeronautics and Space Administration, NASA History Office: For sale by the U.S. G.P.O., Superintendent of Documents, 1999. Description: ix, 219 p.: ill. (some col.); 31 cm. Notes: Includes bibliographical references (p. 191-192) and index. Subjects: Goddard Space Flight Center--History. Series: The NASA history series LC Classification: TL862.G6 W35 1999 Dewey Class No.: 629.4/0973 21 Govt. Doc. No.: NAS 1.21: SP-4312

Wallace, Lane E., 1961- Flights of discovery: 50 years at the NASA Dryden Flight Research Center / by Lane E. Wallace. Published/Created: Washington, DC: NASA History Office: For sale by the U.S. G.P.O., Superintendent of Documents, 1996. Related Authors: United States. National Aeronautics and Space Administration. Description: 198 p.: ill. (some col.); 32 cm. Notes: "National Aeronautics and Space Administration." Includes bibliographical references (p. 173-179) and index. Subjects: NASA Dryden Flight Research Center--History. Series: The NASA history series NASA SP; 4309 LC Classification: TL521.312 .W354 1996 Dewey Class No.: 629.1/0720794/88 20

Waltman, Gene L., 1935- Black magic and gremlins: analog flight simulations at NASA's Flight Research Center / by Gene L. Waltman. Published/Created: Washington, D.C.: National Aeronautics and Space Administration, Office of Policy and Plans, NASA History Division, 2000. Related Authors: United States. National Aeronautics and Space Administration. History Office. Description: x, 232 p.: ill.; 28 cm. Notes: Includes bibliographical references (p. 223-230) and index. Subjects: Flight Research Center (U.S.)--History. Flight simulators--History. Series: NASA SP; 2000-4520 Monographs in aerospace history; no. 20 LC Classification: TL712.5 .W35 2000 Dewey Class No.: 629.132/52/078 21

Warner, Jeffrey, 1939- Apollo 12 lunar-sample information. Published/Created: Washington, National Aeronautics and Space Administration, 1970. Description: xiv, 377 p. illus. 27 cm. Notes: Cover title. Bibliography: p. 64. Subjects: Project Apollo (U.S.) Lunar petrology. Series: United States. National Aeronautics and Space

Administration. NASA technical report, NASA TR R-353 LC Classification: TL521 .A3312 no. 353 Dewey Class No.: 552/.1/09991

Warwick, James W., 1924- Particles and fields near Jupiter, by James W. Warwick. Published/Created: Washington, National Aeronautics and Space Administration; [for sale by the Clearinghouse for Federal Scientific and Technical Information, Springfield, Va.] 1970. Related Authors: Jet Propulsion Laboratory (U.S.) United States. National Aeronautics and Space Administration. Description: v, 123 p. illus. 27 cm. Notes: Cover title. Prepared under contract no. NAS 7-100 by Jet Propulsion Laboratory, California Institute of Technology for National Aeronautics and Space Administration. Bibliography: p. 118-123. Subjects: Interstellar matter. Magnetism. Jupiter (Planet) Series: NASA contractor report; NASA CR-1685. LC Classification: TL521.3.C6 A3 no. 1685 QB661 Dewey Class No.: 629.1/08 s 523.4/5/2

Wells, Helen T. Origins of NASA names / Helen T. Wells, Susan H. Whiteley, and Carrie E. Karegeannes. Published/Created: Washington: Scientific and Technical Information Office, National Aeronautics and Space Administration: for sale by the Supt. of Docs., U.S. Govt. Print. Off., 1976. Related Authors: Whiteley, Susan H., joint author. Karegeannes, Carrie E., joint author. Related Titles: NASA names. Description: x, 227 p.: ill.; 25 cm. Notes: Includes bibliographical references and index. Subjects: United States. National Aeronautics and Space Administration. Astronautics--United States. Aeronautics--United States.

Series: NASA SP; 4402. NASA history series. LC Classification: TL521.312 .W45 Dewey Class No.: 629.4/0973 Govt. Doc. No.: NAS 1.21:4402

Wells, Helen T. The publications of Dr. Hugh L. Dryden, compiled by Helen T. Wells. Edition Information: Comment ed. Published/Created: Washington, Historical Office, National Aeronautics and Space Administration, 1966. Description: 15 l. 27 cm. Subjects: Dryden, Hugh Latimer, 1898-1965 -- Bibliography. Series: [United States. National Aeronautics and Space Administration. Historical Office] HHN-59 LC Classification: Z8243.85 .W45

Whitnah, A. M. Space shuttle wind tunnel testing program summary / A.M. Whitnah and E.R. Hillje. Published/Created: Washington, D.C.: National Aeronautics and Space Administration, Scientific and Technical Information Branch; [Springfield, Va.: For sale by the National Technical Information Service], 1984. Related Authors: Hillje, E. R. United States. National Aeronautics and Space Administration. Scientific and Technical Information Branch. Description: iv, 159, 258, 67 p.: ill.; 28 cm. Notes: "November 1984"--Cover. Item 830-H-11 Bibliography: p. 49-50 (1st group) Subjects: Space shuttles--Testing. Series: NASA reference publication; 1125 LC Classification: TL795.5 .W48 1984 Dewey Class No.: 629.44/1/0287 19 Govt. Doc. No.: NAS 1.61:1125

Wiksten, David B. Dynamic environment of the Ranger Spacecraft: I through IX; final report [by] David B. Wiksten. Published/Created: Pasadena, Jet Propulsion Laboratory, California Institute of Technology, 1966.

Description: vi, 60 p. illus. 28 cm. Notes: "Prepared under contract no. NAS 7-100, National Aeronautics & Space Administration." Bibliography: p. 60. Subjects: Project Ranger. Space vehicles--Design and construction. Series: California Institute of Technology, Pasadena. Jet Propulsion Laboratory. JPL technical report no. 32-909. LC Classification: TL789.8.U6 R355 Dewey Class No.: 629.46

Wilhelms, Don E. Stratigraphy of part of the lunar near side / by Don E. Wilhelms; prepared on behalf of the National Aeronautics and Space Administration. Published/Created: Washington: U.S. Govt. Print. Off.: for sale by the Supt. of Docs., GPO, 1980. Related Authors: United States. National Aeronautics and Space Administration. Description: iv, 71 p.: ill.; 29 cm. Notes: Bibliography: p. 66-71. Subjects: Project Apollo (U.S.) Lunar stratigraphy. Series: Apollo 15-17 orbital investigations Geological Survey professional paper; 1046-A. LC Classification: QB592 .W52 Dewey Class No.: 551.7/00999/1 Govt. Doc. No.: I 19-16:1046-A

Wilson, James H. Two over Mars: Mariner VI and Mariner VII, February to August 1969 [by] James H. Wilson. Published/Created: Pasadena, Jet Propulsion Laboratory, California Institute of Technology; [for sale by the Supt. of Docs., U.S. Govt. Print. Off., Washington, 1971] Related Authors: Jet Propulsion Laboratory (U.S.) United States National Aeronautics and Space Administration. Description: 39 p. illus. 28 cm. Notes: Prepared for National Aeronautics and Space Administration under contract no. NAS7-100. Includes bibliographical references. Subjects: Project Mariner (U.S.) Space flight to Mars. LC Classification: TL789.8.U6 M386 Dewey Class No.: 629.45/5/3 Govt. Doc. No.: NAS 1.19:90.

Wind energy developments in the 20th century. Edition Information: Rev. 1981. Published/Created: Cleveland, Ohio: National Aeronautics and Space Administration, Lewis Research Center; Washington, D.C.: For sale by the Supt. of Docs., U.S. G.P.O., 1981. Related Authors: Lewis Research Center. Related Titles: Wind energy in the twentieth century. Description: 32 p.: ill.; 28 cm. Notes: S/N 033-000-00819-8 Item 830-C (microfiche) Includes bibliographical references. Subjects: Wind power. Wind turbines. LC Classification: TJ823 .W56 1981 Dewey Class No.: 333.9/2 19 Govt. Doc. No.: NAS 1.2:W 72/981

Young, Richard S. (Richard Stuart) An analysis of the extraterrestrial life detection problem [by] Richard S. Young, Robert B. Painter and Richard D. Johnson. Published/Created: Washington, Scientific and Technical Information Division, National Aeronautics and Space Administration; [for sale by the Clearinghouse for Federal Scientific and Technical Information, Springfield, Va.] 1965. Related Authors: Painter, Robert B., joint author. Johnson, Richard D., joint author. Description: 33 p. illus. 27 cm. Notes: Bibliography: p. 32-33. Subjects: Life on other planets. Mars (Planet) Series: NASA SP; 75 LC Classification: QB54 .Y6

WEBSITES

A Blueprint for New Beginnings: Section 33 - National Aeronautics and Space Administration Tues. Jun 12, 2001 News Space Shuttles Space Stations Space History Space Agencies Missions/Launches Science/Astronomy SETI: Search for Life Photos/Videos SpaceTV Space News Business Technology Science Fiction Entertainment Games Spa http://www.space.com/news/spaceagencies/funding_2002_010228.html

About.com: http://www.hq.nasa.gov/office/pao/History/spaceact.html > DVD > Find a Movie Review > Movie http://space.about.com/gi/dynamic/offsite.htm?site=http%3A%2F%2Fwww.hq.nasa.gov%2Foffice%2Fpao%2FH..

Aeronautics and Space Engineering Board Read reports about upgrading the space shuttle, protecting the shuttle from asteroids and debris, and space technology for the new century. http://www.nas.edu/cets/aseb

CAPPP Field Placements - National Aeronautics and Space Administration Center for American Politics and Public Policy Field Placement Description National Aeronautics and Space Administration Alums Jonathan Lopez Research Paper: "Computer Privacy: Has Legislation done an Adequate Job in Protecting an Individual's Privacy… http://www.sscnet.ucla.edu/issr/cappp/dc/nasa.html

Cato Handbook for Congress: National Aeronautics and Space Administration 37. National Aeronautics and Space Administration Congress should shut down the National Aeronautics and Space Administration (NASA). To that end, it should scrap plans to build a space station or, failing http://www.cato.org/pubs/handbook/hb105-37.html

Earth Science Enterprise Search for data gathered by remote sensing missions from spacecraft and aircraft. Examine science mission info or browse kids' sections. http://www.earth.nasa.gov/

Encyclopedia Smithsonian: Aeronautics and Space History Smithsonian aeronautics, space history, aircrafts, airplanes, astronomy, and aviation links. http://www.si.edu/resource/faq/nmah/nasm.htm

EO 12856 Facilities by Department: National Aeronautics and Space Administration (NASA) EO 12856 Facilities by Department: National Aeronautics and Space Administration

(NASA) Ames Research Center Moffett Field, CA 94035 George C. Marshall Space Flight Center Madison County Marshall Space Flight Center, AL 35812 John C. Stennis http://es.epa.gov/program/exec/nasafac.html

Goddard Space Flight Center NASA's main center studying the earth as a global environmental system. With news and mission overviews. http://www.gsfc.nasa.gov/

http://ccf.arc.nasa.gov/jf/jff/personnel/barskey.html X.500 Directory Service move to Ames Research Center National Aeronautics and Space Administration US The World Claire E Barskey, Ames Research Center, National Aeronautics and Space Administration, US Street Address Building: 19, Room: 1027 Tele http://ccf.arc.nasa.gov/jf/jff/personnel/barskey.html

http://x500web.arc.nasa.gov/ X.500 Directory Service move to National Aeronautics and Space Administration US The World Ames Research Center, National Aeronautics and Space Administration, US http://x500web.arc.nasa.gov/

Managing for Results: The National Aeronautics and Space Administration Managing for Results: The National Aeronautics and Space Administration. Federal agency Inspectors General and the GAO highlight the top ten management programs faced by NASA. http://www.freedom.gov/results/ig/nasa.asp

Memorandums of Understanding - List of Agencies/National Aeronautics and Space Administration http://www.usgs.gov/mou/nasa.html

Mundt Inventory - National Aeronautics & Space Administration Karl E. Mundt Historical & Educational Foundation and Archives Madison, South Dakota 57042 Return to Contents RG II Independent Agencies - National Aeronautics & Space Administration 1962-1972 1 Hollinger document box (#275) This series relates to http://www.departments.dsu.edu/library/archive/naerspa.html

Mundt Inventory - National Aeronautics & Space Administration Karl E. Mundt Historical & Educational Foundation and Archives Madison, South Dakota 57042 Return to Contents RG II Independent Agencies - National Aeronautics & Space Administration 1962-1972 1 Hollinger document box (#275) This series http://www.dsu.edu/departments/library/archive/naerspa.html

NACA Technical Report Server National Advisory Committee for Aeronautics Chartered in 1915, operational from 1917-1958. The National Aeronautics and Space Act of 1958 created NASA from NACA. Keyword Searching Abstract Browsing (AND, OR, and NOT available) 1917 1918 1919 1920 http://naca.larc.nasa.gov/

NASA (National Aeronautics & Space Administration) NASA (National Aeronautics & Space Administration) NASA Ames Research Center Dryden Flight Research Center Glenn Research Center at Lewis Field Goddard Institute for Space Studies Goddard Space Flight Center Independent Validation & Verific http://www.ssec.wisc.edu/library/NASA.htm

NASA Education Programs NASA's Implementation Plan Why is NASA involved in education? A guide to NASA's education programs NASA Education Information and

Services NASA Education Contacts Educator Resource Center Network NASA Strategic Enterprises NASA State Directory NASA http://www.education.nasa.gov/

NASA Headquarters Contains HQ-specific information, links to information servers at HQ, as well as some external links relevant to HQ and NASA. http://www.hq.nasa.gov/

NASA Home Page June 27, 2001 Text Only Version NASA's Vision (Flash movie) "NASA is deeply committed to spreading the unique knowledge that flows from its aeronautics and space research...." Read NASA Administrator Daniel S. Goldin's welcome letter, bio and speec http://hypatia.gsfc.nasa.gov/NASA_homepage.html

NASA Human Space Flight Read space news, real-time data and space data related to National Aeronautics and Space Administration programs. Includes an image gallery. http://spaceflight.nasa.gov/

NASA Space Grant Learn about the history and strategic plans of the National Space Grant College and Fellowship Program. http://calspace.ucsd.edu/spacegrant

NASA Space Science News Features news and images of space phenomena, lesson plans and profiles of the National Aeronautics and Space Administration's research efforts. http://www.science.nasa.gov/

NASA Spacelink - An Aeronautics & Space Resource for Education An Aeronautics and Space Resource for Education. NASA Spacelink is a service of the Education Division of the National Aeronautics and Space Administration (NASA). http://spacelink.msfc.nasa.gov/.index.html

NASA Spacelink Current NASA programs and educational materials offers articles on everything from living in space to the big bang theory. http://spacelink.nasa.gov/

NASA Administration, education, and technology development are among its many resources, plus to find out how to become an astronaut. http://www.nasa.gov/

National Aeronautics & Space Admin (NASA) DIVISION TYPE DEADLINE ALL GRANTS/CONTRACTS OPEN NATIONAL AERONAUTICS AND SPACE ADMINISTRATION (NASA) NASA Headquarters Chief Resources, Management http://www.umanitoba.ca/academic_support/research_admin/ors/funding/pages/index_mp/rz678a.htm

National Aeronautics and Space Act of 1958 (As Amended) National Aeronautics and Space Act of 1958 (As Amended) [Editorial headnote: "National Aeronautics and Space Act of 1958," Public Law #85-568, 72 Stat., 426. Signed by the President on July 29, 1958, Record Group 255, National Archives and Record http://www1.umn.edu/scitech/nasa1958.htm

National Aeronautics and Space Act of 1958 (As Amended) National Aeronautics and Space Administration National Aeronautics and Space Act of 1958 (As Amended) [Editorial headnote: "National Aeronautics and Space Act of 1958," Public Law #85-568, 72 Stat., 426. Signed by the President on July 29, 1958,

http://www.42cs.au.af.mil/au/awc/awcgate/amendact.htm

National Aeronautics and Space Administration - encyclopedia article from Britannica.com National Aeronautics and Space Administration - (nasa), independent U.S. governmental agency established in 1958 for the research and development of vehicles and activities for the exploration of space within and outside of Earth's atmosphere.
http://www.britannica.com/seo/n/national-aeronautics-and-space-administration/

National Aeronautics and Space Administration (menu = nasa) National Aeronautics and Space Administration (menu = nasa) Mars Pathfinder / Sojourner Mission Other Information NASA Information Services (WWW) NASA Earth Observatory (WWW) Events from the last shuttle mission (WWW) Marshall Space Center (WWW)
http://www.seflin.org/astron/nasa.mnu.html

National Aeronautics and Space Administration (NASA) form National Aeronautics and Space Administration (NASA) form Download nasabud.wk4 NASABudget Page--Lotus 1-2-3 for Windows-Release 5.0 Return to AGENCY FORMS page Return to DSR home page Return to Vanderbilt University home page
http://www.vanderbilt.edu/SponsoredResearch/nasa.html

National Aeronautics and Space Administration (NASA) privatized.... Political satire: (Sep 15, 2001) New Republican President sells U.S. Space Agency (NASA) to Japanese manga publishers, as step to reduce Federal Government size.
http://www.users.cloud9.com/~bradmcc/nasa.html

NATIONAL AERONAUTICS AND SPACE ADMINISTRATION (NASA) SUPPORT NATIONAL AERONAUTICS AND SPACE ADMINISTRATION (NASA) SUPPORT Revision: February 1999 CAPT Jack McCorkle (703) 428-1585 What this chapter is about Process Description (In Detail) Process Outputs Process
http://www.dcma.mil/onebook/4.0/4.4/4.4.3/NASA.htm

National Aeronautics and Space Administration (NASA) Achieving the Goals--Goal 5 - First in the World in Math and Science Technology Resources - November 1996 National Aeronautics and Space Administration (NASA) World Wide Web: www.nasa.gov/ Links to NASA news, subjects of public interest, NASA...http://www.ed.gov/pubs/AchGoal5/nasa-net.html

National Aeronautics and Space Administration (NASA) At Kennedy Space Center, Cape Canaveral, ... Few external marks remain of the ... NASA Homepage Technicians at the Deep Space Network ... Mission control specialists from the National ... Cape Canaveral Jet Propulsion Laboratory Alan Shepard
http://www.comptons.com/encyclopedia/ARTICLES/0125/01293568_A.html

National Aeronautics and Space Administration (NASA) ENC Directory of Federal Programs, Guidebook to Excellence http://watt.enc.org/gbook/4013_20.htm

National Aeronautics and Space Administration (NASA) Information and Extensive Data Available From CPI Electronic Publishing http://www.citation.com/hpage/nasa2.html

National Aeronautics and Space Administration (NASA) National Aeronautics and Space Administration (NASA) Improved design and simulation of aerospace vehicles and increased ability of scientists to model the Earth's climate and to forecast global environmental trends are NASA's strategic goals withi http://www.ccic.gov/pubs/blue94/section.4.4.html

National Aeronautics and Space Administration Information Lehman College, Grants Office, CUNY, Faculty Funding Information, faculty research, Bronx, New York. Features Grants News, Guidelines for the use of Human Subjects, Application Forms, NIH Submission Schedule, RF/CUNY Information, Links http://www.lehman.cuny.edu/grants/nasainf.html

National Aeronautics and Space Administration Records Schedules National Archives and Records Administration National Aeronautics and Space Administration The National Aeronautics and Space Administration has placed its Records Retention Schedule on the World Wide Web at http://www.sti.nasa.gov/nasarrs/. http://ardor.nara.gov/nasa

National Aeronautics and Space Administration .... Last updated March 1999 New Data >> << Merit Systems Protection Board Executive Branch?Independent Executive Agencies National Archives and Records Administration >> National Aeronautics and Space Administration (NASA) 300 E Street SW...http://www.polisci.com/exec/agency/01440.htm

National Aeronautics and Space Administration > Help Feedback Search for: Lycos Network: Tripod.com | Angelfire.com | MailCity.com | WhoWhere.com | HotBot.com All Sites... Lycos Home Computers Web Guide Science News Shop for Computer Software Lycos Download Library Scientists Weather... http://infoplease.lycos.com/ce5/ce036411.html

National Aeronautics and Space Administration > Home > National Aeronautics and Space Administration National Aeronautics and Space Administration (NASA), civilian agency of the U.S. federal government with the mission of conducting research and developing operational programs in... http://lycos.factmonster.com/ce5/ce036411.html

NATIONAL AERONAUTICS AND SPACE ADMINISTRATION Created for the World-Wide Web: November 20, 1996 BRANCH: EXECUTIVE NATIONAL AERONAUTICS AND SPACE ADMINISTRATION Location Position Title Name of Incumbent Pay Plan Type of Appt. Level Grade or Pay Tenure Expires OFFICE OF SPACE FLIGHT.... http://www.access.gpo.gov/plumbook/ag202.html

National Aeronautics and Space Administration FedStats provides easy access to statistics and information produced by more than 100 U.S. Federal Government agencies http://www.fedstats.gov/key_stats/NASAkey.html

National Aeronautics and Space Administration National Aeronautics and Space Administration news http://www.newsterrier.com/nasa.shtml

National Aeronautics and Space Administration National Aeronautics and Space Administration 300 E. St. SW Washington, DC 20546 Web: http://www.nasa.gov Phone: 202-358-0000 Fax: 202-358-3047 From the people who developed technology used in everything... http://capsules.hoovers.powerize-asap.com/info/com.powerize_co_capsule_4_0_2163_54834_00_html.html..

National Aeronautics and Space Administration National Aeronautics and Space Administration When the United States Congress created the National Aeronautics and Space Administration in 1958, it directed NASA to do three things: Conduct aeronautical and space "activities". http://www.opengroup.org/public/member/mlists/nasa.htm

National Aeronautics and Space Administration National Aeronautics and Space Administration If Uncle Sam charged admission to NASA's website, the U.S. government might be out of debt soon. Full justice cannot be done to NASA in these pages--or via pre-21st-century web technology, even with a 28 http://www.uncle-sam.com/nasa.html

National Aeronautics and Space Administration National Aeronautics and Space Administration Fiscal Year 2001 Estimates (IN Millions of Real Year Dollars) The FY 2001 multi-year budget estimate is submitted in accordance with the NASA FY 1989 Authorization Law (P.L. 100-685). 1999 OPLAN 1 http://www.ifmp.nasa.gov/codeb/budget2001/HTML/fy01_myb.htm

National Aeronautics and Space Administration National Aeronautics and Space Administration Mission and Life Science Programs NASA's life science research programs explore the role of gravity in fundamental biological processes. These programs focus on: gravitational biology, including molecular... http://www.faseb.org/opar/consensus98/nasa.htm

National Aeronautics and Space Administration National Aeronautics and Space Administration Areas Of Global Change Research The National Aeronautics and Space Administration (NASA) is responsible for Earth science research from space, including those studies of broad scientific scope that study http://www.gcrio.org/USGCRP/nasa.html

National Aeronautics and Space Administration National Aeronautics and Space Administration NASA brings to environment and natural resources research the ability to view the Earth in its entirety from space. This unique position has led NASA to focus on the study of the Earth as an integrated system http://www.nnic.noaa.gov/CENR/nasanew.html

National Aeronautics and Space Administration National Aeronautics and Space Administration (All Installations) Unsolicited Proposals NASA's OSS Proposal Processing Site (Electronic Submission) NASA OSS Research Opportunities - On-Line Earth Science Research Announcements Office of Life... http://www.virginia.edu/~opra/nasa/nasa.html

NATIONAL AERONAUTICS AND SPACE ADMINISTRATION NATIONAL AERONAUTICS AND SPACE ADMINISTRATION MARSHALL SPACE FLIGHT CENTER SHUTTLE PROJECT OFFICE SSME Quality- Andrew Braxton 688-1422 CR55 4995 S&E Engineering- Larry Pigott 688-2841 EE23 4995 Ronnie Rigney 688-2796 EE23 4995 Updated 12/00 http://www.ssc.nasa.gov/~telecom/phone/nasamsfc.htm

National Aeronautics and Space Administration National Aeronautics and Space Administration "Our greatest strength is our workforce. In all we do, we operate according to a set of values that form the bedrock of our efforts. As individuals and as a team, we strive to uphold these values... http://www.employeesurvey.gov/2000-NASA.asp

National Aeronautics and Space Administration, an Encarta Encyclopedia Article Titled "National Aeronautics and Space Ad National Aeronautics and Space Administration (NASA), agency of the United States government responsible for developing space exploration and research initiatives, as well as coordinating various communications-related projects. NASA also...
http://encarta.msn.com/index/conciseindex/45/04502000.htm?z=1&amp%3Bpg=2&amp%3Bbr=1

National Aeronautics and Space Administration: Oral Histories and Publications U.S. NATIONAL AERONAUTICS AND SPACE ADMINISTRATION. HISTORICAL OFFICE: Oral histories and publications, 1972-93 (less than one foot) Copies of material concerning the early Space Shuttle program, including NASA publications, near-print items... http://www.ford.utexas.edu/library/findaid/nasa0.htm

NetworkForChange NASA advances science and technology in space.
http://envirolink.netforchange.com/frame.html?page=content.html%3Fitemid%3D199%26thissite%3Denviro..

NetworkForChange NASA advances science and technology in space.
http://library.envirolink.org/frame.html?page=content.html%3Fitemid%3D199%26thissite%3Dlibrary%26c..

NGA - Eyewitness to Space, Paintings and Drawings Done for the National Aeronautics and Space Administration (03/1965) Eyewitness to Space, Paintings and Drawings Done for the National Aeronautics and Space Administration 14 March-23 May 1965 Overview: 70 paintings and drawings created by 14 American artists for the NASA Art Program were on view. http://www.nga.gov/past/data/exh259.htm

NSSDC News National Space Science Data Center News is published quarterly by the National Aeronautics and Space Administration. Read full-text articles.
http://nssdc.gsfc.nasa.gov/nssdc_news/toc.html

Records of the National Aeronautics and Space Administration Guide to Federal Records in the National Archives of the United States http://www.nara.gov/guide/rg255.html

Resources for Middle School Science--11.67-- National Aeronautics and Space Administration, NASA Marshall Space Flight C Previous Entry Next Entry 11.67 National Aeronautics and Space Administration, NASA Marshall Space Flight Center, Mail Code CL-01, Huntsville, AL 35812-0001 (205) 961-1225 http://spacelink.nasa.gov One of 9 NASA field centers serving multistate are
http://www.nap.edu/readingroom/books/rtmss/11.67.html

Sevilleta LTER: Collaboration: National Aeronautics and Space Administration Collaboration: National Aeronautics and Space Administration Participations with NASA: Later Modlers Project Sun Photometer ADAR Multispectral Imagery Note: Web sites external to the Sevilleta LTER will appear in another browser window.
http://sevilleta.unm.edu/collaboration/nasa

Subcommittee on Space and Aeronautics Subcommittee on Space and Aeronautics U.S. House of Representatives Committee on Science B-374 Rayburn House Office Building Washington, DC 20515 (202) 225-7858 (202) 225-6415 FAX NASA FY2001 Budget Request for Human Spaceflight 2318 Rayburn...http://www.house.gov/science/space_witness_031600.htm

The National Aeronautics and Space Administration: An Overview With FY1997 and FY1998 Budget Summaries "To improve the scientific basis for making decisions on environmental issues." The National Aeronautics and Space Administration: An Overview With FY1997 and FY1998 Budget Summaries Full document: http://www.cnie.org/nle/st-15.htmlJuly 1, 1999...http://www.cnie.org/CRSsearch/st15sum.htm

The National Aeronautics and Space AdministrationPresents to The National Aeronautics and Space Administration Presents to Tom Rogers the NASA Distinguished Public Service Medal "For sustained and exemplary leadership in advancing the commercialization of space, encouraging the development of reusable space...http://www.spacetransportation.org/nasaaward.htm

AUTHOR INDEX

A

Akens, David S., 129
Aldrin, Buzz. Collins, Michael, 137
Alexander, Kent, 129
Allaway, Howard, 129
Allen, William Hubert, 151
American Telephone and Telegraph Commpany., 214
Ames Research Center., 195
Anderson, Frank Walter, 131
Anderson, Frank Walter. Orders of magnitude., 141
Anderson, John David, 131
Anderton, David A., 132
Anna, Henry John, 132
Apollo 11 (Spacecraft), 133
Armstrong, Neil, 137
Aroesty, Jerome, 137
Ashmore, Teresa C., 225
Association for the Study of Afro-American Life and History., 142
Atkinson, Joseph D., 138
Atwill, William D., 138

B

Baker, David, 138
Baker, Wendy, 139
Bates, James R., 139
Batson, Raymond M, 139
Batson, Raymond M., 139
Belew, Leland F., 140, 204
Bell Telephone Laboratories, inc., 210
Belton, M. J. S., 149, 168
Benedict, Howard, 140
Benson, Charles D., 140, 146
Berg, Robert Alvin, 149
Bergreen, Laurence, 141
Bilstein, Roger E., 141
Black, David C., 196
Booth, Robert E. (Robert Edmond), 130
Brand, D. O., 142
Bridges, P. M., joint author. Inge, Jay L., 139
Bromberg, Joan Lisa., 142
Brooke, Charles W., 143
Brooks, Courtney G., 143
Bruno, Leonard C., joint author. Rosholt, Robert L., 222
Burgess, Eric., 143, 152
Burrough, Bryan, 143
Burrows, William E., 143
Butrica, Andrew J., 141, 144
Byrnes, Mark E., 144

C

California. University, Santa Cruz. Manned Spacecraft Center (U.S.), 127
Carle, Glenn C., 1936- Schwartz, Deborah E. Huntington, Judith L., 155
Carpenter, M. Scott (Malcolm Scott), 199
Carroll, Roger D., 222
Carroll, William F., 144
Chambers, Joseph R., 144
Chandler, Margaret K., 200
Chapman, Richard L., 145
Chappell, Russell E., 145
Clark, Lenwood G., 217
Clemmons, Dewey L., 145
Coler, Myron A. (Myron Abraham), 162
Colin, Lawrence. Burgess, Eric., 157
Collins, Stewart A., 145
Compton, William David, 146
Cook, Howard., 225
Cooke, Hereward Lester. Dean, James D., 155
Corliss, William R., 146, 147, 148
Cortright, Edgar M., 135
Cote, Charles (Charles E.) Taylor, Ralph E., 1923- Gilbert, Eugene (Eugene L.), 190
Coyle, G. G., 148
Coyle, G. G., joint author. Von Delden, H., 144
Cozic, Charles P., 206
Crabill, Norman L., 179
Crouch, Tom D., 148
Cunningham, Walter, 149

D

Daniels, Lynne L., 149
Davies, Merton E., 149, 150
Dawson, Virginia P. (Virginia Parker), 150
Dayhoff, Margaret O., 219
Deerwester, Jerry M., 150
Dickson, Katherine Murphy., 151
Doctors, Samuel I., 151
Dole, Stephen H., 150
Donn, Bertram., 151

Dugan, Daniel C. Giulianetti, Demo J., 176
Dunar, Andrew J., 151
Dunne, James A., 152

E

Eddy, John A., 152
El-Baz, Farouk. Warner, Delia M., 136
Ertel, Ivan D., 189, 214
Ethell, Jeffrey L., 154
Etzioni, Amitai., 154
Evans, David Donald, 154
Ezell, Edward Clinton., 155
Ezell, Linda Neuman, joint author., 156
Ezell, Linda Neuman., 155

F

Faherty, William Barnaby, 140, 141
Fallaci, Oriana., 156
Faller, Kenneth H., 156
Fimmel, Richard O., 156, 157
Fisher, Allan C., 158
Fleming, William A., 160
Flynn, Barbara J., 197
Ford, J. P. Jet Propulsion Laboratory (U.S.), 208
Ford, John P., 164
French, Bevan M., 146, 158
Frierson, J. Q., 158
Fries, Sylvia Doughty., 158
Froehlich, Walter, 159

G

Garba, J. A., 209
Gawdiak, Ihor, 160, 161
GCA Corporation. United States. National Aeronautics and Space Administration., 213
George C. Marshall Space Flight Center., 201
Glennan, Thomas Keith, 161
Goldstein, Richard M., 161
Gloersen, Per., 136
Golovin, Nicholas E. (Nicholas Erasmus), 162

Goddard Space Flight Center, 161, 196
Goddard Space Flight Center. Lyndon B. Johnson Space Center., 161
Gorn, Michael H., 165
Green, Constance McLaughlin, 163
Grimwood, James M., 143, 163, 164, 213
Grimwood, James M., joint author. Alexander, Charles C., 213
Grimwood, James M., joint author. Swenson, Loyd S., 143
Groudle, T. A., joint author. Mattson, R. F., 154
Groudle, Thomas A., joint author. Kelley, James Howard, 200
Gulkis, S. (Samuel) Stetson, D. S. Stofan, Ellen Renee, 181
Gump, David., 164
Gurman, Joseph Bearak., 187

H

Hacker, Barton C., 164
Hacker, Barton C., 1935- joint author. Vorzimmer, Peter J., 163
Hakkila, Jon Eric, 164
Hall, Harvey, 164
Hall, R. Cargill., 164
Hallion, Richard., 165
Hallman, Theodore M., ed. Lubarsky, Bernard, 185
Hanover, Paul N., 165
Hansen, James R., 165
Harris, Gordon L., 165
Hartman, H. Lawless, J. G. (James G.) Morrison, Philip., 201
Hartmann, William K., 166
Harvey, Edmund H., joint author., 170
Hawkins, Gerald S., ed. 180
Henry, James Paget, 1914- ed. Mosely, John D., 198
Heppenheimer, T. A., 166
Herring, Harvey W., 189
Herring, Mack R., 166
Herskowitz, Mickey, 149
Hillje, E. R., 227

Hirsch, Richard, 167
Hitchcock, Barbara. Riva, Peter,, 202
Hoban, Francis T., 167
Holme, Molly., 167
Holmes, Dyer Brainerd., 177
Hoyt, Edwin Palmer., 168
Hunley, J. D., 161, 172, 221
Hunten, Donald M., 215

I

Ise, Rein. George C., 153

J

Janson, Bette R., 169
Jenkins, Dennis R., 169
Jet Propulsion Laboratory (U.S.), 145, 158, 160, 176, 197, 219, 224, 225, 227, 228
Jet Propulsion Laboratory (U.S.) United States National Aeronautics and Space Administration., 228
Jet Propulsion Laboratory (U.S.) United States. National Aeronautics and Space Administration., 225, 227
John F. Kennedy Space Center., 130
Johnston, Richard S. Dietlein, Lawrence F., 142
Johnston, Richard S. Dietlein, Lawrence F. Berry, Charles A. (Charles Alden), 142
Jones, Robert Lewis, 169
Jordan, Raymond, joint author. Larson, Kathleen B., 139

K

Kalamaras, Mary., 143
Kaplan, Judith., 170
Katayama, Frank Y., 1924- Roth, James A., 150
Kay, W. D., 170
Kennan, Erlend A., 170
Kerrigan, Evans E., 170
Kerrod, Robin., 170
Kibler, James F., 170
King, Michael D., 154

Kloman, Erasmus H., 171
Koskela, P. E. Cutts, James A., 178
Kraemer, Robert S., 171
Kraft, Christopher C., 171
Kranz, Gene., 171, 172

L

Lambright, W. Henry, 172
Langley Research Center., 175, 176
Larson, Wiley J. Pranke, Linda K., 168
Lauderdale, W. W. Kernaghan, Harold., 139
Launius, Roger D., 172, 173
Lawbaugh, William M. Hoffman, Edward J. (Edward Jay), 167
Levine, Arnold S., 173
Levine, Arthur L., 173
Levine, Sol, 173, 174
Lewis Research Center., 228
Lewis, D. W., 220
Lewis, Richard S., 174
Logsdon, John M., 1937- Tindall, Howard W., 177
Lomask, Milton, joint author., 163
Lord, Douglas R., 175
Ludwig, George H., 175
Lundquist, Charles A., 205
Lyndon B. Johnson Space Center., 204

M

Macauley, Molly K. Sponberg, Brant, 191
Machell, Reginald M., 211
Mack, Pamela Etter., 160
Maisel, Martin D., 176
Malik, P. W., 177
Mallove, Eugene F., 128
Malone, Myrtle Davidson., 140
Masursky, Harold., 179
Masursky, Harold, 1922- Colton, George Willis, 1920- El-Baz, Farouk. Doyle, Frederick J., 135
Maurer, Alvin, joint author. Mora, Raul Mina, 212

Maurer, Alvin. Weigle, Oscar. Mora, Raul Mina, ill. Hoss, Norman, 212
McCurdy, Howard E., 179
McRoberts, Joseph J., 180
Mesnard, Donald C., joint author., 196
Mirabito, Michael M., 180
Morgan, Clay, 181
Morgan, Norman Evan, 181
Miró, Ramón, 1968- Stueland, Sam, 160
Moore, H. J. (Henry J.), 193
Mora, Raul Mina, illus., 213
Morath, W. D., 181
Morrison, David, 181, 182
Morrison, Philip. Billingham, John. Wolfe, John, 218
Moser, Diane, 208
Mudgway, Douglas J., 182
Muenger, Elizabeth A., 182
Muniz, Robert., 170
Mutch, Thomas A., 217

N

National Aeronautics and Space Administration., 133, 169
National space program., 154
Naugle, John E., 188
Naumann, Robert J., 189
Neal, Valerie., 189
Neugebauer, Marcia., 206
Newkirk, Roland W., 189
Nicks, Oran W., 190
Nicogossian, Arnauld E., 214
Norman, Susan M., 151
North Carolina State University. Dept. of Mechanical and Aerospace Engineering. Langley Research Center., 205

O

Oakley, Eric Gilbert, 190
Orloff, Richard W., 190

P

Pace, George., 191

Pace, Scott., 191, 215
Page, Lou Williams., 191
Page, Thornton,, 192
Paine, Thomas O., 192
Painter, Robert B., joint author. Johnson, Richard D., 228
Parkinson, Claire L., 137
Pearcy, Arthur., 193
Piccard, Don, 1926- Papers. Piccard, Jean Felix, 1884-1963. Papers. Piccard, Jeannette, 193
Poole, Gray Johnson, 195
Poole, Lynn., 195
Portree, David S. F., 195

Q

Quimby, Freeman Henry, 196

R

Ragsdale, George C., 196
Rahe, Jürgen., 151
Ramaty, R., ed. Stone, Robert G. (Robert Gilbert), 166
Raper, Odell, 166
Renetzky, Alvin, 183
Renetzky, Alvin., 197
Richardson, Adele, 164
Ride, Sally., 199
Ritchie, Eleanor H., 169
Roland, Alex, 199
Rosen, Alan, 155
Rosenthal, Alfred., 200
Rosholt, Robert L., 200
Roth, Ladislav E. Wall, Stephen D., 216
Rumerman, Judy A. Garber, Stephen J., 145

S

Samonski, Frank H., 200
Samz, Jane., 181
Saunders, Joseph F., 216
Sayles, Leonard R., 200
Schilling, G. F., 222
Schirra, Wally., 199
Schmitz, Bruce W., 200
Schultz, James., 201
Seamans, Robert C., 201
Sekanina, Zdenek., 215

Shafritz, Jay M., 138
Sharpless, Jack., 202
Shetler, Charles, 161
Short, Nicholas M., 180
Short, Nicholas M., 202
Siddiqi, Asif A., 202
Simmons, Gene, 203
Sjogren, W. L., 203
Smetana, Frederick O., 205
Smith, Howard Everett, 205
Smull, T L K, 205
Souris, G. A., 177
Souza, Kenneth A. Hogan, Robert (Robert P.) Ballard, Rodney, 174
Spangenburg, Ray, 208
Spehalski, Richard J., 208
Sperling, F., 209
Spilker, Linda J., 193
Stein, John P., 209
Steinberg, Florence S., 210
Stillwell, Wendell H, 210
Strasser, Bruce E., 210
Stuart, Locke M., 202
Stuhlinger, Ernst, 140
Sturms, F. M., 210
Summerlin, Lee R., 204
Sutton, Felix., 212, 213
Swanson, Glen E., 140
Swenson, Loyd S., 213
Swindell, William, joint author. Burgess, Eric, 157

T

Tang, Wen, 213
Taylor, Robert, 213
Thomas, Richard Nelson, 219
Thompson, Milton O., 220
Thornton, T. H., 220
Thostesen, T. O., 220
Tomayko, J. E. (James E.), 220
Torr, Douglas, 143
Trask, N. J. (Newell Jefferson), 221
Trento, Joseph John, 167
Trento, Joseph John, 221
Troika Multimedia, Inc., 187
Tucker, Wallace H., 221
Tyner, Richard L., 222

U

Ulrich, Bertram, 173
United States. Aeronautics and Space Administration, 197
United States. Air Force Ballistic Missile Division, 127
United States. Congress. House. Committee on Science and Astronautics. Subcommittee on NASA Oversight, 218
United States. Congress. House. Committee on Science and Technology, 188
United States. Congress. Senate. Committee on Aeronautical and Space Sciences, 153, 187, 188, 206
United States. Congress. Senate. Committee on Commerce, Science, and Transportation, 207
United States. Congress. Senate. Committee on Governmental Affairs, 169
United States. National Aeronautics and Space Administration Officials and employees, 156
United States. National Aeronautics and Space Administration, 128, 129, 132, 134, 136, 137, 138, 142, 143, 145, 146, 148, 151, 152, 153, 156, 158, 159, 161, 165, 167, 168, 172, 175, 178, 179, 180, 182, 183, 186, 187, 189, 190, 191, 197, 199, 201, 202, 203, 204, 207, 208, 210, 216, 219, 220, 221, 222, 226, 228
United States. National Aeronautics and Space Administration. Educational Programs Division, 129, 204
United States. National Aeronautics and Space Administration. George C. Marshall Space Flight Center, 208
United States. National Aeronautics and Space Administration. History Office, 167, 226
United States. National Aeronautics and Space Administration. International Affairs Division, 222
United States. National Aeronautics and Space Administration. Jet Propulsion Laboratory (U.S.), 219
United States. National Aeronautics and Space Administration. NASA Historical Staff, 151
United States. National Aeronautics and Space Administration. Office of Public Affairs, 132, 145, 159
United States. National Aeronautics and Space Administration. Public Affairs Division, 180, 210
United States. National Aeronautics and Space Administration. Scientific and Technical Information Branch, 182
United States. National Aeronautics and Space Administration. Scientific and Technical Information Division, 137, 138, 146, 172, 186
United States. National Aeronautics and Space Administration. Scientific and Technical Information Office, 179
United States. National Aeronautics and Space Administration. Scientific and Technical Information Program, 152, 158
United States. National Aeronautics and Space Administration. Space Station Task Force, 207
United States. National Aeronautics and Space Administration., 128, 129, 132, 134, 136, 137, 138, 142, 143, 145, 146, 148, 151, 152, 153, 158, 159, 161, 165, 167, 168, 172, 175, 178, 179, 180, 182, 183, 186, 187, 189, 190, 191,

197, 199, 201, 202, 203, 204, 207, 208, 210, 216, 219, 220, 221, 222, 226, 228
United States. Office of Manned Space Flight, 135, 153
United States. Office of Space Flight, 206
United States. Office of Space Science and Applications, 218
United States. Presidential Commission on the Space Shuttle Challenger Accident, 198

V

Van Nimmen, Jane, 1937- Bruno, Leonard C. Ezell, Linda Neuman, 183
Van Nimmen, Jane, 222
Vaughan, Diane., 222
Vogt, Gregory., 223
Von Braun, Wernher, 223

W

Wagner, Richard., 225
Wall, Stephen D., 225
Wallace, Harold D., 225
Wallace, Lane E., 226
Waltman, Gene L., 226
Waring, Stephen P, 152
Warner, Jeffrey, 226
Warwick, James W., 227
Weaver, William L, 180
Wells, Helen T., 227
Wheelock, Harold J., comp, 178
Whiteley, Susan H., joint author. Karegeannes, Carrie E., joint author, 227
Whitnah, A. M., 227
Wiksten, David B., 227
Wilcove, Raymond. United States. Congress. House. Committee on Science and Astronautics, 173
Wilhelms, Don E., 228
Wilson, James H., 228
Wolf, Charles, 1924- joint author, 210
Wolfinger, Jarritus. United States. National Aeronautics and Space Council, 195

X

Yeates, C. M., 160
Young, Carolynn, 217
Young, Richard S. (Richard Stuart), 228

Z

Zimmerman, R. Logan,, 137

TITLE INDEX

#

1958 NASA/USAF space probes (Able-1), 127
1967 Summer Study of Lunar Science and Exploration, 127

A

A bibliography of adult aerospace books and materials, 127
A bibliography of aerospace books and teaching aids for elementary school pupils and teachers., 127
A Bibliography on the search for extraterrestrial intelligence, 128
A catalog of NASA special publications., 128
A catalog of selected Viking Orbiter images, 222
A new sun
 the solar results from Skylab, 152
A preliminary control net of Mars, 149
A review of the Mariner IV results, 190
A study of NACA and NASA published information of pertinence in the design of light aircraft, 205
A treatise on the Surveyor lunar landing dynamics and an evaluation of pertinent telemetry data returned by Surveyor I, 209
A twenty-fifth anniversary album of NASA, 223
A walk in space, 128
Aboard the space shuttle, 210
Aeronautical and astronautical events., 128
Aeronautics, 128, 132
Aeronautics and astronautics. Astronautical and aeronautical events, 138
Aeronautics and space bibliography for elementary grades, 127
Aeronautics and space bibliography of adult aerospace books and materials, 128
Aeronautics in NACA and NASA, 128
Aerospace bibliography, 129

Aiming at targets
 the autobiography of Robert C. Seamans, Jr., 201
Airborne trailblazer, 226
A.L.S.E.P. termination report, 139
ALSEP termination report, 139
Amending the National aeronautics and space act of 1958 with respect to property rights in inventions;, 129
Amending the National aeronautics and space act of 1958., 130
Amending various sections of the NASA act of 1958., 130
Amendment to the National aeronautics and space act of 1958., 130
America in space, 130, 139
America in space the first five years, 130
America's spaceport
 John F. Kennedy Space Center, 130
An administrative history of NASA,, 200
An analysis of the extraterrestrial life detection problem, 228
An annotated bibliography of the Apollo program, 172
An Earth-Mars mission-analysis program, 170
An index to NASA tech briefs (briefs 63-10003 through 64-10211) a keyword-in-context index (KWIC), 130
Analysis of Apollo 8, 131
Analysis of Apollo 10, 131
Analysis of Surveyor 3 material and photographs returned by Apollo 12., 131
Apollo, 132, 133, 134, 135, 136, 145, 159, 190, 191, 226
Apollo 11
 preliminary science report., 133
Apollo 11 mission report., 133
Apollo 11--man on the moon, 133
Apollo 12 lunar-sample information., 226
Apollo 12 preliminary science report., 133
Apollo 13

Apollo 14
   preliminary science report., 134
   science at Fra Mauro., 159
Apollo 15
   preliminary science report., 134
Apollo 15 at Hadley Base., 134
Apollo 16
   preliminary science report., 134
Apollo 16 at Descartes, 159
Apollo 16: expedition to Descartes., 134
Apollo 17
   preliminary science report, 134
Apollo 17 at Taurus Littrow, 132
Apollo 8
   man around the moon., 135
Apollo by the numbers
   a statistical reference, 190
Apollo earth landmark maps., 135
Apollo expeditions to the moon, 135
Apollo program summary report., 135
Apollo over the moon: a view from orbit, 135
Apollo program, 153
Apollo Soyuz, 135, 159
Apollo Soyuz Test Project
   first international manned space flight, July 15-24, 1975., 135
Apollo-Soyuz Test Project preliminary science report., 136
Apollo terminology. Prepared for Office of Manned Space Flight., 135
Apollo translunar/transearth trajectory plotting chart, 136
Apollo-Soyuz pamphlet[s], 191
Apollo-Soyuz Test Project
   summary science report, 136
Appointment on the moon, 174
Arctic and Antarctic sea ice, 1978-1987: satellite passive-microwave observations and analysis, 136
Arctic sea ice, 1973-1976
   satellite passive-microwave observations, 137
Ariel I
   the first international satellite, 137
Astronautics and aeronautics, 137
Astronautics and aeronautics, 1979-1984
   a chronology, 169
Astronautics and aeronautics, 1986-1990
   a chronology, 160
Astronautics and aeronautics, 1991-1995
   a chronology, 161
Astronomy in space, 138
Atlas of Comet Halley 1910 II, 151
Atlas of Mars, the 1
   5,000,000 map series, 139

Atlas of Surveyor 5 television data, 139

B

Before this decade is out--
   personal reflections on the Apollo Program, 139
Beyond the ionosphere
   fifty years of satellite communication, 141
Beyond the moon
   a golden age of planetary exploration, 1971-1978, 171
Biomedical results from Skylab, 142
Biomedical results of Apollo, 142
Black Americans in aerospace., 142
Black magic and gremlins
   analog flight simulations at NASA's Flight Research Center, 226

C

Can democracies fly in space?
   the challenge of revitalizing the U.S. space program, 170
Catalog of NASA's Skylab 2 photography, 203
Challenge to Apollo
   the Soviet Union and the space race, 1945-1974, 202
Chariots for Apollo
   a history of manned lunar spacecraft, 143
Chronology of space shuttle flights, 1981-2000, 145
Compositions of major and minor minerals in five Apollo 12 crystalline rocks, 145
Computers in space
   journeys with NASA, 220
Concepts for detection of extraterrestrial life, 196
Conclusion of Viking Lander Imaging Investigation
   picture catalog of experiment data record, 225
Conquest of the moon., 212
Control net of Mars
   February 1978, 150
Coordinates of features on the Mariner 6 and 7 pictures of Mars, 150
Countdown to glory
   NASA's trials and triumphs in space, 129
Cruise to Jupiter
   contributions of the Galileo Imaging Science Team, 149

D

Daring the unknown
   a history of NASA, 205

Data bases and data base systems, related to NASA's Aerospace Program
   a bibliography with indexes., 149
Data policy issues and barriers to using commercial resources for Mission to Planet Earth, 191
Design, test, and performance of the Mariner V temperature control reference, 144
Designs on space
   blueprints for 21st century space exploration, 225
Development of a hydrogen-oxygen internal combustion engine space power system, 181
Development of an electrochemical energy source for the Mariner II spacecraft, 143
Development of the post-injection propulsion system for the Mariner C spacecraft, 200
Development of the Ranger Block III spacecraft propulsion system, 154
Dictionary of technical terms for aerospace use., 151
Domestic and international implications of the space race, 154
Dragonfly
   NASA and the crisis aboard the MIR, 143
Dreams, hopes, realities
   NASA's Goddard space flight center the first forty years, 226
Dynamic environment of the Ranger Spacecraft I through IX, 227

E

Earth observations and global change decision making
   a special bibliography, 152
Earth photographs from Gemini III, IV, and V., 152
Earth photographs from Gemini VI through XII., 152
Educational horizons, 153
Electrical power management standard., 153
Energy-related research and development, 153
Engineer in charge
   a history of the Langley Aeronautical Laboratory, 1917-1958, 165
Engineering challenges to the long-term operation of the International Space Station, 153
Engines and innovation
   Lewis laboratory and American propulsion technology, 150
EOS science plan
   the state of science in the EOS Program, 154
Establishment of the National space program, 154
Evaluation and comparison of three space suit assemblies, 169
Exobiology in solar system exploratio, 155
Explorer vi and Pioneer v data., 155

Exploring the moon and planets, 146
Exploring the universe with the Hubble Space Telescope, 189

F

Facilities data, 156
Failure is not an option
   mission control from Mercury to Apollo 13 and beyond, 171, 172
Faster, better, cheaper
   low-cost innovation in the U.S. space program, 179
Fire and power
   the American space program as postmodern narrative, 138
First among equals
   the selection of NASA space science experiments, 188
First five years of NASA, 167
Flight
   my life in mission control, 171
Flight research
   problems encountered and what they should teach us, 220
Flights of discovery
   50 years at the NASA Dryden Flight Research Center, 226
Flying the frontiers
   NACA and NASA experimental aircraft, 193
Forecasts and appraisals for management evaluation., 158
From engineering science to big science
   the NACA and NASA Collier Trophy research project winners, 160
Fuel economy in aviation, 154
Future aeronautics and space opportunities., 160
Future National space objectives, 160
Future of planetary astronomy, 190

G

Galileo
   exploration of Jupiter's system, 160
Galileo to Jupiter
   probing the planet and mapping its moons., 160
Gateway to the moon
   building the Kennedy Space Center launch complex, 140
Gemini extravehicular activity, 211

Goddard Space Flight Center/Johnson Space Center LACIE interface control document LANDSAT data/LACIE 2., 161
Governmental organization for space activities, 163
Guide to Magellan image interpretation, 163

H

Heat Capacity Mapping Mission (H.C.M.M.) anthology, 202
High energy phenomena on the sun, 166
Highlights of the year ..., 166
Historical origins of the George C. Marshall Space Flight Center., 129
Historical origins of the National Aeronautics and Space Administration., 167
Historical sketch of NASA., 167
History at NASA, 167
History of aeronautics and astronautics, 151
History of a physicist
 joys and woes, 164
Hubble Space Telescope, 189
Human spaceflight
 mission analysis and design, 168
Human support issues and systems for the space exploration initiative
 results from Project Outreach, 137
Humans to Mars
 fifty years of mission planning, 1950-2000, 195
Hypersonics before the shuttle
 a concise history of the X-15 research airplane, 169

I

If the sun dies, 156
Improved coordinates of features in the vicinity of the Viking 1 lander site on Mars, 150
Improved reporting needed on National Aeronautics and Space Administration projects
 report to the Congress, 168
In orbit at Jupiter
 contributions of the Galileo Imaging Science Team, 168
Index of NACA technical publications, 168
Information security
 many NASA mission-critical systems face serious risks, 169
Injection accuracy characteristics for lunar missions, 220
Inside NASA
 high technology and organizational change in the U.S. space program, 179
Into the thermosphere
 the atmosphere explorers, 143
Issues concerning the future operation of the space transportation system
 study, 169

L

Large space structures & systems in the space station era
 a bibliography with indexes., 172
Leadership and America's future in space
 a report to the Administrator, 199
Life into space
 space life sciences experiments, NASA Ames Research Center, 1965-1990, 174
Linking man and spacecraft, 146
List of selected references on NASA programs, 174
Living and working in space
 a history of Skylab, 146
Lunar impact
 a history of Project Ranger, 164
Lunar Orbiter II, 175
Lunar orbiter II
 photography., 175
Lunar Orbiter III, 175
Lunar Orbiter III, 176
Lunar Orbiter IV, 176
Lunar orbit rendezvous, 175

M

Magellan
 revealing the face of Venus., 176
Maintaining U.S. leadership in aeronautics
 scenario-based strategic planning for NASA's aeronautics enterprise, 176
Man in space, 132, 159
Man in space., 159
Managing large systems
 organizations for the future, 200
Managing NASA in the Apollo era, 173
Managing the moon program
 lessons learned from Project Apollo, 177
Manned space flight, 1963, 177
Mariner IV mechanical operations, 208
Mariner IV science platform structure and actuator design, development and flight performance, 148
Mariner Mars 1964 mechanical configuration, 208
Mariner Mars 1964 Project report

Mission operations, 178
Mariner Mars 1964 temperature control hardware, 144
Mariner Mars 1971 television picture catalog, 178
Mariner Mars absorptivity standard, 220
Mariner mission to Venus, 178
Mariner Venus 67 dynamic flight data, 142
Mariner Venus 67 guidance and control system, 191
Mariner Venus 67 spacecraft environmental test results, 196
Mariner-Venus, 1962, 178
Mariner-Venus, 1967, 178
Mariner-Mars 1969
   a preliminary report, 178
Mars
   the Viking discoveries, 158
Mars Observer mission, 158
Materials processing in space
   early experiments, 189
Meteor orbits and dust, 180
Mission to earth
   Landsat views the world, 180
Mission to the moon, 170
Mission to the solar system
   exploration and discovery
      a mission and technology roadmap, 181
Model research
   the National Advisory Committee for Aeronautics, 1915-1958, 199
Models of Mars atmosphere, 1967., 181
Moon launch!
   a history of the Saturn-Apollo launch operations, 140
Moonport
   a history of Apollo launch facilities and operations, 140

N

NASA, 139, 140, 142, 146, 147, 158, 164, 170, 172, 173, 182, 183, 184, 185, 186, 187, 197, 222
   a history of the U.S. civil space program, 173
   a quarter century of space achievement, 140
   America in space, 139
   visions of space
      capturing the history of NASA, 170
N.A.S.A. space communications program, 185
NASA & the exploration of space
   with works from the NASA art collection, 172
NASA 1958-1983
   remembered images., 182
NASA accident and injury data., 182
NASA and the space industry, 142

NASA astronauts., 183
NASA day, April 27, 1962., 183
NASA engineers and the age of Apollo, 158
NASA factbook
   guide to National Aeronautics and Space Administration programs and activities., 183
NASA historical data book, 1958-1968, 222
NASA historical data book., 183
NASA magazine [microform]., 183
NASA magazine., 184
NASA mishap and injury data., 184
NASA names, 227
NASA patent abstracts bibliography., 184
NASA photography from five years of space., 184
NASA pocket statistics., 184
NASA PPMI lessons learned [computer file], 185
NASA reactor facility hazards summary., 185
NASA reports required by Congress, 185
NASA sounding rockets, 1958-1968, 146
NASA space communications program
   report, 185
NASA space plans and scenarios to 2000 and beyond, 186
NASA spacecraft, 147
NASA space vehicle design criteria
   environment, 181
NASA specifications and standards., 186
NASA tech briefs, 130
NASA thesaurus., 186
NASA, the first 25 years, 1958-1983
   a resource for teachers
      a curriculum project., 187
NASA's Solar Maximum Mission
   a look at a new sun, 187
NASA--the 25th year [computer file]., 187
NASA visitor information facilities, 187
NASA/USAF space probes, 127
National aeronautics and space act of 1958, 130
National aeronautics and space act of 1958 as amended through October 6, 1961 (with footnotes), 187
National aeronautics and space act of 1958, as amended, and related legislation, 188
National Aeronautics and Space Act of 1958, as amended, and related legislation, 188
National Aeronautics and Space Administration factbook, 183
Near-earth phase operations plan for Mariner Mars 1969 project., 189
New horizons, 189
Nimbus 6 Random access measuring system
   applications experiments, 190

O

On Mars
 exploration of the Red Planet, 1958-1978, 155
On the frontier
 flight research at Dryden, 1946-1981, 165
 flight research at Dryden, 1946-1999, 165
On the moon with Apollo 15, 203
On the moon with Apollo 17, 203
On the moon with Apollo 16, 203
On the shoulders of Titans
 a history of Project Gemini, 164
Orders of magnitude
 a history of NACA and NASA, 1915-1976, 131
 a history of the NACA and NASA, 1915-1990, 141
Origins of NASA names, 227
Other worlds from earth
 the future of planetary astronomy, 190
Outlook for space, 191
Oversight reports during the 95th Congress
 report, 191

P

Papers of Nicholas E. Golovin, 1943-1970 (bulk 1958-1969), 162
Papers of Thomas O. Paine, 1931-1992 (bulk 1960-1982), 192
Papers of Wernher Von Braun, 1796-1970 (bulk 1950-1970), 223
Particles and fields near Jupiter, 227
Partners in freedom, 144
Passage to a ringed world
 the Cassini-Huygens mission to Saturn and Titan, 193
Physical properties of the surface materials at the Viking landing sites on Mars, 193
Piccard family papers, ca. 1470-1983 (bulk 1926-1983), 193
Pioneer odyssey, 156, 157
 encounter with a giant, 157
Pioneer Venus, 157
Pioneer, first to Jupiter, Saturn, and beyond, 157
Pioneering Venus
 a planet unveiled, 157
Planetary exploration, 147
Pocket statistics history., 195
Politics and space
 image making by NASA, 144
Portable life support systems, 195
Power to explore
 a history of Marshall Space Flight Center, 1960-1990, 151
Powering Apollo
 James E. Webb of NASA, 172
Preliminary inventory of the records of the National Aeronautics and Space Council, record group 220, 195
Prescription for disaster, 221
Pricing policies for the space shuttle, 209
Project Gemini, 177
Project Gemini
 technology and operations, 163
Project Gemini., 196
Project management in NASA, 145
Project Mercury, 163
Project Mercury
 man-in-space program of the National Aeronautics and Space Administration, 196
Project Orion
 a design study of a system for detecting extrasolar planets, 196
Project Telstar, 190
Putting satellites to work, 147

R

Ranger VII, 197
Ranger VII lunar charts., 197
Reference system characteristics for manned stopover missions to Mars and Venus, 150
Remembered images, 182
Remote sensing of oceanic parameters during the Skylab/Gamefish Experiment, 156
Report from Mars
 Mariner IV, 1964-1965., 197
Report on the coordination of proposed NASA/DOD FY 1967 facilities, 197
Report to the Administrator, National Aeronautics and Space Administration., 198
Review of recent launch failures, 199
Research program on pyroelectric detection techniques and materials, phase II., 198
Results of the Project Mercury ballistic and orbital chimpanzee flights, 198
Results of the second U.S. Manned orbital space flight, May 24, 1962., 198
Results of the third U.S. manned orbital space flight, October 3, 1962., 199
Return to the red planet
 the Mars Observer mission., 158
Review of the National Aeronautics and Space Act of 1958, as amended
 report together with additional views, 199

Rocketed into history
  NASA claims a paradise, 158

S

Saturn SA-1., 200
Science in orbit
  the shuttle & spacelab experience 1981-1986, 201
Scientific American inventions from outer space
  everyday uses for NASA technology, 138
Scientific satellites, 147
Scientists who work with astronauts, 195
Search for the universal ancestors, 201
Searching the horizon
  a history of Ames Research Center, 1940-1976, 182
Selling to NASA., 201
Selling Uncle Sam, 165
Shuttle atmospheric lidar research program
  final report of Atmospheric Lidar Working Group., 202
Shuttle-Mir = [Mir-shattl]
  the U.S. and Russia share history's highest stage, 181
Sightseeing
  a space panorama
    84 photographs from the NASA archives, 202
Significant NASA inventions., 202
Sixty years of aeronautical research, 1917-1977, 132
Skylab
  a chronology, 189
  a guidebook, 140
Skylab 2 photography catalog
  multispectral, earth terrain, Hasselblad, Nikon cameras., 203
Skylab EREP investigations summary, 204
Skylab experiments, 204
Skylab explores the Earth, 204
Skylab, classroom in space, 204
Skylab, our first space station, 204
Skylab's astronomy and space sciences, 204
Small business participation in the NASA research and development program, 205
Some aspects of the atmospheric circulation on Mars,, 213
Sounds from space. [Sound recording], 210
Space benefits
  the secondary application of aerospace technology in other sectors of the economy, 206
Space colonies, 192
Space enterprise
  beyond NASA, 164

Space exploration
  opposing viewpoints, 206
Space flight
  the first 30 years., 206
Spacelab, 175
Space mission patches, 223
Space missions to comets
  a conference, 206
Space patches
  from Mercury to the space shuttle, 170
Space physics and astronomy, 147
Space scientists and engineers
  selected biographical and bibliographical listing, 1957-1961., 206
Space shuttle
  human capital and safety upgrade challenges require continued attention, 207
Space shuttle wind tunnel testing program summary, 227
Space station
  key to the future., 207
  the next logical step, 159
Space Station Program
  description, applications, and opportunities, 207
Space telescope, 180
Spaceborne radar observations
  a guide for Magellan radar-image analysis, 208
Spacecraft power, 147
Spacecraft tracking, 148
Spaceflight revolution
  NASA Langley Research Center from Sputnik to Apollo, 165
Spacelab 1., 208
Spacelab, an international short-stay orbiting laboratory., 159
Spacelab, an international success story, 175
Stages to Saturn
  a technological history of the Apollo/Saturn launch vehicles, 141
Status and issues pertaining to the proposed development of the space telescope project, 209
Status of actions taken on recommendations of the Apollo 204 Accident Review Board., 209
Stellar atmospheric structural patterns, 219
Stratigraphy of part of the lunar near side, 228
Structural systems and program decisions., 210
Summary of Gemini extravehicular activity, 211
Surveyor I, 211
Surveyor I mission report., 211
Surveyor III, a preliminary report., 211
Surveyor V, 211
Surveyor program results., 211
Surveyor VI

a preliminary report., 212
Surveyor VII, 212

T

Task groups and linkages in complex organizations a case study of NASA, 132
Technical facilities catalog., 214
Technical history of the environmental control system for Project Mercury, 200
Telstar I., 214
The all-American boys, 149
The Apollo-Soyuz Test Project medical report, 214
The Apollo spacecraft: a chronology., 214
The birth of NASA
  the diary of T. Keith Glennan, 161
The Challenger launch decision
  risky technology, culture, and deviance at NASA, 222
The Comet Halley archive summary volume, 215
The contributions of Ranger photographs to understanding the geology of the Moon, 221
The early years,, 215
The earth below
  purchasing science data and the role of public-private partnerships, 215
The earthbound observer
  a personal look at the people of NASA and the space shuttle effort, 202
The Echo I inflation system,, 145
The evolving universe
  structure and evolution of the universe roadmap 2000-2020., 215
The Experiments of Biosatellite II, 216
The exploration of outer space with cameras
  a history of the NASA unmanned spacecraft missions, 180
The face of Venus
  the Magellan radar-mapping mission, 216
The fall of NASA's Skylab--the communications problem
  fourteenth report, 216
The Final Skylab mission
  man at home and at work in space., 216
The first, 216
The first lunar landing
  20th anniversary, 137
The future of the U.S. space program, 173
The great venture, 158
The Heat Capacity Mapping Mission (HCMM) anthology, 202
The history of NASA, 208
The history of the XV-15 tilt rotor research aircraft
  from concept to flight, 176
The how and why wonder book of the moon, 212, 213
The how and why wonder book of the moon., 213
The infinite journey
  eyewitness accounts of NASA and the age of space, 143
The interplanetary pioneers, 148
The Long duration exposure facility (LDEF) mission 1 experiments, 217
The Magellan Venus explorer's guide, 217
The Mariner 6 and 7 pictures of Mars, 145
The Martian landscape, 217
The moon-doggle, 154
The NASA program for particles and fields research in space., 175
The NASA scientific and technical information system
  its scope and coverage, 217
The NASA scientific and technical information system and how to use it, 217
The NASA technology transfer program, 151
The National Aeronautics and Space Administration, 148, 167
The National space program
  its values and benefits, 218
The nature and scope of the NASA university program, 205
The new Mars
  the discoveries of Mariner 9
    prepared for the NASA Office of Space Science, 166
The next 100 years [sound recording]
  your future on earth & in space, 165
The partnership
  a history of the Apollo-Soyuz test project, 155
The pioneer mission to Jupiter. Prepared for the Office of Space Science and Applications., 218
The atmosphere of Titan, 215
The publications of Dr. Hugh L. Dryden, 227
The Ranger III flight path and its determination from tracking data, 203
The real stuff
  a history of NASA's astronaut recruitment program, 138
The role of federal agencies in technology transfer, 151
The Search for extraterrestrial intelligence, SETI, 218
The six orbits of Sigma 7
  Walter M. Schirra's space flight, October 3, 1962., 218
The space dealers, 168

The space shuttle, 129, 166, 213
The space shuttle at work, 129
The space shuttle decision
    NASA's search for a reusable space vehicle, 166
The space telescope., 218
The star splitters
    the High Energy Astronomy Observatories, 221
The superior conjunction of Mariner IV, 161
The view from Ranger., 219
The Viking mission to Mars, 148
The voyage of Mariner 10
    mission to Venus and Mercury, 152
The Voyager flights to Jupiter and Saturn., 219
Thermodynamic equilibrium in prebiological
    atmospheres of C, H, O, N, P, S, and Cl, 219
This is NASA., 219
This new ocean, 213
This new ocean
    a history of Project Mercury, 213
To see the unseen
    a history of planetary radar astronomy, 144
Toward a better tomorrow with aeronautical and space technology, 220
Toward Mach 2
    the Douglas D-558 program, 221
Trajectory analysis of an Earth-Venus-Mercury mission in 1973, 210
Twenty-five years of NASA international programs, 221
Twenty-five years of N.A.S.A. international programs. 25 years of NASA international programs, January 1, 1983, 222
Two over Mars
    Mariner VI and Mariner VII, February to August 1969, 228

U

Unmanned space project management, 171
Uplink-downlink
    a history of the NASA Deep Space Network, 1957-1997, 182

V

Vanguard, 163
Venture into space

early years of Goddard Space Flight Center, 200
Venus cusp observations during 1969, 222
Viking 1
    early results, 223
Viking site selection and certification, 179
Voyage to Jupiter, 181
Voyage to Mars
    NASA's search for life beyond Earth, 141
Voyager
    journey to the outer planets, 225
Voyager 1 and 2 atlas of six Saturnian satellites, 139
Voyager 1 encounters Saturn, 224
Voyager 1, encounter with Jupiter, 224
Voyager encounters Jupiter, 225
Voyager to Jupiter and Saturn, 225
Voyages to Saturn, 182

W

Wallops Station and the creation of an American space program, 225
Walter M. Schirra's space flight, October 3, 1962, 218
Way station to space
    a history of the John C. Stennis Space Center, 166
Where do you go after you've been to the moon?
    a case study of NASA's pioneer effort at change, 167
Where no man has gone before
    a history of Apollo lunar exploration missions, 146
Wind energy developments in the 20th century, 228
Wind energy in the twentieth century, 228
Winds of change
    expanding the frontiers of flight
        Langley Research Center's 75 years of accomplishment 1917-1992, 201

X

X-15 research results, 210

Y

Your future in NASA, 173, 174

SUBJECT INDEX

#

20th century, 155, 228
21st Century, 53, 118

A

Advanced Research Projects Agency, 162, 163
Advanced Space Transportation, 25
advanced technologies, 19, 21, 32, 47, 48, 105, 122
Advisory Committee on the Future of the U.S, 192
Aéro-Club Suisse, 194
aeronautical engineers, 159
aeronautics, 127-132, 136-138, 141-146-153, 155-164, 166-168, 170-187, 189-197, 199-202, 204-236
aeronautics, commercial, 226
aeronautics, military, 144
aerospace engineering, 160, 194
aerospace engineers, 140, 171, 224
aerospace industries, 140, 143, 168, 182, 184, 222
Aerospace Medicine and Occupational Health Division (AMOHD), 29
aero-space technology (AST), 13, 14, 24-26, 31, 33, 34, 45, 96
African Americans in aeronautics, 142
African Americans in astronautics., 142
Air Force, 10, 29, 41, 53, 55, 57, 58, 60, 89-92, 96, 98, 99, 104, 111, 117, 127, 137, 194, 195, 197, 201
Air Force Space Command, 89-91
Aircraft Energy Efficiency Program (U.S.) Airplanes, 154
airplanes, 144
airplanes, military, 144
airships, 194
Aldrin, Jr., Edwin E. "Buzz", 5, 6, 16, 137, 192

American literature, 138
American Rocket Society, 224
Ames Aeronautical Laboratory, 4
Ames Research Center (ARC), 14, 31, 36-38, 40, 41, 96, 129, 155-157, 174, 182, 191, 195, 215, 218, 230
Apollo 11 (Spacecraft) , 5, 6, 16, 133
Apollo 13 (Spacecraft) Project Apollo (U.S.), 6, 17, 134
Apollo 16 (Spacecraft), 134
Apollo 8, 6
Apollo Lunar Surface Experiments Package., 139
Apollo program, 6, 10, 14, 16, 17, 41
Apollo Soyuz Test Project (ASTP), 6, 17, 135, 136, 156, 159, 192, 214
application satellites, 17, 18
APStar-2 satellite, 63
Arianespace, 60, 61, 62
Ariel I (Artificial satellite), 137
Armstrong, Neil A., 5, 6, 16
Army Air Forces, 194
Army Ballistic Missile Agency, 221
art, American, 155
artificial satellites, 136, 141, 144, 147, 148, 161, 186, 210
artificial satellites in agriculture, 161
artificial satellites in remote sensing., 144
artificial satellites in telecommunication, 141, 186, 210
astronautical charts, 135, 136
astronautics, 130, 134, 137, 138, 154, 160, 161, 167-170, 173, 177, 180, 181, 199, 205, 206, 209, 218, 223
astronautics in astronomy, 134, 138, 181
astronautics, military, 206
astronauts, 128
astronomical unit, 132
astronomy, 136, 138, 147, 190, 204, 205, 221, 229

atlas, 54, 55, 57, 97, 99, 122
atmosphere, 143, 205
atmosphere explorer, 143
atmospheric chemistry, 154
Austria, 83, 100
automated transfer vehicle (ATV), 75, 77, 82, 85
auxiliary power supply, 147
Avezda Service Module, 29
Aviation Operations System, 25
Aviation Safety Program (AvSP), 26

B

B-52, 58
ballistic missile, 3, 54, 55, 62, 63, 66, 71, 79
Balloon Federation of America, 194
ballooning, 194
bathyscaphe, 194
Bechtle, Otto Wolfgang, 224
Belgium, 81, 100
bilateral trade agreement, 61, 67
Boeing, 10, 29, 54, 56-59, 66, 73, 74, 78, 98, 106, 109, 111, 117, 118
Boeing 737 (Jet transport) Research aircraft, 226
Brazil, 29, 69, 74, 82
Brucker, Wilber Marion, 1894-1968, 224
Bush and Clinton Administrations, 65
Bush, President George H., 62, 66, 71, 92

C

Canada, 29, 69, 73, 74, 81, 82, 98, 100, 103, 107, 110, 114, 119
Canadian Space Agency, 81
Cape Canaveral, 21, 40, 96, 104
Carty, Denzil A, 194
Case for Mars Conference, 192
Centaur rocket, 199
Chaffee, Roger B., 6
Challenger, 6, 17, 18, 40, 55, 56, 72, 92, 117
Challenger (Spacecraft), 198, 221, 222
Challenger accident, 18, 55
Chandra X-ray Observatory, 41
chemistry, 194
chemistry, organic, 194
China, 53, 54, 60-64
China Great Wall Industry Corp. (CGWIC), 61-63
Climate Orbiter, 22, 30
climatic changes, 152
Clinton Administration, 54, 55, 64, 79, 80, 108, 123
Clinton, President, 55, 60, 63, 67, 69, 71, 73, 79, 80
Coler, Myron A, 162

Collier Trophy, 160
Commander in Chief of the North American Aerospace Defense Command (CINCNORAD), 89-91
Commander in Chief of U.S. Space Command (CINCSPACE), 89-91
Commerce Department, 63, 65
commercial launch industry, 22
commercial launch services, 54, 60, 66, 67
Commercial Remote Sensing Program (CRSP), 49, 50
Commercial Space Act, 50, 55
Commercial Space Centers, 30
commercial space launch, 53, 55, 57, 60
Commercial Space Launch Act Amendments, 55
commercial technology, 19
complex organizations, 133
Compton Gamma Ray Observatory(CGRO), 20, 32, 97, 102
computer networks, 169
Consolidated Appropriations Act, 65
corporate culture, 179
Corrigan, Daniel, 194
Cosmic Background Explorer (COBE), 18, 32
cosmic physics, 147, 155
cosmic rays, 194
cosmology, 216
Creative Science Program, 162
crew return vehicle (CRV), 54, 70, 75-77, 81, 85, 97, 98, 118
crew transfer vehicle (CTV), 97, 98

D

database management, 149
decision making, 222
Deep Space 1, 21, 32, 98
Deep Space network, 182
Delta, 54, 55, 57, 99, 106, 109, 122
Delta IV, 57
Denmark, 29, 81, 100
Department of Agriculture, 49
Department of Defense (DOD), 3, 15, 18, 22, 25-27, 36, 46, 53-55, 57, 63-65, 67, 68, 87-91, 97, 98, 102, 108
Department of Transportation (DOT), 49, 55, 59, 60, 107
Development Operatons Division., 221
digital-fly-by-wire (DFBW), 10
Discovery program, 32, 107, 115
Discovery Program, 32
Distributive Active Archive Centers (DAACs), 98, 121

Don Piccard Balloons, 194
DREAMTiME, 84
Dryden Flight Research Center (DFRC), 14, 36-39, 98
Dryden Flight Research Facility, 165
Dryden, Hugh Latimer, 1898-1965, 227
Durant, Frederick C, 224

E

Earth, 133, 136, 141, 152, 154, 166, 180, 191, 204, 215, 218, 229, 232-234
Earth Observing System (EOS), 21, 22, 27, 28, 40, 46, 47, 49-51, 95, 99, 100, 107, 119-124
Earth Observing System (Program) Atmospheric physics, 154
Earth Probes, 28, 47, 100, 119, 120, 124
Earth Radiation Budget Experiment, 180
Earth Radiation Budget Satellite, 180
Earth satellite, 4
Earth Science Applications Research Program (ESARP), 49
Earth Science Enterprise (ESE), 25, 27, 42, 45, 46, 48-51
Earth Science Enterprise Program (U.S.) Earth sciences, 215
Earth sciences, 152, 167, 191, 215, 218
Earth System Science Pathfinder (ESSP), 28, 47, 100, 107, 108
education, 153, 194, 204, 230, 231
Edwards Air Force Base, 29, 98, 111
Einstein, Albert, 1879-1955, 194
Eisenhower, President Dwight D., 3
energy budget (geophysics), 180
Engineers Joint Council, 192
engineers., 192
environmental monitoring, 152
Environmental Protection Agency (EPA), 49
EOS Data Information System (EOSDIS), 27, 40, 46, 47, 49, 50, 98, 100, 107, 119-123
Episcopal Church, 194
Europa Orbiter, 30
Europe, 53, 60, 65, 69, 71, 74-77, 81, 85
European Space Agency (ESA), 21, 29, 60, 61, 73, 81, 82, 97-100, 103, 106, 113, 116, 119, 120
evaluation, 169
evolved expendable launch vehicle (EELV), 53, 57, 99
exobiology, 141, 155
expandable space structures., 145
Expedition 2, 69, 70
expendable launch vehicles (ELVs), 18, 53-55, 57, 96, 98-100, 104-107, 112, 114-117, 122
experiments, 216
exploration, 127, 157, 158, 182, 190
Explorer (Artificial satellite) Orbiting geophysical observatories., 143
Explorer (Artificial satellite) Pioneer (Space probes), 155
Explorer 1, 4, 16, 100
explosives, 194
extrasolar planets., 196
extravehicular activity (EVA), 4, 6, 72, 75, 100, 128, 211

F

facilities, 156, 229
Federal Council for Science and Technology (U.S.), 163
Flight Research Center (U.S.), 226
flight simulators, 226
Florida, 130, 140
Focused Technology Programs (FTPs), 26
FORTRAN (Computer program language), 171
France, 21, 29, 61, 65, 81, 82, 83, 98, 100, 120
full-time equivalent (FTE), 19
Functional Cargo Block, 29, 101
Future-X, 26

G

Galileo, 9, 18
Galileo Project, 149, 160, 168
Gartmann, Heinz, 224
gases, 219
Gemini program, 16
General Dynamics Corporation, 224
General Electric Company, 192
General Motors Corporation, 224
geostationary orbit (GEO), 61, 67, 101, 102, 110, 115
Gerhards, Jupps, 224
German Aerospace Center, 28
Germany, 29, 81-83, 100, 223
Giacobini-Zinner comet, 215
Gilruth, Robert R, 194
Glenn Research Center, 14, 36-39
Glennan, Thomas Keith, 161
global climate change, 46, 51
global environmental change, 215
Global Learning and Observations to Benefit the Environment (GLOBE), 48, 101
global warming, 152
Goddard Institute for Space Studies (GISS), 38, 39

Goddard Space Flight Center, 14, 36-39, 102, 121, 161, 187, 190, 196, 200, 206, 226, 230
Goddard, Robert Hutchings, 215
Golovin, Anne Castrodale, 162
Gore, Vice President, 18, 23, 79
Government Performance and Results Act (GPRA), 23, 24
government purchasing, 202
Grosse, Aristid von,, 224
ground support systems (Astronautics), 171, 172, 207
guided missiles, 224

H

Habitation Module (Hab), 70, 76, 77, 84
Hall, Harvey, 164
handbooks, manuals, etc., 215
Heat Capacity Mapping Mission Program., 202
Herbert (John Herbert) Jones, Thomas V, 192
Hercules Powder Company, 194
High Energy Astronomy Observatories, 221
High Performance Computing and Communication (HPCC), 26
High-Energy Astronomy observatory 1 (HEAO 1), 17
high-speed aeronautics, 220, 221
high-speed aeronautics., 220
history, 138, 140-145, 151, 160-162, 164-167, 171, 172, 174, 176, 177, 183, 187, 189, 192, 193, 201, 202, 213, 221, 223, 225, 226, 229
Hornig, Donald F., 162
Hubble Space Telescope (HST), 8, 9, 20, 31, 40, 84, 98, 102, 108, 109, 115, 180, 189, 209, 219
human engineering, 137
Human Exploration and Development of Space (HEDS), 13, 14, 25, 28
human habitation, 29, 30
human spaceflight, 16, 83

I

imaging systems in astronomy, 225
India, 53, 60, 66, 67, 79, 96, 112
Indian Space Research Organization (ISRO), 66
industry and state, 191
information services, 217
information systems, 169
Infrared Astronomical Satellite (IRAS), 18
Inspector General, (IG), 34, 50
intelligence community (IC), 87-91
Intelsat, 62, 63

Interim Control Module (ICM), 75, 77, 78, 80, 85
internal combustion engines, 181
International Astronautical Congress, 162
International Business Machines Corporation, 224
International Geophysical Year (IGY), 3
International Geosphere-Biosphere Program (IGBP), 27, 46, 102, 103, 107, 117
International Space Station (ISS), 8, 10, 13, 20, 28, 29, 36, 40, 46, 56, 69, 70, 71, 73-77, 79-85, 103, 111, 153
International Telecommunications Satellite Organization, 62
International Ultraviolet Explorer (IUE), 17
internet, 9, 26, 84, 98
interplanetary voyages., 151
interstellar communication, 128, 161, 218
interstellar matter, 180, 227
inventions, 138
Iran Nonproliferation Act (INA), 70, 79, 85
Italian Space Agency, 28
Italy, 29, 77, 81, 82, 100

J

Japan, 29, 53, 60, 67, 69, 71, 73, 74, 76, 77, 81, 82, 103, 104, 119, 124, 192
Japanese Experiment Module (JEM), 82, 104
Jet Propulsion Laboratory (JPL), 4, 14, 30, 36-38, 40, 104, 105, 121
Johnson Space Center (JSC), 14, 36-38, 40, 42, 73, 104, 118
Johnson, Lyndon B, 162
Jupiter (Planet), 149, 157, 160, 168, 181, 209, 218, 219, 224, 225, 227
Jupiter probes, 149, 168, 182, 219, 225

K

Kennedy Space Center (KSC), 7, 14, 29, 36-38, 40, 96, 104, 111, 112, 130, 140, 141, 232
Kennedy, President John F, 16, 162
Kölle, Heinz Hermann,, 224
Kraemer, Robert S., 171
Kraft, Christopher C, 171
Kranz, Gene, 171, 172

L

landforms, 179
Landsat program, 11, 17
Landsat system, 18
Langley Aeronautical Laboratory, 165

Langley Research Center, 14, 36-38, 41, 104, 121, 128, 132, 144, 165, 175, 176, 179, 201, 205, 226
Large Launch Vehicle Planning Group, 162
large scale structure (astronomy), 216
large space structures (astronautics), 172
launch complexes (astronautics), 130, 158
launch services, 59, 61
launch vehicle technology, 123
launch vehicles (astronautics) Manned space flight, 162
LDEF (artificial satellite), 217
legal documents, 194
Ley, Willy, 224
life, 128, 137, 142, 155, 174, 195, 196, 200, 204, 218, 228, 229, 234
life on other planets, 128, 155, 196, 218, 228
life support systems (space environment), 137, 195, 200
LightSAR, 105
literature and technology, 138
location, 233
Lockheed Aircraft Corporation, 224
Lockheed Martin, 22, 29, 54, 56-59, 63, 65, 66, 96, 115, 117, 118
Loeser, G., 224
low Earth orbit (LEO), 61, 67, 102, 104, 106, 110, 111, 114, 115
lunar geology, 134, 196, 221
lunar landing sites, 175, 176, 203
lunar mineralogy., 146
Lunar Module, 5
Lunar Orbiter (Artificial satellite), 4, 171, 175, 176
lunar petrology, 146, 226
lunar photography, 175, 176, 219
lunar probes, 127
lunar stratigraphy, 228

M

Magellan (Spacecraft), 164, 176, 216
Magellan and Galileo space probes, 18
magnetism, 227
maintainability (engineering) service life, 154
Manhattan Project, 6
manned space flight, 132, 137, 168, 171, 172, 174, 177, 223
manufacturing processes., 189
maps, 193, 194
Mariner Jupiter/Saturn 1977 Project., 209
Mariner probes, 17
Mars (Planet) , 4, 7-9, 13, 17, 20, 22, 30, 32, 71, 77, 83, 84, 95, 98, 99, 105-107, 113, 117, 139, 141, 145, 148, 150, 158, 166, 178, 179, 181, 192, 193, 213, 217, 222, 223, 225, 228
Mars Global Surveyor, 9, 20, 30, 106
Mars Observer (Spacecraft), 9, 20, 105, 106, 158
Mars Pathfinder, 8, 9, 20, 30, 32, 98, 106, 113
Mars Polar Lander, 22, 23, 30
Mars probes, 141, 158
Marshall Space Flight Center (MSFC), 14, 36-38, 41, 107, 121, 129, 151, 152, 201, 204, 208, 223, 230, 235
mass, 144, 151
mass media, 144
McDivitt, James Alton, 128
McDonnell Douglas, 57, 73, 98, 106, 109
McNairy, Philip F, 194
Medaris, John B, 224
Megallan (Spacecraft), 208
Mercury (Planet), 152
Mercury program, 16
meteors, 180
microgravity research, 30, 38
microwave remote sensing, 136, 137
military spaceplane, 58
Millikan, Robert Andrews, 194
mini-pressurized logistics modules (MPLMs), 81
Mir (Space station) , 8, 20, 42, 69, 74, 78, 82, 99, 106, 110, 111, 114, 143, 181
Missile Technology Control Regime (MTCR), 62, 63, 79, 80
mission control, 40, 42, 104
Mission to Planet Earth (MTPE), 25, 45, 96, 98-101, 107-109, 111, 116, 117, 119, 120, 122, 123, 191
mobile servicing system (MSS), 82, 107, 114
moon, 131, 134, 140, 212, 213, 221
munitions list, 62, 63
Murphy, Charles S, 162

N

NASA Act, 15, 16, 24
NASA Dryden Flight Research Center, 226
NASA tech brief, 130
National Academy of Sciences (U.S.), 189
National Advisory Committee for Aeronautics (NACA), 4, 10, 11, 15, 129, 131, 141, 168, 199, 230
National Advisory Committee for Aeronautics History, 129, 131, 141, 200
National Aeronautics and Space Act, 129, 130, 173, 187, 188
National Aeronautics and Space Act of 1958, 163, 188, 199, 230, 231

National Aeronautics and Space Administration (NASA), 127-236
National Aeronautics and Space Council, 130, 173, 195
National Archives and Records Service, 195
National Bureau of Standards, 162
National Commission on Space, 192
National Imagery and Mapping Agency (NIMA), 28, 48
National Oceanic and Atmospheric Administration (NOAA), 16, 27, 28, 31, 46-48, 101, 106, 108, 218
National Polar-orbiting Operational Environmental Satellite System (NPOESS), 27, 46, 106, 108
National Research Council (U.S.) Aeronautics (NRC), 50, 51, 72, 81, 122, 194
National Science Foundation, 26, 31, 36
National Security Council (NSC), 89, 92
National Space Council, 92
National Space Transportation Policy, 55
Navy, 192
Near Earth Asteroid Rendezvous (NEAR), 32, 98, 107
Neptune, 18, 117
Netherlands, 29, 81, 100
Nets (Geodesy), 150
New Millennium, 21, 48, 98, 99, 107, 108
New Millennium Program (NMP), 32, 48
New York University, 162
Next Generation Space Telescope (NGST), 31, 108
Nimbus (Artificial satellite) Random Access Measurement System, 190
Nimbus (Artificial satellite) Sea ice, 137
NOAA (Weather satellite) Terrestrial radiation, 180
non-governmental organization (NGO), 74, 81, 84
NORAD, 89
Northrop Corporation, 192
Norway, 29, 66, 81, 100, 118
nuclear engineering, 185

O

Ocean Topography Experiment, 21, 27
Office of Aero-Space Technology (OAST), 25
Office of Earth Science (OES), 27, 33
Office of Life and Microgravity Science and Applications (OLMSA), 28, 29, 109
Office of Management and Budget (OMB), 19, 24, 76, 107
Office of Safety and Mission Assurance (OSMA), 34, 109
Office of Science and Technology, 162, 191, 215
Office of Science and Technology Policy (OSTP), 72, 117
Office of Space Flight (OSF), 22, 28, 29, 33, 109
Office of Space Science (OSS), 30, 33, 107-109, 119
Office of the Chief Technologist (OCT), 34
Optus 1, 62
orbital transfer (space flight), 220
orbiter *Columbia*, 7
orbiter *Discovery*, 7
orbiting astronomical observatories, 189, 199, 209, 219
Orbiting Astronomy Observatory (OAO), 17
Orbiting Geophysical Observatory (OGO), 17
Orbiting Solar Observatory (OSO), 16, 17
organizational behavior, 133, 222
organizational effectiveness, 179
organizational sociology, 179
outer planets., 225
outer space, 127, 130, 137, 143, 148, 149, 155-157, 159, 164, 165, 173, 177, 179-181, 189, 191, 192, 196, 197, 206, 208, 225

P

Panama Canal, 6
particles (nuclear physics), 166
patents, 130
patents and government-developed inventions, 130
photographs, 192, 194, 224
physicists, 163, 164
Picard family, 194
Piccard, Auguste, 194
Piccard, Jacques, 194
Piccard, Jules, 194
Pioneer (Space probes), 148, 157, 158, 182
Pioneer 10, 4, 8, 17, 110
Pioneer 11, 8, 17
planetary probes, 13, 14, 17, 18, 30
planets, 212
planning, 162, 207
Pleiades (Aerostat) Calco Chemical Company, 194
Pluto, 13, 14, 18, 110
polar regions, 190
Project Echo, 210
power resources, 153
Presidential Commission on the Space Shuttle Challenger Accident, 198
Presidential Space Advisory Group, 89
pricing, 209
privatization, 56, 60, 122
program budgeting, 162
Project Apollo (U.S.), 4-6, 131-137, 140-143, 145, 146, 153, 155, 158, 159, 162, 168, 170, 172, 175,

177, 190, 192, 196, 198, 203, 209, 210, 212, 214, 224, 226, 228
Project Echo, 145, 210
Project Gemini (U.S.), 4, 5, 128, 163, 164, 177, 196, 211, 224
project management., 145, 171
Project Mariner (U.S.), 132, 142-145, 148, 150, 152, 161, 166, 178, 179, 189-191, 196, 197, 199, 201, 208, 220, 228
Project Mercury (U.S.), 4, 163, 177, 196, 198-200, 213, 218, 224
Project Orion, 196
Project Ranger, 155, 164, 197, 203, 219, 221, 228
Project Telstar, 190, 214
Project Vanguard, 3, 163
propulsion laboratory (U.S.), 145, 150, 152, 158, 160, 164, 176, 197, 208, 215, 219, 224, 225, 227, 228
propulsion module, 70, 75, 76, 77
propulsion system, 25, 38, 75, 85
public officials, 163, 224
public-private sector cooperation, 215
pyrometry, 198

R

radar in astronomy, 144
ranger, 4
Reagan Administration, 61, 83, 92
Reagan, President, 18, 55, 71, 92
Redstone Arsenal (Ala.), 129, 224
religion, 194
remote sensing, 144, 156, 204, 208
remote sensing industry, 21, 49, 50
research, 128, 144, 150, 153, 155, 157, 162, 165, 174-176, 179, 182, 190, 193-195, 198, 201, 205, 210, 215, 218, 219, 222, 226, 228-230, 234
research aircraft, 176, 193
research and development contracts, 202, 205
reusable launch vehicle, 17, 22, 26, 53, 54, 56-59, 110, 114, 117, 118
reusable space vehicles, 207, 210
Ridlen family, 194
Ridlon, John, 194
robotic missions, 4
rockets, 146, 224
rockets, sounding, 146
Roentgen Satellite, 32
Rumsfeld Commission, 87, 88, 90-92
Rumsfeld, Mr., 87
Russia, 8, 20, 28, 29, 53, 54, 60, 61, 65-67, 69-71, 73-75, 77-85, 96, 98, 101, 103, 108, 110, 111, 114, 118, 119, 181

Russian Prime Minister Chernomyrdin, 79
Russian Space Agency, 66
Ryan, Cornelius, 224

S

Salyut, 8, 82, 110, 111, 114
Satellite Export Issues, 60
Satellite Export Licenses, 65
satellites, 147
Saturn (Planet), 139, 182, 193, 224
Saturn launch vehicles, 140, 141, 200
Saturn probes, 182, 193, 219, 224, 225
Saturn Project (U.S.), 140-142
Saturn rockets, 142
Schirra, Wally, 199, 218
science, 127, 138, 149, 154, 162, 166, 168, 173, 177, 185, 188, 191, 199, 201, 205-207, 209, 212, 215, 216, 218, 222, 229, 231-235
science and industry, 191
science and state, 154, 162
scientific probes, 8, 16
scientific satellites, 147
scientists, 195, 233
sea ice, 136
Seamans, Robert C, 192, 201
sea-viewing wide filed sensor (SeaWiFS), 120
Security Assistance Act, 65, 68
Senior Interagency Group/Space (SIG/Space), 92
separation of powers, 162
Shepard Jr., Alan B., 4
Shuttle Atmospheric Lidar Research Program., 202
Shuttle Radar Topography Mission (SRTM), 28, 47
Sikorsky, Igor Ivan, 1889-1972, 224
single-stage-to-orbit (SSTO), 26, 57-59
Skylab, 4, 8, 17, 18, 71, 111, 112, 114
Skylab Program, 140, 142, 146, 153, 156, 189, 190, 203, 204, 216
Skylab Project, 205
Skystreak (Supersonic planes), 221
small business., 205
Small Spacecraft Technology Initiative (SSTI), 21, 96, 97, 105, 114
social sciences and state, 162
software, 233
solar activity, 166
Solar and Heliospheric Observatory (SOHO), 21, 103
solar atmosphere., 187
Solar Maximum Mission (Project) Solar flares, 187
solar system, 14, 30, 32
solar system astronautics in astronomy, 155
solid rocket boosters (SRBs), 6, 28, 114

sources, 133
Soviet Union, 3, 6, 16, 66, 82, 202
space biology, 174, 216
space colonies, 192
Space Corps, 90, 91
Space Council, 92
space exploration, 20, 24
space flight, 133-135, 137, 142, 145, 147, 149, 151, 158, 159, 164, 168, 170, 174, 177, 187, 190, 192, 195, 199, 203, 206, 210, 212, 214, 216-218, 220, 223, 224, 228
Space Flight Operations Contract (SFOC), 29, 56
space flight to Jupiter, 149, 168, 218
space flight to Mars, 147, 151, 158, 170, 195, 228
space flight to Mercury, 210
space flight to the moon, 133-135, 137, 159, 190, 192, 203, 212, 220
space flight to Venus, 164, 216, 217
space industrialization., 164
space law, 173, 187, 199
space maneuver vehicle, 58
space medicine, 142, 198, 214
Space News, 50, 59, 64, 65, 92, 123
space photography, 131, 202
space probes, 147, 171, 180, 225
Space Program United States, 192
space race, 202
Space Science Board History, 189
Space Science Enterprise (SSE), 25, 30, 31, 32, 48
space sciences, 133, 134, 160-162, 171, 173, 182, 189, 205, 216, 217, 223
space ships in art, 173
space shuttle, 4, 6, 7, 9, 10, 17, 18, 22, 28, 29, 39-42, 47, 56, 114, 120, 122, 129, 131, 145, 166, 169, 181, 201, 202, 207, 214, 227
space simulators, 210
Space Station Freedom, 8, 72
Space Station Program (U.S.) Space stations., 207
space stations, 154, 159, 160, 164, 172, 207, 216
Space Studies Board (SSB), 72
space suits, 170
space trajectories, 151, 170, 210
space transportation system (STS), 6, 7, 22, 28, 109, 115
space vehicle accidents, 134, 143, 198, 209, 221
space vehicles, 15, 39, 90, 131, 143, 144, 147, 148, 153, 155, 162, 168, 173, 179, 181, 186, 196, 201, 203, 209, 210, 216, 220, 228
space vehicles in art, 173
Spacelab Program, 160, 175, 208
Spain, 29, 81, 100
Sparkman, John, 224
Special Purpose Dextrous Manipulator, 82

Sputnik 1, 3, 16
stars, 212
State Department, 62-66
Stennis Space Center (SSC), 14, 36-38, 42, 109, 114, 166
Stevens, A, 194
stratosphere, 194
Stratospheric Observatory for Infrared Astronomy (SOFIA), 109, 113
Stuhlinger, Ernst, 140, 224
subject headings, 186
submarine boats, 192
sun, 153, 192, 235
Sun-Earth Connection (SEC), 30, 31, 32
supersonic transport planes, 162
surveyor, 4, 9, 20, 30, 99, 105, 106, 107
Surveyor Program (U.S.), 131, 139, 171, 209, 211, 212
Swann, W, 194
Sweden, 29, 81, 100
Switzerland, 29, 81, 100, 194

T

technology, 130, 132, 138, 142-144, 148, 151, 152, 154, 158, 160-163, 168, 172, 176, 178, 181, 185, 186, 188-191, 196, 199, 200, 203, 206, 208-211, 213-217, 219, 220, 224, 225, 227-229, 232
technology and state, 162
Technology Development, 24, 48
technology transfer, 151, 220
Technology Utilization Division, 130, 151, 220
Technology Utilization Office, 206
television in astronautics., 148
Terrestrial Planet Finder (TPF), 31
terrestrial radiation, 180
thermodynamic equilibrium, 219
Thermosphere, Ionosphere, Mesophere Energetics and Dynamics (TIMED), 31, 115
tilt rotor aircraft., 176
Titan (Satellite), 54, 55, 57, 97, 99, 103, 193, 215
TOPEX/Posideon, 21, 27
Total Ozone Mapping Spectrometer (TOMS), 21, 27, 28, 47, 95, 116, 124
Transhab, 81, 84
Tropical Rainfall Measuring Mission (TRMM), 27, 46, 47, 116, 121, 124
Truax, Robert Collins, 224

U

U.S. Air Force, 26, 31

U.S. Geological Survey (USGS), 28, 47, 48
U.S. Naval Research Lab (NRL), 75
U.S. Trade Representative (USTR), 60, 63
Ukraine, 53, 54, 60, 61, 65-67, 118
United Kingdom, 29, 81, 96, 100
United Space Alliance (USA), 29, 56, 87, 117
United States, 127-146, 148-174, 176-223, 225-228, 234, 235
United States Global Change Research Program (USGCRP), 27, 46, 50, 117, 119, 123
United States National Aeronautics and Space Administration, 127, 171, 172, 228
universe, 14, 30, 32
University Class Explorers, 42
University Earth System Science (UnESS), 28, 47
University Explorers (UNEX), 32, 42, 96, 100, 108, 115, 117
Upper Atmosphere Research Satellite (UARS), 21, 27, 95, 116, 120
Uranus, 18, 117

V

V-2 rocket, 223
Van Allen Radiation Belt, 4
Venus (Planet), 132, 152, 176, 208, 216, 223
Venus probes, , 4, 17, 18, 105, 110157, 158, 164, 217
video recording, 178
Viking Mars Program (U.S.), 148, 150, 155, 158, 179, 217, 222, 225
Viking rocket, 3
Virgil "Gus" Grissom, 6
Voyager Project, 139, 182, 219, 224, 225

W

Wallops Flight Facility, 225
Watkins, Armitage, 224
Webb, James E, 172
Whipple, Fred Lawrence, 224
White, Edward Higgins Jr., 5, 128
White Sands, 29, 38, 42, 104
White Sands Missile Range, 162
White Sands Test Facility (WSTF), 29, 38, 42
Wide-Field Infrared Explorer (WIRE), 22, 109, 112, 118
wind power, 228
wind turbines, 228
women in the Episcopal Church, 194
World Climate Research Program (WCRP), 27, 46, 117
World War II, 3
World War, 1939-1945, 194, 224

X

X-15 (Rocket aircraft), 169, 210
X-ray, 17, 22, 32, 41

Z

Zero Base Review (ZBR), 19
Zielinski, Francis, 194